00368535 7
Y0-CQR-837

POLLUTION PREVENTION FOR CHEMICAL PROCESSES

POLLUTION PREVENTION FOR CHEMICAL PROCESSES

David T. Allen

and

Kirsten Sinclair Rosselot

A WILEY-INTERSCIENCE PUBLICATION

JOHN WILEY & SONS, INC.

New York / Chichester / Brisbane / Toronto / Singapore / Weinheim

This text is printed on acid-free paper.

Library of Congress Cataloging in Publication Data

Allen, David T.
Pollution prevention for chemical processes / David T. Allen and Kirsten Sinclair Rosselot.
p. cm.
Includes index.
ISBN 0-471-11587-8 (alk. paper)
1. Chemical industry—Waste disposal. 2. Pollution prevention.
I. Rosselot, Kirsten Sinclair. II. Title.
TD195.C45A43 1996
660—dc20 96-8438

Printed in the United States of America

10 9 8 7 6 5 4 3 2

CONTENTS

PREFACE

Billions of tons of industrial waste are generated annually in the United States. Managing and legally disposing of these wastes costs American industry hundreds of billions of dollars each year, and these costs have been increasing rapidly over the past decade. The escalation is likely to continue if emission and treatment standards become more stringent. In the face of rising costs and increasingly stringent performance standards, traditional end-of-pipe approaches to waste management have become less attractive, and a strategy for environmental compliance variously known as *waste minimization*, *waste reduction*, or *pollution prevention* has been gaining prominence. The basic premise of this strategy is that avoiding the generation of wastes or pollutants can often be both more cost-effective and better for the environment than controlling or disposing of pollutants once they are formed.

Strategies for pollution prevention have gained widespread acceptance largely through case study reports. Now, after almost a decade of case study development, education in pollution prevention is at a crossroads. One path is a continued reliance on case studies, but as pollution prevention becomes more specialized and technologically sophisticated, case studies that have broad applicability will become rare. The other available path pursues the development of generic design tools for pollution prevention, and it is the first few steps along this path that this text takes, with particular focus on design tools that are relevant for the chemical process industries. Many of the engineering design tools for pollution prevention are at a rudimentary stage in their development, and much more sophisticated tools are likely to appear. It is the authors' hope that the presentation of emerging approaches in this book will enhance their development.

The text begins with a chapter that briefly introduces the concept of pollution prevention and provides definitions of waste management terms. Since many commonly used terms are loosely defined, the range of definitions currently in use are discussed and precise meanings for use in this book are established. The remainder of the text is divided into three major sections on pollution prevention at the (1) macroscale, (2) mesoscale, and (3) microscale. The scope of each of these three scales of pollution prevention is presented in Figure P-1. The first of the major sections consists of Chapters 2–4, where pollution prevention at the most macroscopic scale is examined.

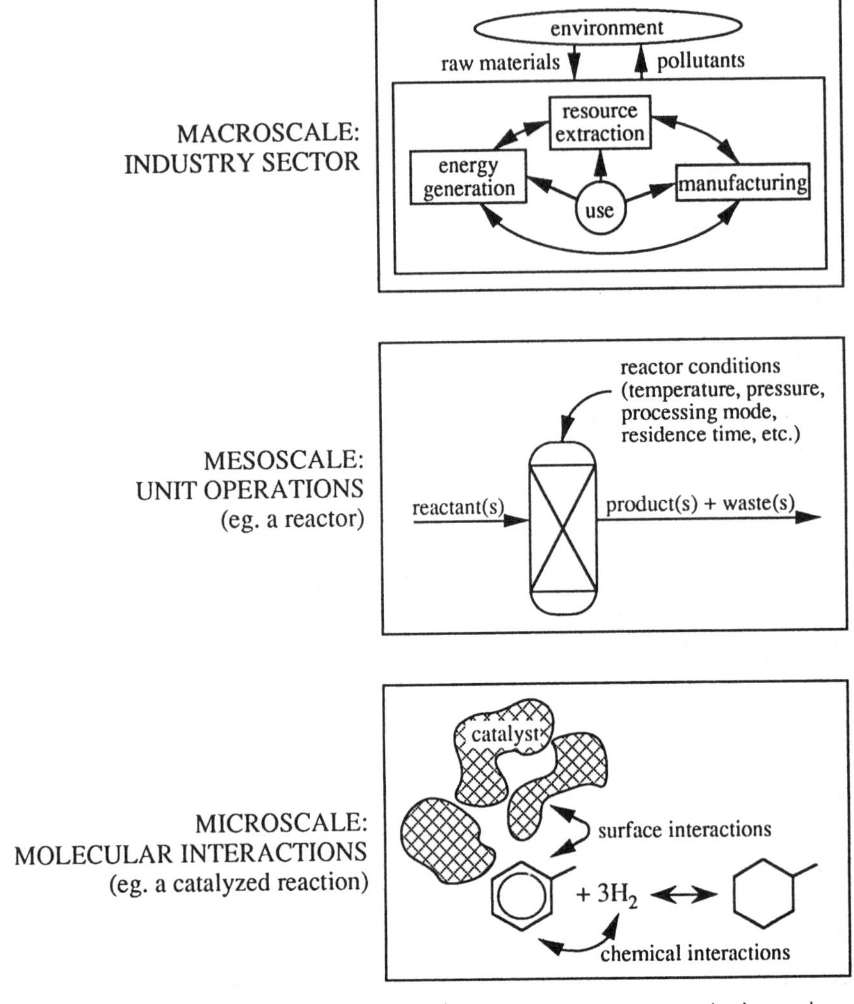

Figure P-1. Pollution prevention can be engineered at macro-, meso-, and microscales.

The flows of materials in our industrial economy, from natural resource extraction to consumer product disposal, are described in this section, beginning with an overview of waste generation and management in the United States. Next, two frameworks for assessing opportunities for pollution prevention at the macroscale are discussed: (1) industrial ecology, where the flows of materials are studied; and (2) product life-cycle assessment, where the environmental impacts due to the life cycles of products are assessed. The second major section of this text consists of Chapters 5–10, where pollution prevention at the level of chemical manufacturing processes is examined. This section on mesoscale pollution prevention begins with a chapter on waste audits and emission inventories. Next, an overview of pollution prevention options for unit operations is presented, and methods for preventing fugitive and secondary emissions are discussed. Then, flowsheeting tools for pollution prevention are described, and methods for quantifying the economic benefits of pollution prevention and prioritizing design options are described. The last chapter in this section consists of in-depth case studies of pollution prevention at the mesoscale. In the last major section, Chapter 11, case studies of pollution prevention at the microscale are presented, including reaction pathway analysis and material design methods.

Draft manuscripts of this text have been used in senior-level engineering electives at the University of Texas at Austin and at the University of California, Los Angeles. In a typical semester, all of the material in the text was presented, augmented by design assignments.

The guidance, support, and assistance of many individuals helped make this book possible. Sheldon Friedlander refined the scope and tone of the book; Mahmoud El-Halwagi and Howard Klee contributed to the case studies and methodologies presented in many chapters. Students at the University of California, Los Angeles and The University of Texas at Austin suffered through early drafts of problems and chapters. Research librarians at UCLA magically produced key references, and our families were patient as we labored to produce the final manuscript. We thank you all.

DAVID T. ALLEN
KIRSTEN SINCLAIR ROSSELOT

Austin, Texas
Calabasas, California

ACRONYMS

AHP analytic hierarchy process
AIRS Aerometric Information Retrieval System
BOD biological oxygen demand
BOOS burners out-of-service
BRS Biennial Report System
BSRP Beavon sulfur removal process
CARB California Air Resources Board
CEMS chemical emissions monitoring system
CERCLA Comprehensive Environmental Response, Compensation, and Liability Act
CFC chlorinated fluorocarbon
CID composition interval diagram
COD chemical oxygen demand
DAF dissolved-air flotation
EC environmental choice
EEP ethyl 3-ethoxypropionate
ELU environmental load unit
EPCRA Emergency Planning and Community Right-to-Know Act
EPS environmental priority strategy(ies)
EV electric(ity-powered) vehicle
FCCU fluidized-bed catalytic cracking unit
FGR flue-gas recirculation
FTIR Fourier transform infrared
GWP global-warming potential
HAZOP HAZard and OPerability (analysis)
HEN heat-exchange network
HHV higher heating value
HVGO heavy-vacuum gas oil
LAER lowest achievable emission rate
LCA life-cycle assessment
LDAR leak detection and repair
LDPE low-density polyethylene
LEA low excess air
LNB low-NO_x burner

ULNB	ultra-low NO_x burner
LQG	large-quantity generator
MDI	methylene diphenylene isocyanate
MEN	mass-exchange network
MGD	million gallons per day
MM	million
MTBE	methyl *tert*-butyl ether
NAAQS	National Ambient Air Quality Standard
NOS	not otherwise specified
OC	operating cost
ODP	ozone-depletion potential
OFA	overfire air
OSC	off-stoichiometric combustion
OSHA	Occupational Safety and Health Act
OVA	organic vapor analyzer
PCB	printed-circuit board
PEL	permissible exposure limit
PEMS	predictive emission monitoring system
PET	polyethyleneterephthalate
POTW	publicly owned (wastewater) treatment works
PSE	polyethylene-sack equivalent
PV	present value
QSAR	quantitative structure–activity relationship
RCRA	Resource Conservation and Recovery Act
SCOT	Shell Claus off-gas treatment
SCR	selective catalytic reduction
SFP	smog-formation potential
SIC	Standard Industrial Classification
SIMS	surface impoundment modeling system
SNCR	selective noncatalytic reduction
SOCMI	synthetic organic chemical manufacturing industry
SQG	small-quantity generator
TCA	1,1,1-trichloroethane
TCE	trichloroethylene
TDI	toluene diisocyanate
TRI	Toxic Chemical Release Inventory
TSDF	treatment, storage, or disposal facility
ULEV	ultra-low-emission vehicle
VOC	volatile organic compound
WSPA	Western States Petroleum Association

POLLUTION PREVENTION FOR CHEMICAL PROCESSES

1

INTRODUCTORY TERMS AND CONCEPTS

In this chapter, the vocabulary of pollution prevention is introduced. To provide a framework for definitions, a waste management hierarchy is described, which explicitly recognizes that some waste manage-alternatives, such as prevention, are preferable to other alternatives, such as disposal. Along with the hierarchy, a number of related and loosely defined terms, including source reduction, waste reduction, waste minimization, pollution prevention, and toxics use reduction, are presented and described. The range of definitions commonly assigned to these terms is reviewed and more precise definitions for use in this text are established.

1.1 INTRODUCTION

Before beginning a study of engineering design for pollution prevention, it is first necessary to define what is meant by the term *pollution prevention*. In this text, where the focus is on chemical manufacturing processes, pollution prevention means more efficient use of raw materials and energy and avoiding the use or generation of hazardous materials. While this engineering definition of pollution prevention is relatively simple, a definition of pollution prevention that addresses both engineering and regulatory issues is far more complex. Raw materials or byproducts might or might not be classified as hazardous wastes or toxic releases (see Chapter 2 for a discussion of these terms), and as demonstrated in the next section, there is a broad gray area between making a process more energy- and mass-efficient and merely doing a better job of managing wastes and emissions once they are formed. The remainder of this chapter addresses these ambiguities and presents the vocabulary of pollution prevention that will be employed in this text.

1.2 THE WASTE MANAGEMENT HIERARCHY

A logical place to begin a presentation of the vocabulary of pollution prevention is with a discussion of the waste management hierarchy, because an understanding of the hierarchy is a prerequisite to understanding the definitions of many waste management terms. It is well accepted that a wide range of solutions to environmental problems exists and that some solutions are preferable to others, so that there is a hierarchy of waste management alternatives. In the Pollution Prevention Act of 1990 (42 U.S.C. §§13101-13109), the waste management hierarchy is defined as follows:

> The Congress hereby declares it to be the national policy of the United States that pollution should be prevented or reduced at the source whenever feasible; pollution that cannot be prevented should be recycled in an environmentally safe manner, whenever feasible; pollution that cannot be prevented or recycled should be treated in an environmentally safe manner whenever feasible; and disposal or other release into the environment should be employed only as a last resort and should be conducted in an environmentally safe manner.

Based on this and similar descriptions, the elements of the waste management hierarchy can be placed in the following order of preference:

1. Source reduction
2. In-process recycling
3. On-site recycling
4. Off-site recycling
5. Waste treatment to render the waste less hazardous
6. Secure disposal
7. Direct release to the environment

The distinctions between these seven elements for a conceptual process (in this case, a reactor) are illustrated in Figure 1-1. The elements of the hierarchy, as shown in this figure, are

1. *Source reduction*—the reactor is modified so that less waste is created.
2. *In-process recycling*—the output of the reactor is sent through a separator and any unreacted feed is sent back to the reactor.
3. *On-site recycling*—waste from the reactor is converted in a second reactor to a salable product.

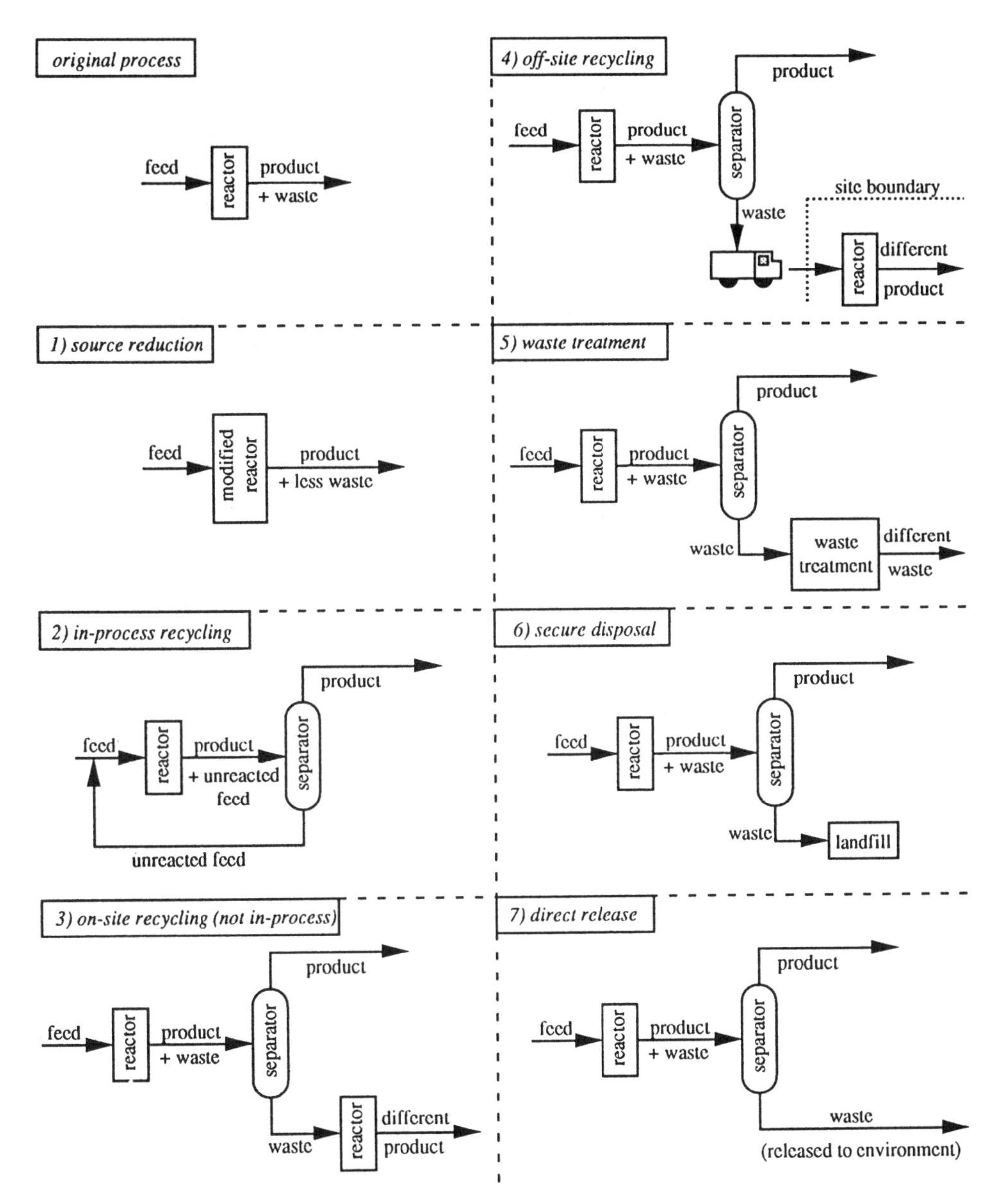

Figure 1-1. Examples of waste management modifications for a simple reactor, classified according to the elements of the waste management hierarchy.

4. *Off-site recycling*—waste from the reactor is sent off-site, where it is converted in a second reactor to a salable product.
5. *Waste treatment*—waste from the reactor is treated so that it is less hazardous.
6. *Secure disposal*—waste from the reactor is sent to a secure landfill.
7. *Direct release*—waste from the reactor is released directly to the environment.

In some cases, there could be a cascade of waste management alternatives applied to the same waste. For example, the waste might be treated to make it less hazardous, then sent to secure disposal. In the examples of Figure 1-1, there is no ambiguity about the element of the hierarchy to which each process modification belongs, but as shown in this chapter, it is difficult to distinguish between the elements in many situations.

1.3 CLOSELY RELATED TERMS AND COMPETING DEFINITIONS

The waste management hierarchy introduces a number of terms that require definition. Precisely defining terms such as source reduction, waste minimization, waste reduction, and pollution prevention is difficult. The definitions are made more complex when regulatory definitions are added to an operational or technical definition.

The focus of discussion in this section is on four legislatively defined terms: source reduction, waste reduction, pollution prevention, and toxics use reduction. A review of the definitions of these terms provides illustration of the subtle differences in definitions between terms and substantive differences in the use of these terms by various regulatory agencies.

1.3.1 Source Reduction

The term *source reduction* has been used in a number of state statutes and has been defined in the federal Pollution Prevention Act of 1990. The federal definition states that

A. The term "source reduction" means any practice that
   1. Reduces the amount of any hazardous substance, pollutant, or contaminant entering any waste stream or otherwise released into the environment (including fugitive emissions) prior to recycling, treatment, or disposal
   2. Reduces the hazards to public health and the environment associated with the release of such substances, pollutants, or contaminants

   The term includes equipment or technology modifications, process or procedure modifications, reformulation or redesign of products, substitution of raw materials, and improvements in housekeeping, maintenance, training, or inventory control.

B. The term "source reduction" does not include any practice which alters the physical, chemical, or biological characteristics or the volume of a hazardous substance, pollutant, or contaminant through a process or activity which itself is not integral to and necessary for the production of a product or the providing of a service.

The federal definition of source reduction is controversial, primarily because of the restrictions imposed by part B. The reasons for the controversy are described later in this chapter, but for now note the key phrase "not integral to and necessary for."

Next, consider some of the definitions of source reduction that appear in state legislation. In California, source reduction is defined as "input change, operational improvement, production process change or product reformulation." In the California definition, as in the federal definition, certain activities are explicitly identified as not being part of source reduction:

Source reduction is not
- Actions after generation, concentration to reduce volume or dilution to reduce toxicity, shifting to other media, or treatment

Vermont's definition is quite similar:

Source reduction is
- Any action causing a net reduction in hazardous waste generation

Source reduction methods include
- Input change to avoid, reduce, or eliminate waste generation and hazardous releases
- Operational improvement
- Production process change, including reuse of materials or their components

Source reduction is not
- Action taken after wastes are generated
- Concentration to reduce volume or dilution to reduce toxicity
- Media shifting
- Treatment

In contrast to these definitions, which are at times exclusive, is the position offered by the Bureau of Pollution Prevention in New York (from Foecke, 1992):

Source reduction is the in-plant practices used to reduce, avoid, or eliminate the generation of waste, including input substitution, technology modification, good housekeeping practices and product reformulation.

Additional definitions of source reduction are described elsewhere (WRITAR, 1991); the general consensus appears to be that process or product modifications that reduce waste can be classified as source reduction. As shown later, this definition requires a definition for what constitutes a process and what constitutes a waste.

1.3.2 Waste Reduction

The term *waste reduction* and the related term *waste minimization* have, in some ways, broader meanings than source reduction. Waste reduction and waste minimization generally incorporate both source reduction and on-site recycling, the first two or three elements of the waste management hierarchy. For example, Louisiana law states that

A. Waste reduction is
   1. In-plant practices reducing, avoiding, or eliminating generation of solid and hazardous waste
   2. Recycling within process shall be considered reduction
B. Waste reduction is not
   1. Recycling outside the process or after generation
   2. Dilution or concentration
   3. Changes to composition or concentration that do not change degree of hazard

In a few cases, the terms waste reduction and particularly waste minimization include even more elements in the hierarchy. An extreme example is the legislation in Delaware where

Waste minimization is
1. The reduction or elimination of the generation of waste, or the recovery, reuse, recycling, or sound treatment and disposal of generated wastes
2. A process used by a facility to analyze production processes for waste minimization opportunities

The extreme example provided by Delaware is not common, and waste reduction and waste minimization generally include only source reduction and on-site recycling.

While waste reduction and waste minimization are more expansive terms than source reduction in some ways, in other ways they are more restrictive. While source reduction can be used in the context of gaseous, liquid, or solid wastes, the terms waste minimization and waste reduction are generally used to refer to solid and liquid waste management activities.

1.3.3 Toxics Use Reduction

In 1989 Oregon created the Toxics Use Reduction and Hazardous Waste Reduction Act, according to which

> Toxics use reduction is
>
> In-plant production changes, or raw materials that reduce, avoid or eliminate the use or production of toxic substances without shifting risk, including
>
> 1. Input substitution
> 2. Product reformulation
> 3. Production process redesign
> 4. Production modernization
> 5. Improved operation and maintenance
> 6. Recycling or reuse within production processes
> 7. Proportionate changes in use of particular toxic substances.

In 1990, Massachusetts passed a related piece of legislation, the Toxics Use Reduction Act. According to the Massachusetts Act:

> Toxics use reduction is
>
> 1. Front-end substitution
> 2. Product reformulation
> 3. Process modernization
> 4. Improved operation and maintenance controls
> 5. Changes in production processes or raw materials that avoid, reduce, or eliminate the use of toxic or hazardous byproducts per unit product to reduce risk without shifting it
>
> Toxics use reduction is not
>
> 1. Incineration
> 2. Media shifting
> 3. Off-site or out-of-process recycling
> 4. End-of-pipe treatment

These definitions of toxics use reduction are in some ways reminiscent of the definitions of waste reduction in that they include process changes and in-process recycling as possible strategies. In underlying principle, however, toxics use reduction is fundamentally different from waste reduction. With waste reduction, the focus is reducing the burden of toxics exposure to the worker and the environment, and the use of hazardous chemicals is not questioned if they are used safely. In contrast, the underlying principle of toxics use reduction is to question any use of toxics. As a consequence, the relatively new concept of toxics reduction is highly controversial. As this concept attracts more attention, competing definitions are certain to appear.

1.3.4 Pollution Prevention

Of all the terms described in this chapter, *pollution prevention* is the most difficult to define with any rigor. Even the federal Pollution Prevention Act of 1990 fails to define this elusive term, despite using it in the Act's citation. In May 1992, however, the EPA announced a formal definition (Habicht, 1992):

> Pollution prevention means "source reduction," as defined under the Pollution Prevention Act, and other practices that reduce or eliminate the creation of pollutants through
>
> - Increased efficiency in the use of raw materials, energy, water, or other resources, or
> - Protection of natural resources by conservation

The announcement goes on to say that recycling is not included within the definition of pollution prevention, and that

> Drawing an absolute line between prevention and recycling can be difficult. "Prevention" includes what is commonly called "in-process recycling," but not "out-of-process recycling." Recycling conducted in an environmentally sound manner shares many of the advantages of prevention, such as energy and resource conservation, and reducing the need for end-of-pipe treatment or waste containment. . . . Some practices commonly described as "in-process recycling" may qualify as pollution prevention.

The EPA therefore considers pollution prevention to encompass only the first two elements of the waste management hierarchy: source reduction and in-process recycling. Even though the EPA has issued a formal definition of pollution prevention, the concept remains ambiguous. If "[p]ollution prevention means . . . practices that reduce or eliminate the creation of pollutants through . . . increased efficiency in the use of raw materials, energy, water, or other resources," as outlined in the EPA definition, then out-of-process and off-site recycling would in many cases fall within the scope of the definition of pollution prevention, except that they are expressly excluded.

Some states have also formulated a definition of pollution prevention. At least 23 state laws that approach the problem of defining pollution prevention have been identified (Foecke, 1992). Within state legislation, there is some consensus concerning the elements of the definition. For example, 17 laws exclude off-site recycling and 14 exclude treatment or incineration. Indiana's definition is typical:

> Pollution prevention is
>
> Reduction of use of toxic materials or reduction of environmental health

hazard without dilution or concentration prior to release, handling, storage, transport, treatment or disposal. It includes changes in production:

1. Technology
2. Materials
3. Processes
4. Operations and procedures
5. In-process, in-line, or closed-loop recycling

Pollution prevention is not

Practice applied after waste is generated or after waste exits the facility. It does not promote or require

1. Burning waste for energy recovery
2. Shifting waste from one environmental medium to another or to the work place or the product
3. Off-site recycling
4. Any other end-of-pipe waste management, including waste exchanges or the incorporation of waste into products or byproducts

The common definition of pollution prevention that appears in the states is therefore quite similar to the common definition of waste minimization or waste reduction, except that the activities are not restricted to solid and liquid wastes. Instead, the focus is on pollution in gas, liquid, or solid form. The general pattern of consensus among state definitions of pollution prevention is not duplicated elsewhere, however. Some definitions are more expansive than the common state definition. The Pollution Prevention Task Force of the American Petroleum Institute defines pollution prevention as (API, 1993) "a multi-media concept that reduces or eliminates pollutant discharges to air, water, or land and includes the development of more environmentally acceptable products, changes in processes and practices, source reduction, beneficial use and environmentally sound recycling." Many more definitions, some more expansive, some more restrictive, exist. A report from the California Environmental Protection Agency Task Force on Pollution Prevention (California EPA, 1991) contains examples of the wide range of definitions of pollution prevention in use. The task force included representatives from government, industry, academia, and public interest groups, who found that "Pollution Prevention almost defies precise description or definition." Members of the task force suggested a broad range of definitions, from

> Include every strategy/management practice/technology in the hazardous waste hierarchy as pollution prevention but give priority to those highest on the list, consistent with overall risk reduction.

to

Define pollution prevention as a combination of toxic substance use reduction, source reduction, and waste minimization.

To summarize, a precise definition of pollution prevention remains elusive. Within the limited context of state legislation, however, there appears to be a consensus that pollution prevention encompasses the first two or three elements of the waste management hierarchy for all wastes and emissions.

1.4 DISTINGUISHING BETWEEN ELEMENTS OF THE WASTE MANAGEMENT HIERARCHY

As discussed in the previous section, one source of confusion in the language of pollution prevention is the use of terms that are almost synonymous, or that have competing definitions. A second source of confusion is that the boundaries between the elements of the waste management hierarchy can be blurred. In the examples of Figure 1-1, there is no ambiguity about the classifications. In practical applications, however, making distinctions between some of the elements of the waste management hierarchy is difficult. Exactly what can be considered to be source reduction? How is the site defined when differentiating between on- and off-site recycling? The questions are endless and because of legislative uses of these terms, the controversy cannot be ignored. The degree of complexity in these definitions is illustrated in this section with examples that show that the distinction between in-process and out-of-process recycling and between recycling and waste treatment is not always clear.

1.4.1 Definition of a Process

The definition of a "process" can impact the classification of specific process modifications. Potential sources of ambiguity are examined in this section using a hypothetical petrochemical complex. Waste generating industries in general, and petrochemical complexes in particular, are continually searching for new uses for current waste streams. Complex refining and chemical manufacturing facilities typically consist of dozens of individual process units. These units are tightly integrated, so that the byproduct (waste) from one unit is frequently a raw material or an intermediate for another process. In refineries, for example, sludges from wastewater treatment can sometimes be used in the production of coke and hydrogen sulfide extracted from refinery wastewaters is used to produce elemental sulfur. Whether these activities are defined as in-process or out-of-process recycling depends on the definition of the process. If the process is broadly defined as an entire petrochemical manufacturing facility, then exchange of material between process units would be in-process recycling. In contrast, if the process is defined narrowly, as an individual process unit designed to manufacture a particular chemical, then exchange of streams between process units would be defined as out-of-process recycling. This issue is shown conceptually for an integrated chemical manufacturing facility in Figure 1-2. In this example, the modification is in-process recycling if the coupled system is viewed as the process. If the processes are considered individually, however, the modification is not in-process recycling.

In practical engineering terms, these definitions are irrelevant. In the examples of Figure 1-2, waste is being reduced and materials are being put to productive use. However, in current legislation, once a material becomes defined as a waste, overhead costs associated with using it are substantial. Whether a material is defined as a waste going from one process to another or as an intermediate material can have a major impact on process economics.

1.4.2 Distinguishing between Recycling and Waste Treatment

A second critical issue of definitions arises in attempting to distinguish recycling from waste treatment. Figure 1-3 shows a process that has been modified to make more effective use of its raw materials. A waste component that formerly had gone to disposal is converted back into reactant and a byproduct that can be recycled. This process modification is in the poorly defined gray area between waste treatment and recycling. If treatment is defined as any alteration of the physical, chemical, or biological characteristics of a waste unless that alteration is integral to and necessary for the production process, as implied in the Pollution Prevention Act of 1990, then this process modification is not recycling, even though the end result is reduced raw material and waste disposal requirements. In this case the key definition concerns what is integral to and necessary for the process. It is reasonably clear that the example of Figure 1-3 would not be viewed as source reduction. It is still unclear, however, whether it would be viewed as waste reduction or waste treatment.

1.5 THE DEFINITION OF POLLUTION PREVENTION USED IN THIS TEXT

The definition of pollution prevention adopted in this text is equivalent to the federal definition, without the exclusion of recycling. Figure 1-4 depicts the entire waste management hierarchy and the elements considered to be pollution prevention for the purposes of this text. Using this definition, a design modification that results in reduced environmental risks or reduced material and energy consumption is pollution prevention.

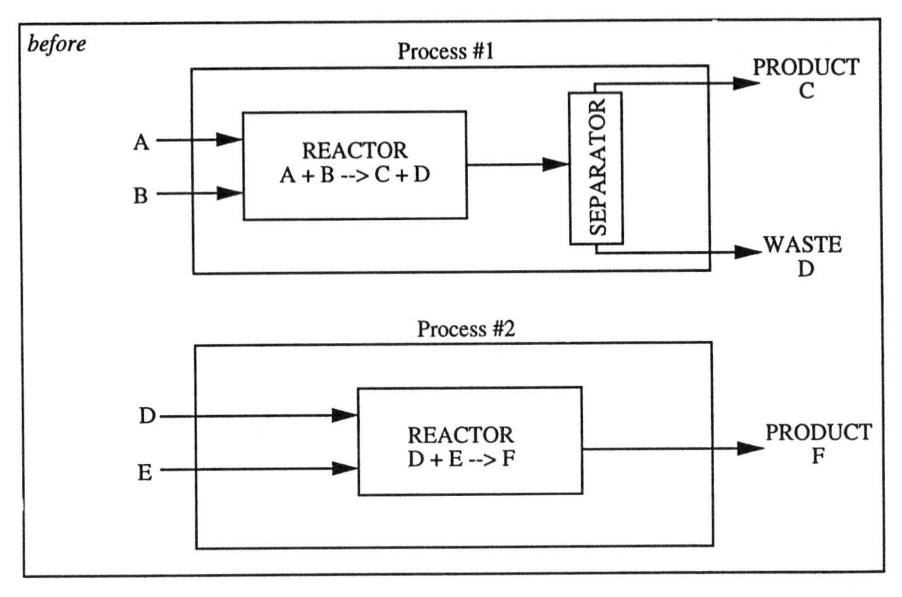

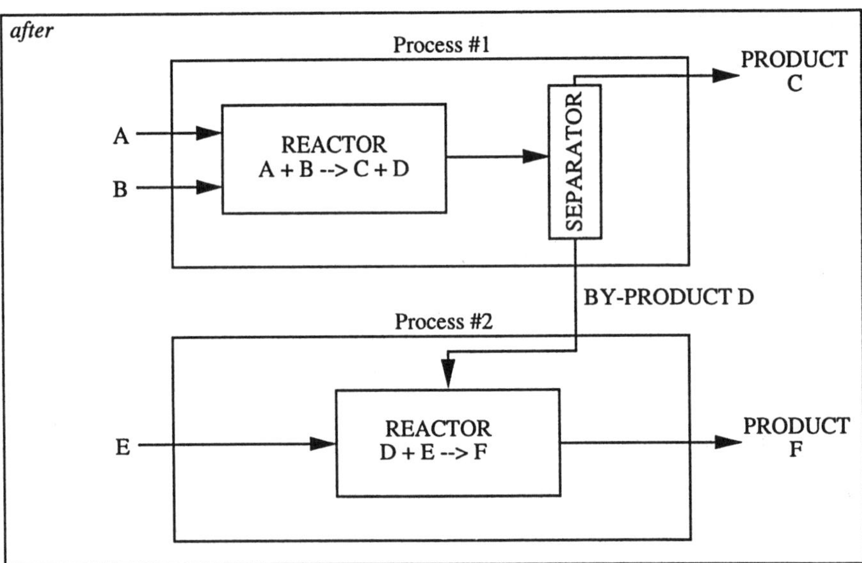

Figure 1-2. A generic process modification: Is it in-process recycling? Is it out-of-process recycling? Assume that Process#1 and Process#2 are part of the same industrial facility.

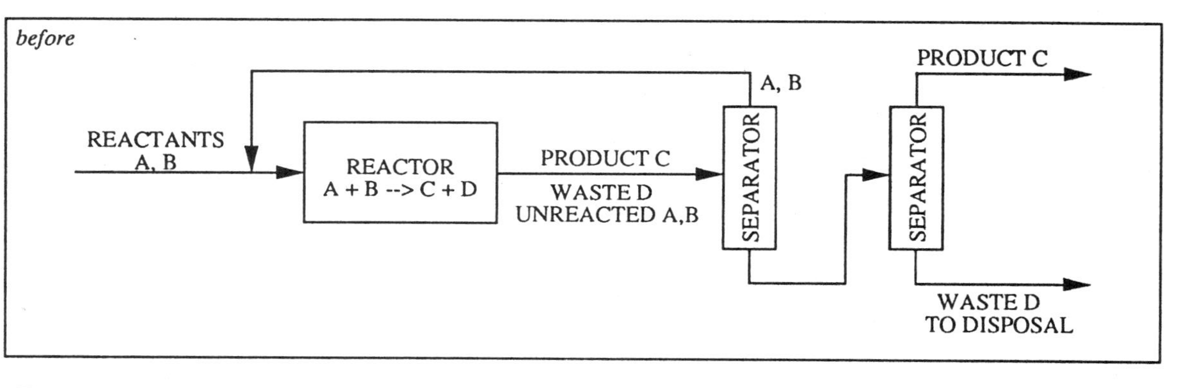

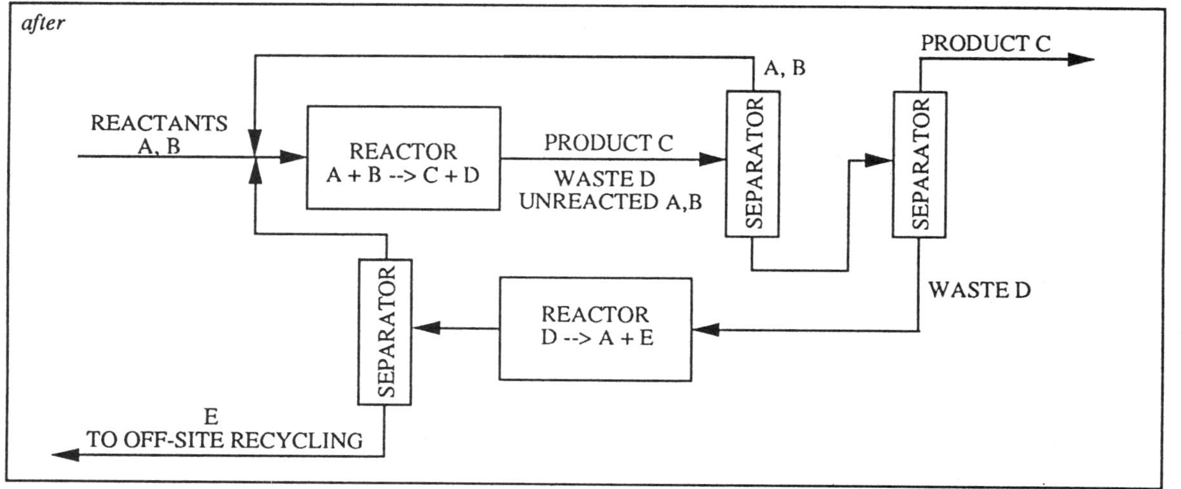

Figure 1-3. A generic process modification: Is it recycling? Is it treatment?

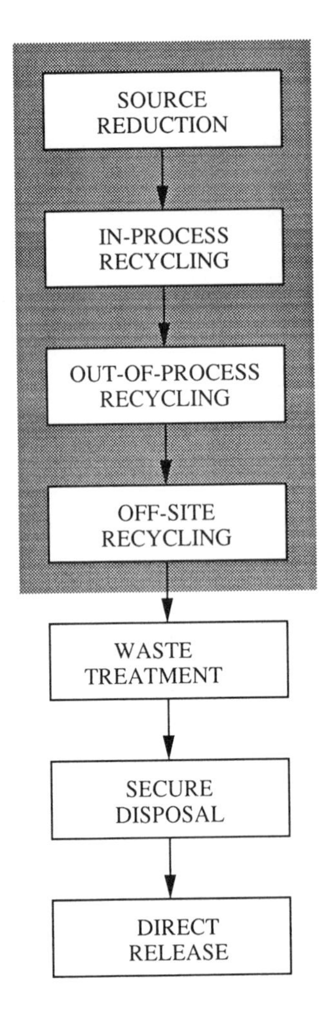

Figure 1-4. The elements of the waste management hierarchy that are included in the definition of pollution prevention in this text.

1.6 SUMMARY

Understanding the waste management hierarchy, in which waste management alternatives are ranked in order of preference, is essential to understanding many of the terms in the vocabulary of waste management. These terms, which include waste minimization, waste reduction, source reduction, and pollution prevention, are almost—but not quite—synonymous. Distinguishing between the terms can be ambiguous and tedious. However, one of the reasons for presenting detailed and competing definitions of terms such as pollution prevention is to highlight the difficulty of making legislative definitions. Source reduction, waste reduction, and pollution prevention involve changing manufacturing processes, which are almost infinitely variable and complex. Thus, attempting to precisely define these terms is difficult. Succinct definitions are open to broad interpretation. Detailed definitions can limit the uncertainty in the use of terminology, but they cannot be comprehensive in describing all possible modifications to all possible processes. Thus, a clear, unambiguous definition for these terms is probably beyond reach. What is possible, in principle, is consistent definitions by state and federal regulatory bodies. It is counterproductive to have definitions of terms, such as those described in this chapter, vary across state lines.

In this text, a more inclusive rather than exclusive definition of pollution prevention is adopted that includes the first four elements of the waste management hierarchy: source reduction and in-process, on-site, and off-site recycling. A more inclusive definition of pollution prevention allows the examination of more alternatives for reducing industrial impacts on the environment and eliminating adverse health effects that are due to manufacturing.

QUESTIONS FOR DISCUSSION

1. Write a one-paragraph definition of pollution prevention.
2. Discuss the relative merits of strategies such as pollution prevention, which focus on reducing releases of hazardous materials to the environment, to toxics use reduction strategies, which seek to eliminate the use of hazardous materials.

REFERENCES

American Petroleum Institute (API), "Environmental Design Considerations for Petroleum Refining Crude Processing Units," Publication 311, Feb. 1993.

California Environmental Protection Agency (California EPA), "Report of the 90 Day External Program Review of California's Toxic Substances Control Program," 1991.

Foecke, T. "Defining Pollution Prevention and Related Terms," *Pollution Prevention Review*, **2**(1), 103–112, Winter 1991/92.

Habicht, F. H. memo to all EPA personnel, May 28, 1992.

Waste Reduction Institute for Training and Applications Research, Inc. (WRITAR), "State Legislation Relating to Pollution Prevention," Minneapolis, MN, 1991.

I

MACROSCALE POLLUTION PREVENTION

Following the flow of materials in our industrial economy, from raw-material acquisition to product and waste disposal, provides a perspective that is essential for the most effective practice of pollution prevention. Such analyses can help to identify whether materials currently regarded as wastes in one industrial sector could be viewed as raw materials in another sector. These studies also reveal what types of processes and products are responsible for waste generation, and identification of the sources of waste is the first step toward prevention.

In this section, material flows are examined from three perspectives. First, in Chapter 2, an overview of waste generation and management is presented. This inventory of wastes helps to identify processes and products that may benefit from pollution prevention, but a mere listing of wastes and emissions ignores the complex interdependencies of many processes and products. Two related approaches are used to study the complex systems used to convert raw materials to products. One approach, described in Chapter 3, involves selecting a particular raw material, such as lead, and following it as it flows through processes and into products. The transformations and pathways followed by the raw material are known as its industrial metabolism. A second approach, called life-cycle assessment, starts with a particular product and identifies all the precursors that were required for the product's manufacture, use, and disposal. This approach is the subject of Chapter 4.

Pollution prevention at the macroscale utilizes these three elements: waste inventories, industrial metabolism, and life-cycle assessment.

2

WASTES AND EMISSIONS IN THE UNITED STATES

More than 12 billion tons of industrial waste are generated annually in the United States. This is equivalent to more than 40 tons of waste for every man, woman, and child in the country. The sheer magnitude of these numbers is cause for concern and drives us to identify the characteristics of the wastes, the industrial operations that are generating the waste, and the manner in which the wastes are being managed. The first portion of this chapter is devoted to an assessment of the strengths and limitations of data on wastes and emissions in the United States. Then, the sources of industrial wastes and emissions and the methods used to manage them are examined. Finally, the ways in which waste generation and management data are being used to evaluate environmental performance and to assess the potential for pollution prevention at industrial facilities are reviewed.

2.1 AN OVERVIEW OF WASTE FLOWS AND EMISSION RATES

There are a number of reasons to review waste generation and emissions from a national perspective. For an engineer confronted with a new waste stream, national inventories may be helpful in revealing how other facilities or industries have managed similar wastes. Data on waste generation also provide the public its view of the waste problem. The public, regulatory agencies, and advocacy groups are increasingly using public-domain databases to judge environmental performance [see, e.g., the West Virginia Discharge Reduction Scorecard (1990) and Table 2-7 later in this chapter]. It is therefore essential that anyone concerned with the engineering design of chemical processes have a working knowledge of publicly available waste and emission inventories. This section begins with a review of data resources on waste flows and emissions. A discussion of the magnitudes of these flows is next, followed by an analysis of the ways in which wastes in the United States are currently managed and what trends are being observed.

2.1.1 National Waste and Emission Inventories

Dozens of national data sources provide partial pictures of waste generation and management and pollutant emissions in the United States. These data are collected by public agencies such as the Environmental Protection Agency, the Department of the Interior, and the Department of Energy, and by private-industry groups such as the Chemical Manufacturers Association and the American Petroleum Institute. Each of these data sources focuses on a particular aspect of wastes and emissions; none provide a global view. In this subsection, important inventories for criteria air pollutants, toxic chemicals, and hazardous wastes are described.

2.1.1.1 ***Criteria Air Pollutants*** National Ambient Air Quality Standards (NAAQSs) have been established through the Clean Air Act for the following six pollutants, referred to as *criteria pollutants*: (1) particulate matter <10 μm in diameter (PM_{10}), (2) sulfur dioxide (SO_2), (3) nitrogen oxides (NO_x), (4) carbon monoxide (CO), (5) ozone (O_3), and (6) lead (Pb). NAAQSs are time-averaged concentrations that cannot be exceeded in the ambient air more than a specified number of times in a year. Attainment areas are those that are in compliance with the NAAQSs; nonattainment areas are those that are not. In nonattainment areas, permitting for a new source or for a major modification of an existing source requires that lowest achievable emission rate (LAER) technology be used. Also, any increase in emissions in a nonattainment area must be offset by decreases in emissions from other sources in the same region. Existing sources in nonattainment areas are also subject to emission control measures. Although volatile organic compounds (VOCs) are not on the list of criteria pollutants, facilities emitting VOCs can be affected by the NAAQS for ozone because of the role VOCs play in the creation of ozone in the lower atmosphere [see, e.g., Seinfeld (1986)]. The same holds true for NO_x: in a region that is an attainment area for NO_x but a nonattainment area for ozone there may be strict controls placed on NO_x emissions.

Information on emission rates of criteria air pollutants is available primarily from the Aerometric Information Retrieval System (AIRS). Information about accessing these data and obtaining other sources of information on criteria air pollutants is available from the US EPA's Office of Air Quality Planning and Standards in Research Triangle Park, North Carolina. In addition to reporting emissions of criteria air pollutants, AIRS includes measured ambient levels of the pollutants at thousands of monitoring sites in the United States.

The intent of the EPA in gathering data on criteria air pollutant emissions is to quantify the important sources of these emissions, including both mobile sources such as automobiles and stationary sources such as chemical manufacturing processes. Particulate emissions from area sources such as dirt roads are estimated. The inclusion of all important sources has recently involved attempts to quantify biogenic emissions (those that are emitted through biological processes), particularly VOCs from plant life. A high level of detail is available in AIRS: entries for each point source (such as a process heater) at each facility are identified separately. However, the data quality is inconsistent. Some of the emission data for point sources are generated from year-round direct measurement of pollutants while biogenic, mobile, and area emissions are impossible to measure directly and are not very well understood. Data quality tends to depend on geography as well. Emissions in nonattainment areas are more carefully monitored than emissions in attainment areas.

2.1.1.2 *Toxic Chemicals* More than 600 chemicals and chemical categories are currently reported in the Toxic Chemical Release Inventory (TRI). Prior to 1995, about 300 chemicals were included in the inventory. These chemicals were chosen for reporting because of their perceived potential to cause adverse human health effects or environmental damage. Some are highly "toxic," while others are relatively benign. The TRI was created because of Section 313 of the Emergency Planning and Community Right-to-Know Act (EPCRA) of 1986. Originally, manufacturing operations (those with Standard Industrial Classification codes between 20 and 39) that had more than 10 employees and handled more than a threshold quantity of chemicals were required to report in the TRI. In August 1993, an Executive Order was signed requiring federal facilities meeting the threshold requirements to report their releases and transfers in the TRI as well.

One of the best ways to become familiar with the data content in the TRI is to study the reporting instructions sent to facilities, which can be obtained free of charge by calling the EPCRA hotline at (800) 535-0202. These instructions contain lists of the chemicals and chemical categories for which TRI data are available. The TRI Reporting Form R for the year 1994 is reproduced in Appendix A of this text.

TRI data are available to the public on-line, and are updated annually, and individual reporting facilities are explicitly identified. Further, the data are in most cases reported by individual compound and can therefore be more easily related to risk than the data in most other waste inventories. However, the accuracy of the data is, at times, questionable. Many of the data reported in the TRI are estimates, not measurements, and as a consequence the information may be subject to considerable errors. The magnitude of the potential errors is demonstrated later in this text, through a discussion of fugitive and secondary emission estimation methods. Nonmanufacturing activities not reported in the TRI may be significant and include agriculture, mining, construction, transportation, and utilities. Small facilities (those with less than 10 employees) are also excluded.

2.1.1.3 *Hazardous Wastes* Prior to the late 1980s, a detailed accounting of the management patterns for the hundreds of millions of tons of hazardous waste generated each year in the United States was unavailable. However, with the data collection provisions enacted under the Superfund reauthorization and the Resource Conservation and Recovery Act (RCRA), the legal authority to collect such data was put in place.

Hazardous waste is defined in many different ways. All legal definitions in the United States, however, include wastes defined as hazardous under RCRA. Subtitle C of RCRA governs the "cradle to grave" management and manifesting of wastes that are defined as hazardous. Hazardous waste is defined in RCRA as any waste that (1) exhibits greater than threshold properties of ignitability, corrosivity, reactivity, or toxicity, or (2) is specifically listed as hazardous by compound or by the generating process or industry. Once a waste is defined as hazardous, the costs of managing it skyrocket. Partly because of these high costs, the focus of most pollution prevention efforts in the chemical process industries is on hazardous wastes. Also, while hazardous waste constitutes only a small fraction of the total waste mass generated by U.S. industry, most of the data available on waste generation and management focus on this segment of the waste inventory.

The most important ongoing source of data on hazardous waste generation is the Biennial Report System (BRS). Data from large quantity generators of RCRA hazardous waste and RCRA-permitted treatment, storage, and disposal facilities that accept RCRA hazardous waste are collected every other year. Like the TRI and AIRS, the data are publicly accessible and individual facilities are identified. Waste stream quantities are reported by their characteristic(s) and/or listed name(s). The chemicals present in the waste that are also listed in the TRI must be identified. However, the concentration of individual compounds is not reported and there is no mechanism for determining, given the data, how much of a reported waste stream is composed of water. Treatment and disposal methods applied to the waste are also reported.

There are many exclusions to the reporting requirements in the BRS. Household wastes and many mining and agricultural wastes are excluded, as are small-quantity generators and treatment, storage, and disposal facilities not subject to RCRA permitting.

An important hazardous waste survey was conducted in 1986 that asked for data on concentration of hazardous constituents and that included a wider universe of hazardous waste generators. This National Hazardous Waste Survey (US EPA, 1988c, 1991) has two basic components: a generator survey focusing on waste characterization and a survey of treatment, storage, disposal, and recycling facilities that focuses on waste treatment and disposal.

2.1.1.4 *Nonhazardous Wastes* Nonhazardous wastes fall under the provisions of Subtitle D of RCRA. Compared to hazardous wastes, there are relatively few sources of data on nonhazardous waste generation. As noted in a report prepared for the Department of Energy (US DOE, 1991):

> Information on non-hazardous waste is not collected on a regular basis, either by the Federal government or by industry groups. Most of the information that has been gathered reflects only a snapshot of conditions at a given time.

This creates particular problems since nonhazardous wastes represent over 90% of all industrial waste mass. In addition, not all types of nonhazardous wastes are included in the studies, most notably agricultural and construction wastes. The main sources of data on nonhazardous waste are a series of EPA reports to Congress (EPA, 1988a, 1988b).

2.1.1.5 *Integrating Information in Waste and Emission Inventories* As shown in Table 2-1, there are a number of waste and emission data sources other than the ones just discussed. Although this table lists more than 20 sources of data, developing a comprehensive and coherent picture of waste generation and management in the United States remains problematic for a number of reasons (US DOE, 1991). First, the definition of waste is not consistent across the data sources. Definitions of hazardous waste and reporting limits vary from inventory to inventory and from state to state. Some waste inventories account for only hazardous waste managed in facilities permitted through RCRA. Others account for any waste defined as hazardous under RCRA, regardless of the permitting status of the facility in which the waste was managed. There are almost as many definitions of hazardous materials or wastes as there are inventories, and the same wastes may appear in more than one inventory, but under a different name or category. A waste might be reported in one inventory as part of a waste stream and in another inventory in compound-specific fashion, but there is no formal way to link information found in the different inventories. Also, each inventory focuses on a specific industry sector and/or a specific type of waste, and measures for coordinating the data sources on a large scale are not in place. For example, most, but not all, of the data sources classify the wastes by the Standard Industrial Classification (SIC) code of the industry generating the waste (US DOE, 1991), and different data sources have different criteria for determining the appropriate SIC code.

A graphic depiction of the universe of wastes and emissions and some of the corresponding inventories is given in Figure 2-1. It is important to note that no attempt has been made to scale the components of this figure with waste and emission quantities. The outside box includes all wastes and emissions, human-produced or not, in all physical states. Inside that box is a subset of wastes that have been defined as criteria, hazardous, or toxic

Table 2–1. Sources of industrial waste, economic, and energy data

Document/Database Title and Contact Information
Industrial Waste Data Sources
Toxic Chemical Release Inventory (TRI);[a,b] available through National Library of Medicine, Bethesda, Maryland and RTK NET, Washington, DC
Generation and Management of Residual Materials: Petroleum Refining Performance (replaces *The Generation and Management of Wastes and Secondary Materials* series);[b] American Petroleum Institute, Washington, DC
Preventing Pollution in the Chemical Industry: Five Years of Progress (replaces the *CMA Hazardous Waste Survey series*);[b] Chemical Manufacturers Association, Washington, DC
National Biennial Report of Hazardous Waste Treatment, Storage, and Disposal Facilities Regulated Under RCRA;[c] US EPA Office of Solid Waste, Washington, DC
National Survey of Hazardous Waste Generators and Treatment, Storage, Disposal, and Recycling Facilities in 1986; available through NTIS as PB92-123025
Biennial Report System[a,c] (BRS); available through RTK NET, Washington, DC
National Air Pollutant Emission Estimates;[b] US EPA Office of Air Quality Planning and Standards, Research Triangle Park, NC
Aerometric Information Retrieval System[a,d] (AIRS); US EPA Office of Air Quality Planning and Standards, Research Triangle Park, NC
Permit Compliance System;[a,d] US EPA Office of Water Enforcement and Permits, Washington, DC
Toxic Release Inventory: Public Data Release (replaces *Toxics in the Community: National and Local Perspectives*);[b] EPCRA hotline (800) 535-0202
Report to Congress on Special Wastes from Mineral Processing; US EPA Office of Solid Waste, Washington, DC
Report to Congress: Management of Wastes from the Exploration, Development, and Production of Crude Oil, Natural Gas, and Geothermal Energy, Vol. 1, *Oil and Gas*; US EPA Office of Solid Waste, Washington, DC
Databases on Economic Aspects of Waste Generation
Manufacturers' Pollution Abatement Capital Expenditures and Operating Costs;[c] Department of Commerce, Bureau of the Census, Washington, DC
Minerals Yearbook, Vol. 1, *Metals and Minerals*;[b] Department of the Interior, Bureau of Mines, Washington, DC
Census Series: *Agriculture*, *Construction Industries*, *Manufacturers-Industry*, *Mineral Industries*;[e] Department of Commerce, Bureau of the Census, Washington, DC
Industrial Energy Consumption Database
Manufacturing Energy Consumption Survey: Consumption of Energy;[f] US DOE Office of Energy Markets and End Use, Washington, DC

[a]On-line database.
[b]Data are collected annually.
[c]Data are collected every 2 years.
[d]Data are updated frequently.
[e]Data are collected every 5 years.
[f]Data are collected every 3 years.
Source: US DOE (1991), with updates.

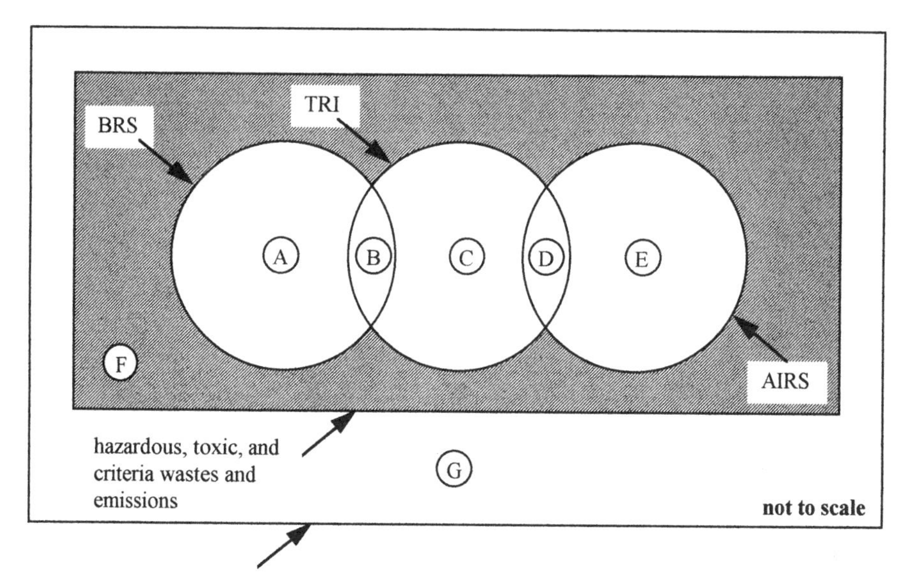

Figure 2-1. Existing waste and emission inventories overlap and are not comprehensive. A, wastes that are reported in the BRS only, e.g. the water in wastewaters; B, wastes reported in the TRI and BRS, e.g. benzene-contaminated wastewaters (TRI would only report benzene content); C, releases, transfers, and generation that fall under TRI reporting requirements only, e.g. air emissions of nickel that are not PM10; D, air emissions that fall under TRI and AIRS reporting requirements, e.g. VOCs on the TRI list; E, air emissions that fall under reporting requirements of AIRS only, e.g. mobile sources; F, hazardous wastes and toxic emissions that do not fall under the reporting requirements of BRS, TRI, or AIRS, e.g. many mining and oil exploration-related wastes; G, all other wastes and emissions, e.g. nonhazardous wastes.

wastes. A subset of the wastes in this box are reported under RCRA in the BRS as hazardous. Another subset is reported in the TRI, and another subset is reported under the Clean Air Act in AIRS. Some of the other data sources listed in Table 2-1 could be used to fill in waste and emissions areas not covered by the three databases given in the figure. Take a moment to review this figure and understand where the data sources overlap and where data may be unavailable.

One of the most frequently heard criticisms of available waste and emission data is that they are too inaccurate to be useful. Waste and emission quantities are often estimated by facility personnel. The quality of the data therefore varies from facility to facility, depending on the expertise of the personnel and the resources available to them.

The limitations of available waste data cannot be ignored when the infor-

that the inventories do not have an important place in pollution prevention efforts. On the contrary, the information they contain is vital to just and productive policymaking, and is certainly complete and accurate enough to base assessments on the overall level of concern that should be directed toward preventing pollution in our country. It must be remembered that the databases are all relatively young and are evolving rapidly to address concerns about their utility.

2.1.2 Waste Generation and Emission Rates

Given the shortcomings in the available data, it is clear that there are significant uncertainties in national estimates of industrial waste generation. Nevertheless, it is worthwhile to cautiously examine the information provided in existing waste and emission inventories. Because there is no single source of data on waste generation that is truly comprehensive, trying to develop a national picture of industrial waste generation is much like trying to develop a picture of an elephant by having a dozen blindfolded people touch different parts of the animal. Some waste databases touch the tail, others the trunk, only a few touch the main body of the elephant.

The best estimates available from national waste generation data indicate that more than 12 billion tons of industrial waste are generated in the United States annually. Only 750 million tons, or less than 10% of this waste mass, are defined as hazardous under RCRA. Based on EPA data from sources such as the Biennial Report System, the Toxic Chemical Release Inventory, and the National Survey of Hazardous Waste Generators, more than 20,000 generators of hazardous waste can be identified. The precise number of generators is subject to considerable uncertainty, due to definitions of hazardous waste and reporting limits that vary from inventory to inventory and from state to state. Despite the uncertainties, it is clear that there are tens of thousands of industrial hazardous waste generators in the United States producing 250–750 million tons of hazardous waste annually. The large uncertainty in the amount of waste generated is again due to differing definitions.

Thus, even for issues as basic as the number of generators and the total mass of waste generated, there is considerable disagreement among various inventories. The uncertainty is significant; estimates of these quantities vary by factors of at least 2 or 3. While these different pictures of waste generation can be confusing, there is general agreement among the data sources on some issues at the national level. One issue on which there is consensus is the physical state of the waste: the vast majority of the hazardous waste mass, as reported in virtually all of the data sources, is in the form of wastewaters.

Two other issues on which there appears to be consensus among the hazardous waste databases are the distribution of waste types and the distri-

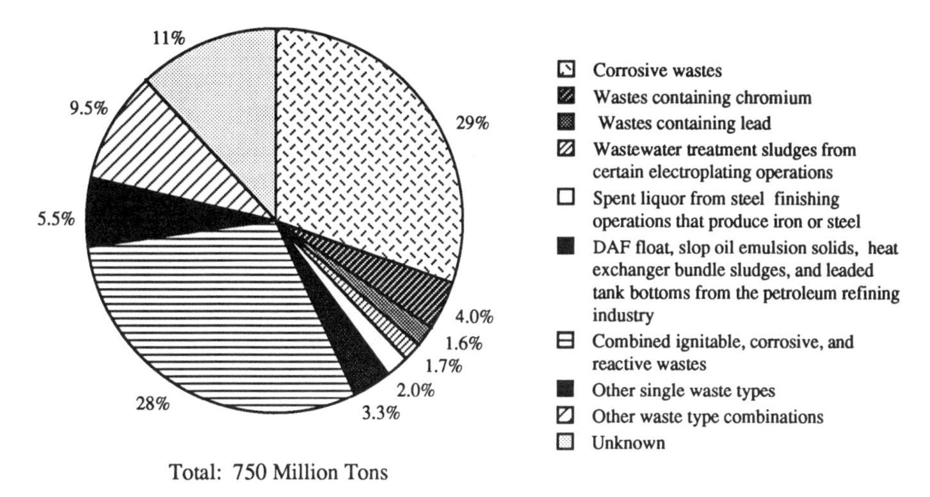

Figure 2-2. Quantity of hazardous waste generated in 1986 by RCRA waste code (Baker and Warren, 1992). (Reprinted with permission of Mary Ann Liebert, Inc., publishers.)

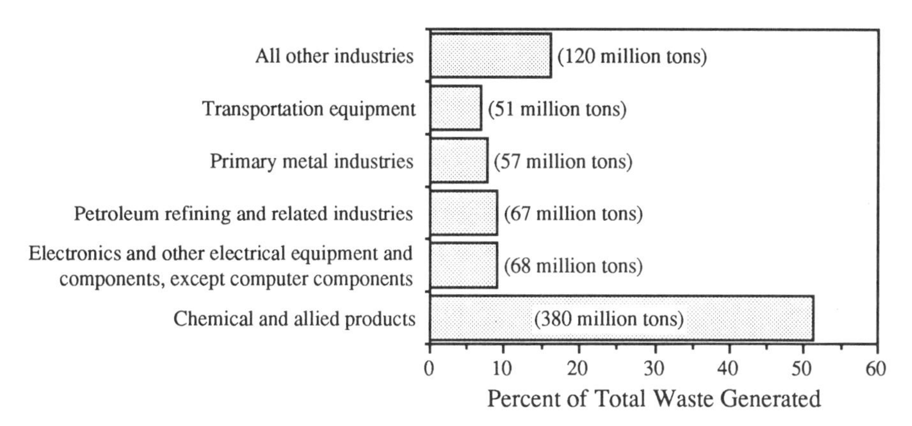

Figure 2-3. Largest general industries by the quantity of hazardous waste generated (Baker and Warren, 1992). (Reprinted with permission of Mary Ann Liebert, Inc., publishers.)

bution of waste generation among industrial sectors. Figures 2-2 and 2-3 report results from the National Hazardous Waste Survey on these two issues (Baker and Warren, 1992).

Wastes in the National Hazardous Waste Survey are described by characteristic(s) (e.g., ignitable, corrosive, or reactive), and if applicable by listed waste (specified wastes and waste containing specific materials). Waste generators reported single waste streams using as many as five different codes, and because these codes were entered in the database in alphabetic order rather than in an order that best described the waste type, results showing waste generation by type of waste must be viewed with caution. It is clear in Figure 2-2, however, that corrosive wastes (DOO2) are the most frequently generated category of waste.

Reporting the distribution of managed waste by industry sector involves fewer uncertainties than reporting the distribution by type of waste managed. As shown in Figure 2-3, the chemical manufacturing industry is the dominant generator of hazardous waste in the United States.

As the geographic area that is surveyed is reduced, conclusions regarding waste codes and waste generators can be questioned. For example, in three separate studies, the State of Illinois found very different distributions of industrial waste generation (Thomas and Miller, 1992). One study reported that chemical manufacturing produced the largest quantity of waste, while two others reported that blast furnaces and steelmills generated the largest quantity of waste. The studies employed different definitions and were conducted at different times, so the differences can perhaps be rationalized, but this example illustrates how different analyses of available hazardous waste surveys can produce very different results.

A summary of releases and transfers reported in the 1994 TRI, which totaled 6100 million pounds (i.e., 6100×10^6 lb), are given by Standard Industrial Classification code in Table 2-2. Of the total reported releases and transfers, 37% (2300 million lb/yr) were direct releases of listed chemicals to the environment. Note that the releases listed here are given on a dry-mass basis and are not waste stream flow rates. The great majority (92%) of releases and transfers reported in the TRI for this year are from facilities that identify themselves with only one major class of manufacturing activity, but there are also a few pairs of manufacturing classifications with significant releases. Manufacturers of chemicals and allied products are the single largest contributor to TRI-reported releases and transfers, at 1800 million lb/yr and 30% of the total. Primary metal manufacturing is the next highest industry group at 1400 million lb/yr and 24% of the total.

Direct releases to the environment are important because they are untreated and unsecured and have the most potential to cause adverse health and environmental effects. Air emissions in particular have been shown to be logical targets for toxics reduction in terms of risk reduction, cost, and technical feasibility, even though air emission quantities represent only a tiny fraction of the total waste stream flow. It must be remembered that most of the reported hazardous waste flows are in the form of wastewaters, and the mass of the water is reported along with the hazardous constituents of the wastewater. Thus, only a fraction of the 12 billion tons of waste reported under RCRA consists of hazardous constituents. In contrast, when atmospheric emissions are reported, only the mass of the contaminant is counted. In 1994, TRI-reported atmospheric releases of approximately 300 chemicals and chemical categories by large manufacturing operations totaled 780,000 tons (TRI94, 1996). Criteria air pollutant emissions from industrial processes are given in Figure 2-4.

Table 2-2. Sources of releases and transfers reported in the Toxic Chemical Release Inventory for 1994

Industry	Standard Industrial Classification Code	Total Releases and Offsite Transfers	
		Millions of Pounds	% of Total
Single manufacturing classifications			
Food and kindred products	20	62	1.0
Tobacco products	21	1.5	0.025
Textile mill products	22	26	0.44
Apparel, etc.[a]	23	1.8	0.030
Lumber and wood products, nec[b]	24	43	0.70
Furniture and fixtures	25	63	1.0
Paper and allied products	26	310	5.1
Printing, publishing, and allied industries	27	45	0.74
Chemicals and allied products	28	1800	30
Petroleum refining and related industries	29	200	3.3
Rubber and miscellaneous plastic products	30	170	2.7
Leather and leather products	31	8.7	0.14
Stone, clay, glass, and concrete products	32	35	0.58
Primary metal industries	33	1400	24
Fabricated metal products, nec[b]	34	410	6.9
Industrial and commercial machinery and computer equ't	35	93	1.5
Electronic and other electrical equ't and components, nec[b]	36	420	6.9
Transportation equipment	37	320	5.3
Measuring instruments, optical goods, watches[c]	38	41	0.67
Miscellaneous manufacturing industries	39	38	0.63
Multiple manufacturing classifications			
Primary metals/fabricated metals	33 and 34	79	1.3
Food products/chemical products	20 and 28	35	0.59
Chemical products/petroleum refining	28 and 29	26	0.43
Primary metals/electronic equ't	33 and 36	24	0.40
Chemical products/rubber, plastic products	28 and 30	21	0.35
Fabricated metals/electronic equ't	34 and 36	16	0.27
Fabricated metals/machinery	34 and 35	13	0.21
Fabricated metals/transportation equipment	34 and 37	12	0.20

Table 2-2. (*Continued*)

Industry	Standard Industrial Classification Code	Total Releases and Offsite Transfers	
		Millions of Pounds	% of total
Paper products/chemical products	26 and 28	11	0.19
Primary metals/machinery	33 and 35	11	0.18
Other manufacturing industry combinations	20 to 39	174	2.9
Manufacturing combined with nonmanufacturing	(20 to 39) and (00 to 19 or 40 to 99)	51	0.84
Nonmanufacturing	00 to 19 or 40 to 99	23	0.38
No data		4.7	0.078
Total		6100	100

[a]*Full description is*: Apparel and other finished products made from fabrics and other similar materials.
[b]nec: not elsewhere classified.
[c]*Full description is*: Measuring, analyzing, and controlling instruments; photographic, medical, and optical goods; watches and clocks.
Source: TRI94 (1996).

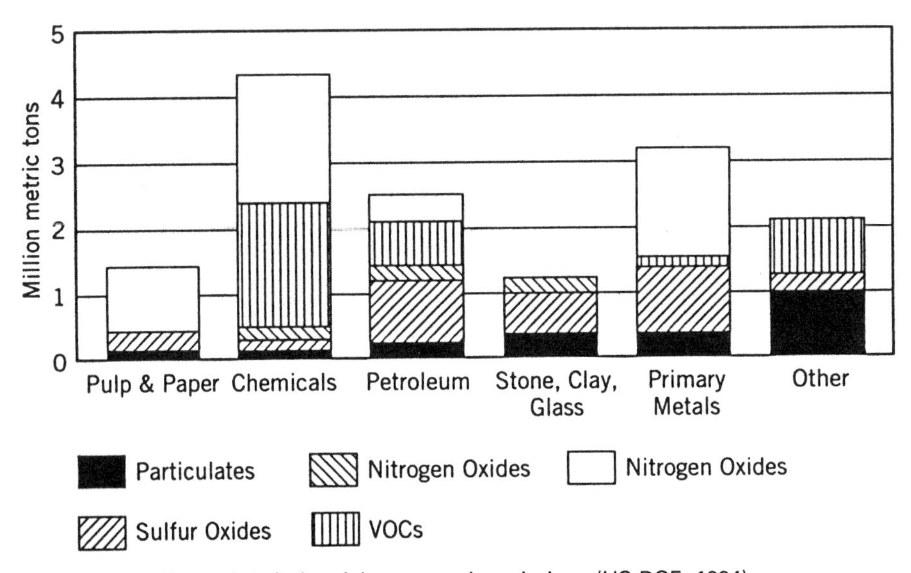

Figure 2-4. Industrial process air emissions (US DOE, 1994).

Table 2-3. Nonhazardous (RCRA Subtitle D) waste generation

Waste Category	Estimated Annual Generation Rate (million tons)
Industrial nonhazardous waste[a,b]	7600
Oil and gas waste[c,e]	
Drilling waste[d]	129–871
Produced waters[f]	1966–2738
Mining waste[c,g]	>1400
Municipal waste[b]	158
Household hazardous waste	0.002–0.56
Municipal waste combustion ash[h]	3.2–8.1
Utility waste[c,i]	
Ash	69
Flue-gas desulfurization waste	16
Construction and demolition waste[j]	31.5
Municipal sludge[b]	
Wastewater treatment	6.9
Water treatment	3.5
Very-small-quantity[k] generator hazardous waste ($<$ 100 kg/mo)[b,e]	0.2
Waste tires[g]	240 million tires
Infectious waste[e,l]	2.1
Agricultural waste	Unknown
Approximate total	>11,387

[a]Not including industrial waste that is recycled or disposed of off-site.
[b]These estimates are derived from 1986 data.
[c]See SAIC (1985).
[d]Converted to tons from barrels: 42 gal = 1 barrel, ~17 lb/gal.
[e]These estimates are derived from 1985 data.
[f]Converted to tons from barrels: 42 gals = 1 barrel, ~8 lb/gal.
[g]These estimates are derived from 1983 data.
[h]This estimate is derived from 1988 data.
[i]These estimates are derived from 1984 data.
[j]This estimate is derived from 1970 data.
[k]Small-quantity generators [100–1000 kg/month waste] have been regulated under RCRA, Subtitle C, since October 1986. Before then, approximately 830,000 tons of small-quantity generator hazardous wastes were disposed of in Subtitle D facilities every year.
[l]Includes only infectious *hospital* waste.

Source: US EPA (1988a).

The types of waste important in nonhazardous waste generation are summarized in Table 2-3. Extractive and commodity material processing industries are the major sources of nonhazardous wastes in the United States. Pulp and paper manufacturing is the single most important source of nonhazardous wastes from manufacturing, with about 2.5 billion tons of annual waste generation. Chemical products, primary metals, and stone, clay, and glass are next with approximately 1.5, 1.5, and 0.5 billion tons generated, respectively. As shown in Table 2-3, mining, petroleum exploration, and electric power generation are important nonmanufacturing sources of nonhazardous waste. Again, it is important to note that these waste flow rates are reported on a wet basis and include substantial quantities of water.

2.1.3 Waste Management

The previous section outlined the quantity of waste generated annually in the United States. In this section methods of waste management are briefly examined. For hazardous wastes, there are five primary methods for management:

1. Recycling materials off-site or to other on-site processes
2. Reuse as fuel
3. Incineration
4. Physical, chemical, and biological treatment of aqueous wastes
5. Land treatment and disposal.

Even though hazardous wastes constitute less than 10% of the total waste flow, the costs associated with treating and disposing of these wastes are a substantial fraction of national waste management expenditures. One of the reasons for the high costs associated with hazardous wastes is the level of treatment required. For example, the hazardous constituents in an organic waste stream must typically be 99%, 99.99%, or 99.9999% destroyed before the waste stream is considered successfully treated. Specified levels of destruction depend on the toxicity of the hazardous constituent present. Achieving such high levels of destruction is expensive and typically involves the use of sequential treatment technologies.

One of the most recent national data sources available on the relative magnitudes of wastes handled by each of these management methods is the 1986 National Hazardous Waste Survey (Baker et al., 1992). The categories of hazardous waste management technologies reported in the National Hazardous Waste Survey, along with the quantities of waste managed and the number of facilities, are listed in Table 2-4. As shown in Table 2-5, the chemical manufacturing industry is the dominant generator of hazardous waste sent to most types of waste management. One logical exception to this pattern is the primary metals industry, which is a disproportionately large

Table 2-4. Hazardous waste generation for each management technology

Management Method	Quantity Managed in 1986[a] (million tons)	Number of Facilities
Metal recovery	1.4	330
Solvent recovery	1.2	1500
Other recycling	0.96	240
Fuel blending	0.75	180
Reuse as fuel	1.4	300
Incineration	1.1	200
Solidification	0.77	120
Land treatment	0.38	58
Wastewater treatment	730	4400
Disposal impoundment[b]	4.6	70
Surface impoundment[c]	230	300
Landfill	3.2	120
Waste pile	0.68	71
Underground injection	29	63
Storage (RCRA permitted)	190	1800
Other treatment	2.0	130

[a]Quantities reported were obtained using the TSDR section of the National Hazardous Waste Survey. Total waste generated in 1986 was approximately 750 million tons; note that some wastes were managed in multiple treatment technologies and that wastes can be sent to and removed from storage.

[b]Surface impoundments used for disposing of hazardous waste.

[c]Includes waste entering surface impoundments for disposal, treatment and storage.

Source: Baker et al. (1992). Reprinted with permission of Mary Ann Liebert, Inc., publishers.

user of metal recovery operations. The geographic distribution of hazardous waste management, shown in Figure 2-5, closely resembles the geographic distribution of waste generation, because 96% of hazardous waste is managed on-site. Incineration, reuse as fuel, land treatment, and underground injection are management technologies that occur mostly in EPA Region VI (Texas, Oklahoma, Louisiana, and Arkansas). Regions VII (Nebraska, Kansas, Iowa, and Missouri) and VIII (Montana, Wyoming, Utah, Colorado, and the Dakotas) are notable for their lack of hazardous waste management activity.

A flow diagram providing an overview of hazardous waste management practices is given in Figure 2-6. The mass of waste managed in each of 13 different types of technologies (storage is not included here, and surface impoundment and disposal impoundment are integrated as one method) is given in this figure and the management patterns and approximate amounts of industrial waste streams regulated under RCRA that are processed through various treatment and disposal routes are shown. The flow rates in Figure 2-6 are totals for approximately 40,000 industrial waste streams generated

Table 2-5. Hazardous waste management technology usage by industrial sector

Management Method	Major Contributor	% Contribution by Major Contributor
Metal recovery	Primary metal	34
Solvent recovery	Chemical manufacturing	21
Other recovery	Nonelectrical machinery	38
Fuel blending	Chemical manufacturing	46
Reuse as fuel	Chemical manufacturing	53
Incineration	Chemical manufacturing	83
Solidification	Chemical manufacturing	29
Land treatment	Petroleum industry	83
Wastewater treatment	Chemical manufacturing	46
Disposal Impoundment	Chemical manufacturing	79
Surface impoundment	Chemical manufacturing	86
Landfill	Metal fabrication	20
Waste pile	Transportation equipment	30
Underground injection	Chemical manufacturing	60

Source: Baker et al. (1992). Reprinted with permission of Mary Ann Liebert, Inc., publishers.

from all U.S. industry regulated under RCRA in 1986. This figure provides useful information about the relative amounts of the wastes managed by different techniques. For example, it shows that by mass, only a small fraction of the waste streams flow through recycling loops. The total mass involved in solvent, metal, and other recycling is about 5 million tons per year. The largest single stream in terms of total mass flow, approximately 700 million tons per year (more than 90% of the total waste mass flow), is hazardous wastewater. However, because most of this stream is water, the mass of the chemically hazardous component of this stream is on the order of the components being recycled. A third set of waste streams, about 4 million tons per year, is sent to various thermal treatment technologies that include direct incineration, fuel blending, and reuse as fuel. Although incineration destroys less than 1% of the hazardous waste mass currently generated, these incinerated wastes generally contain moderate to high concentrations of regulated substances. Land treatment and disposal operations received 30 million tons of waste.

From Figure 2-6 it is clear that waste management frequently involves multiple technologies. Figure 2-6 also reveals the interdependencies of many waste management technologies. For example, roughly 1 million tons of waste were incinerated in 1986. Scrubbers in the air pollution control equipment of these incinerators generated 40 million tons of wastewaters that are legally defined as hazardous and that must be treated as hazardous wastewaters. That is why the waste flow diagram shows one million tons of waste entering and 40 million tons leaving the incineration step. Other interdepen-

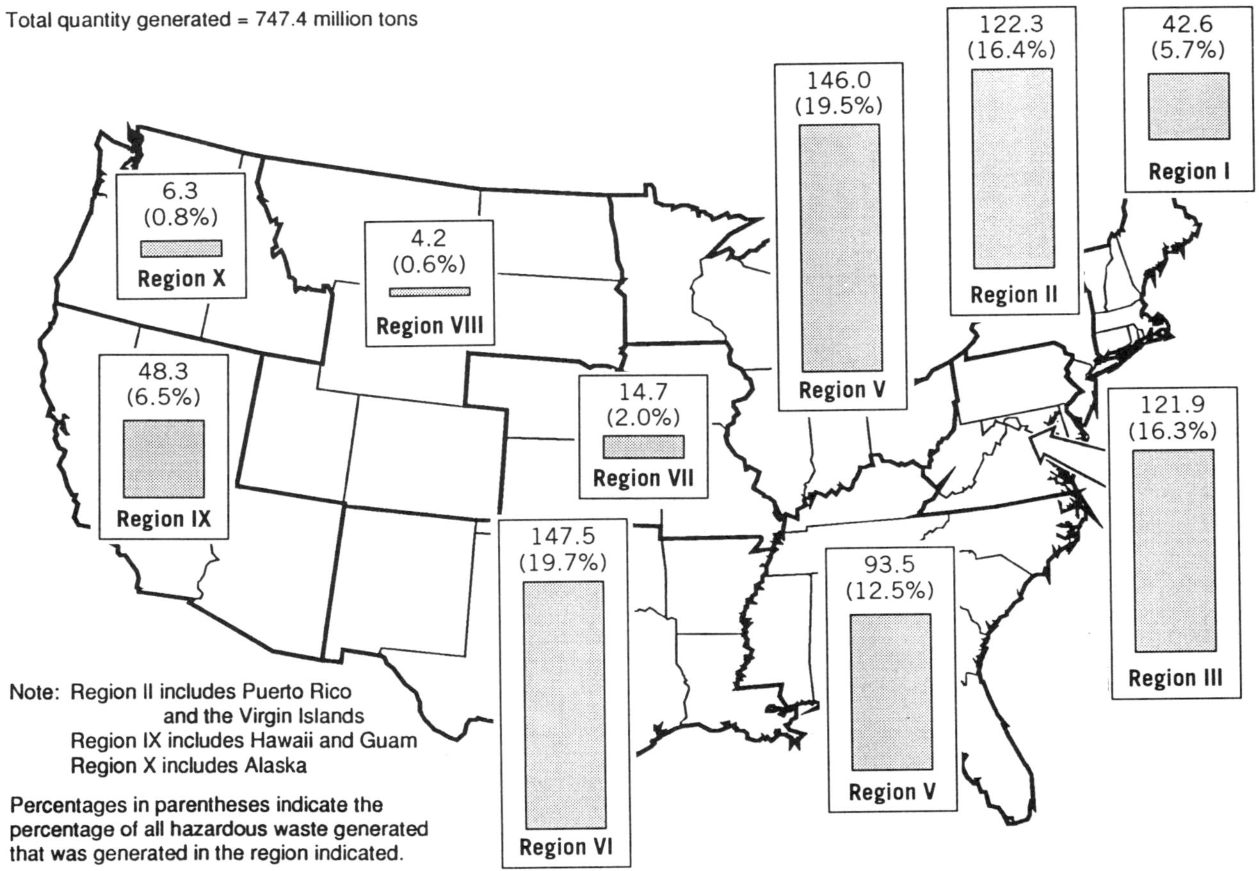

Total quantity generated = 747.4 million tons

6.3 (0.8%) Region X

4.2 (0.6%) Region VIII

48.3 (6.5%) Region IX

14.7 (2.0%) Region VII

147.5 (19.7%) Region VI

146.0 (19.5%) Region V

93.5 (12.5%) Region V

122.3 (16.4%) Region II

42.6 (5.7%) Region I

121.9 (16.3%) Region III

Note: Region II includes Puerto Rico and the Virgin Islands
Region IX includes Hawaii and Guam
Region X includes Alaska

Percentages in parentheses indicate the percentage of all hazardous waste generated that was generated in the region indicated.

Figure 2-5. The distribution by EPA region of hazardous waste management in 1986 in millions of tons (Baker and Warren, 1992). (Reprinted with permission of Mary Ann Liebert, Inc., publishers.)

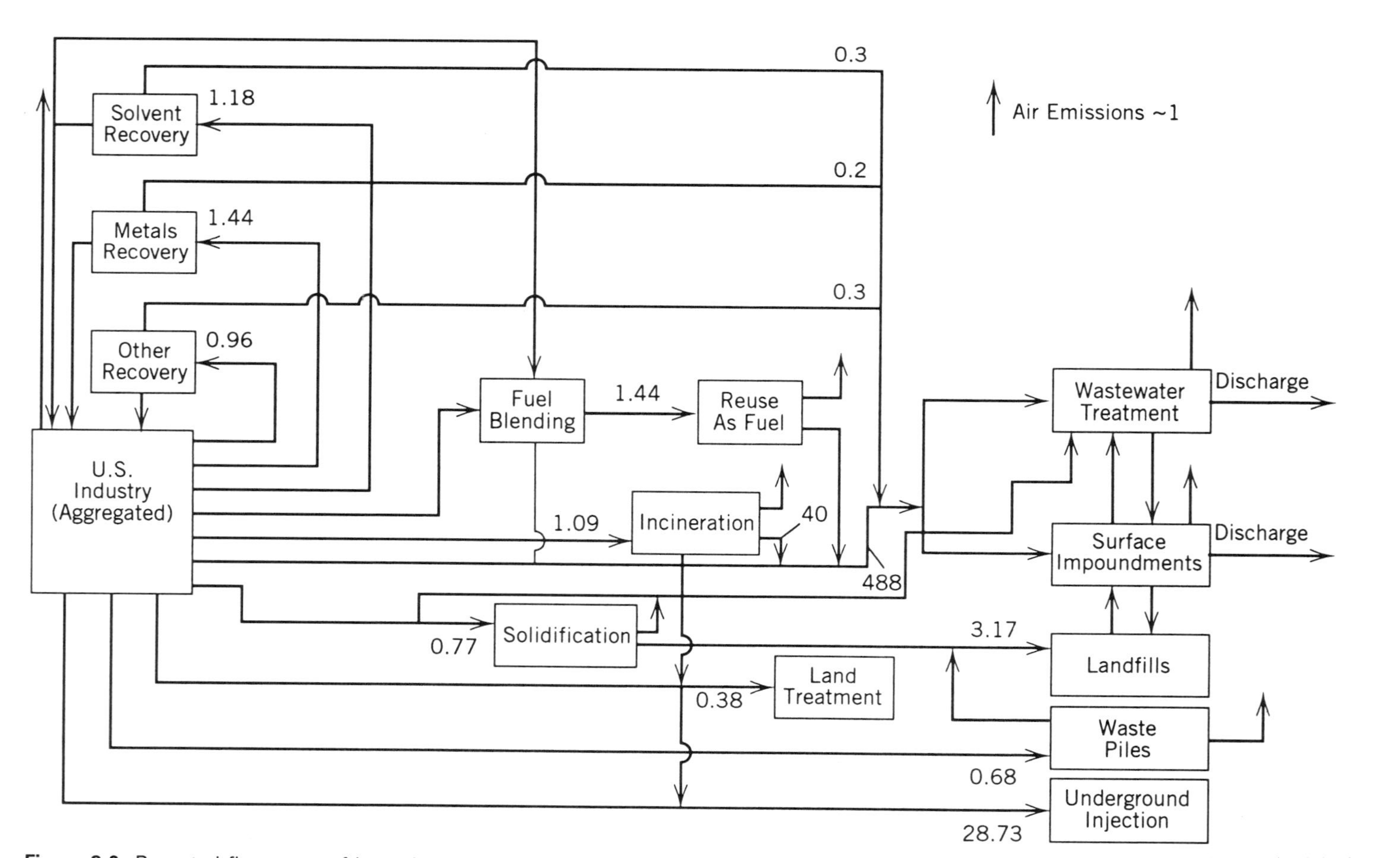

Figure 2-6. Reported flow rates of hazardous waste to treatment technologies: 1986 data in millions of tons per year (Baker et al, 1992; US EPA, 1991). (Reprinted with permission of Mary Ann Liebert, Inc., publishers.)

dent management technologies include wastewater treatment and land treatment, because biological sludges from hazardous wastewater treatment plants are sent to solidification and hazardous waste landfills. Also, material from waste piles and surface impoundments are sometimes sent to landfills and solvent and metal recovery operations generate hazardous wastewaters.

The National Hazardous Waste Survey can be used to determine patterns of usage for the hazardous waste management technologies listed in Table

Table 2-6. Estimated number of nonhazardous waste management units

	Unit Type				
Waste Category	Landfills[a]	Surface[b] Impoundments	Land[c] Application Units	Waste Piles	Total
Municipal solid waste	6,584[e]	—[f]	—[f]	—[f]	6,584
Industrial waste	2,757[g]	15,253[g]	4,308[g]	5,355g	27,654
Municipal sewage sludge	—[h]	1,938	11,937	—[f]	13,875
Oil and gas waste[i]	—[h]	125,074	726	—[f]	125,800
Agricultural waste	—[h]	17,159	—[f]	—[f]	17,159
Mining waste[i]	—[h]	19,813	—[f]	—[h]	19,813
Municipal runoff	—[f]	488	—[f]	—[f]	488
Construction and demolition debris	2,591	—[f]	—[f]	—[h]	2,591
Miscellaneous waste	1,030	11,118	621	—[h]	12,769
Appropriate total number of units	12,962	190,843	17,592	5,335	226,732

[a]A *landfill* is a facility where waste is placed in or on land for permanent disposal.
[b]A *surface impoundment* is a natural topographic depression, human-made excavation, or diked area formed primarily of earthen materials (although it may be lined with human-made materials) that is designed to hold an accumulation of liquid wastes or wastes containing free liquids for treatment, storage, or disposal. Also called *pits*, *ponds*, or *lagoons*. Does not include tanks.
[c]A *land application unit* is an area where waste is applied onto or incorporated into the soil surface for the purpose of beneficial use or waste treatment and disposal. Also called *land treatment*, *landfarming*, or *landspreading*. Manure spreading operations are excluded.
[d]A *waste pile* is a mass of solid nonflowing waste. Functions as treatment or storage.
[e]The results of a previous census of the States indicated 9300 municipal solid waste landfills. However, the table entry is considered more accurate. It is based on a 1986 survey.
[f]Unknown, none or few thought to exist.
[g]These estimates differ from previously published results from a census of the states. Table entries are considered to be more accurate. They are based on a 1986 industrial survey.
[h]Unknown, some may exist.
[i]See SAIC (1985).
Source: US EPA (1988a).

2-4. The types of wastes that were managed by each waste management technology, the industry sectors that generated the wastes, and the locations of waste management facilities can be determined, as shown in the figures of Appendix B (also see Baker et al., 1992). However, because multiple treatment technologies are frequently used in waste management, examination of usage patterns for individual technologies must be done with care and the data must not be overinterpreted.

As mentioned previously, there are few sources of data on nonhazardous waste. However, it is known that in contrast to hazardous waste, nonhazardous waste is almost entirely disposed of in surface impoundments, landfills, and land application units. Data on the number of units handling RCRA nonhazardous waste are summarized in Table 2-6.

2.1.4 Trends in Waste Generation and Management

Any evaluation of trends in waste generation or waste management must rely on data collected on a regular basis, based on a consistent set of contributors. Such data are exceedingly scarce. For hazardous wastes, the primary sources of information are the Toxic Chemical Release Inventory and the Biennial Report System. These data sources present evidence for stable to slightly declining rates of waste generation during the most recent survey period. A more complete understanding of trends in waste generation will take several more years to develop.

2.2 WASTE INVENTORIES AND POLLUTION PREVENTION

Waste generation and management data are necessary for determining the health effects of pollution on populations, for formulating effective environmental policies, and for modeling global, regional, and local distributions of pollutants. It is important to know what the available data suggest and how the data can be interpreted. In this section, interpretation of waste generation and management data is focussed on the prioritizing of wastes and on the measurement of pollution prevention.

2.2.1 Prioritizing Wastes

While lists of releases like the one in Table 2-2 are useful in identifying important industries to target for toxics reduction, it is important to remember that release quantities represent only a fragment, although an important fragment, of the information necessary to measure the environmental effects of toxic chemicals. Most importantly, Table 2-2 gives the sum of releases for hundreds of different chemicals, some of which are potentially harmful to the environment and to humans and some of which are less detrimental. Therefore, an industry with small releases may be a more important target for reduction if its releases are made up of compounds more harmful than those of another industry with larger releases. However, large release numbers tend to generate the most public alarm. Also, the environmental compartment (air, water, soil, etc.) to which the chemicals are released is important, as is the environmental fate of the chemical. Once released, compounds can transform into more dangerous chemicals or degrade into harmless ones and can be transported from one environmental compartment to another (e.g., air pollutants washed out of the atmosphere by rain enter the soil or surface waters). Confusion also results from combined reporting of direct environmental releases and off-site transfers to publicly owned (wastewater) treatment works (POTWs), and recycling, treatment, storage, or disposal facilities. Lumping together the amount of a compound sent to a treatment facility with the amount released directly to the air or to a waterway for prioritization purposes is not rational.

Although wastewaters dominate waste generation on a mass basis, they are not generally thought of as the highest-priority waste. This is because they are made up of mostly water and are treated for destruction of the hazardous compounds they contain. Air emissions tend to be targeted for waste reduction when risk reduction, cost, and technical feasibility are the criteria for choosing waste reduction projects.

2.2.2 Measuring Pollution Prevention: The Toxic Chemical Release Inventory as an Environmental Report Card

As statutes and public interest increasingly focus on waste generation and waste reduction there is a natural tendency to turn to a single source of waste and emission data. It is becoming increasingly clear that the source of choice for these data will be the Toxic Chemical Release Inventory (TRI). This is because the TRI is publicly available, facilities are specifically identified, and releases and transfers are reported on a per compound basis. However, as a tool for measurement of environmental stewardship, the TRI has a number of flaws. An important one related to measuring pollution prevention is that there is only very limited data on production in the TRI. Production data are important because emissions should be normalized by output. For example, if an ethylene manufacturing facility generates emissions equal to a similar facility, but has only half the production volume, then the facility is not as efficiently managing its emissions as its neighbor. As of reporting year 1991, facilities must provide either a production ratio or an activity index to account for changes in waste quantities that are attributable to changes in production at their facility. Unfortunately these indices cannot be used to compare one facility to another. Another challenge in using the TRI to measure pollution prevention is that it is continually changing. Figure 2-7 shows how the information contained in the TRI was expanded beginning with reporting year 1991. There is continuing pressure to include even more data.

Despite its flaws, the TRI has become the primary data source used to

√ reported in lbs/yr
? not reported

A. Toxic Chemical Release Inventory (EPCRA, 1986)

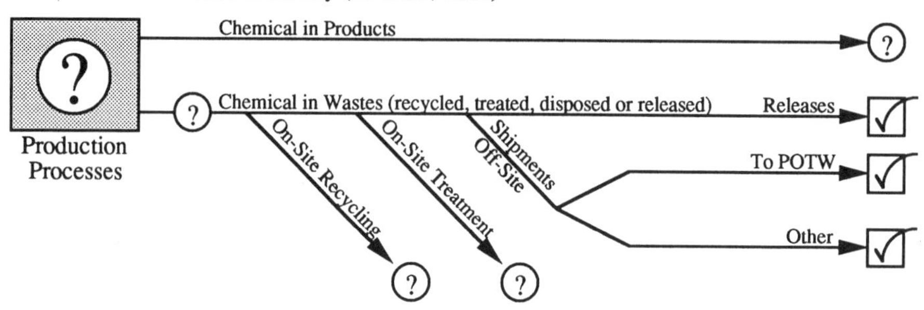

B. Pollution Prevention Act (1990)

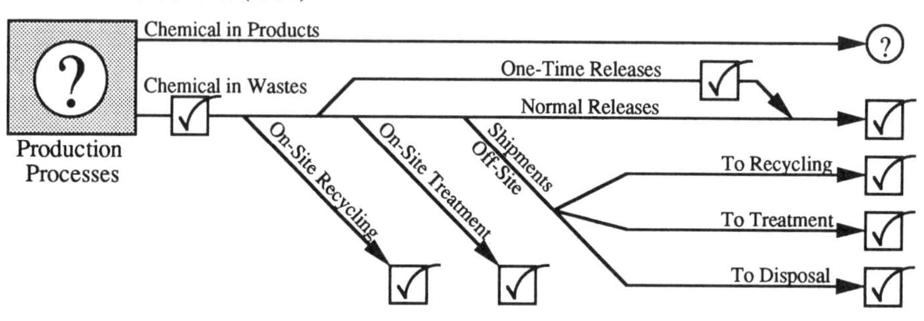

C. Community Right-to-Know-More Act (proposed)

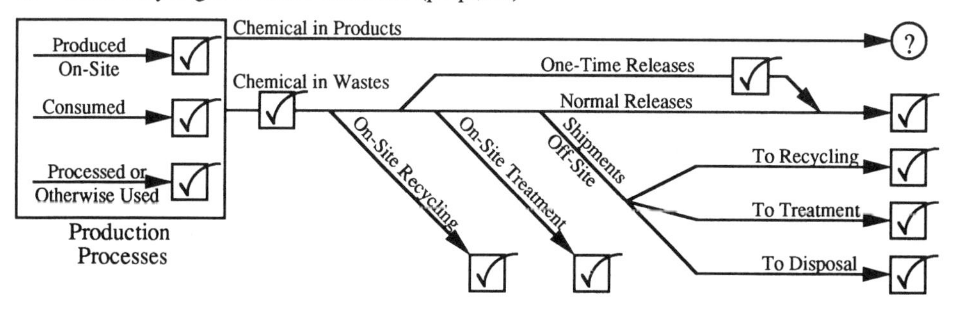

Figure 2-7. Information reported in the TRI: (A) in its original version, (B) as modified by the 1990 Pollution Prevention Act, and (C) including expansions proposed by the "community right-to-know-more movement" (Working Notes, 1991).

evaluate environmental performance by industrial facilities. A representative cross section of the many reports available based on TRI data is listed in Table 2-7. Some of the reports are produced by public-interest groups, while others are created by regulatory agencies. Their geographic focus can be local, regional, or national.

When reported TRI releases decrease for a facility there is a tendency to immediately credit pollution prevention efforts. However, a number of factors other than pollution prevention can lead to an increase or decrease in the reported emissions from a facility. Changes in production levels and alternative emission estimation methods are among these factors. Lumping total releases from a facility can result in an even more complicated source of confusion in a case where large quantities of a relatively benign substance are being used as a substitute for small quantities of a potently harmful substance. Despite such shortcomings and difficulties, attempts are being made to measure pollution prevention, primarily using the TRI. This section briefly describes the five most common metrics by which pollution prevention can be measured:

1. Absolute measures
2. Measures indexed to a production (output)-based quantity
3. Measures indexed to a production (input)-based quantity
4. Measures indexed to throughput
5. Measures indexed to production activity (e.g., the number of batches run in a chemical manufacturing operation)

All except the first of these metrics are based on a ratio of emissions to some measure of business activity.

2.2.2.1 *Absolute Measures* Comparison of the mass of wastes and emissions generated in one year to that generated in another year is an absolute measure of pollution prevention. Such measures of pollution prevention are consistent and easy to understand (a pound of a given waste or emission generated in one year can be compared directly to a pound of the same waste or emission generated in a subsequent year or at a different facility). The major disadvantage of such measures is that they do not reveal direct information about the level of business activity associated with the waste generation. Thus, a 50% reduction in wastes and emissions might be associated with a 50% reduction in production or might indicate a major effort at pollution prevention. The obvious solution to this problem is to index waste and emissions to some measure of production.

A number of alternatives are available for indexing waste and emission generation to production. They can be roughly grouped into output-based indices, input-based indices, and throughput-based indices. These measures are discussed in the following sections.

Table 2-7. A sampling of publicly issued reports based on the Toxic Chemical Release Inventory (TRI)

Organization	Report
California Public Interest Research Group	*Industrial Toxic Pollution in California; An Industry by Industry Analysis of Chemical Releases and Opportunities for Toxics Use Reduction*, July 1989 (1987 data), 48 pp. This report presents California's toxic-release data and outlines a state toxics use reduction program. *The Good, The Bad and The Toxic*, March 1990 (1987/88 data), 50 pp. The study investigates year-to-year changes in TRI releases reported by California's largest emitters.
Citizens for a Better Environment	*Richmond (CA) at Risk: Community Demographics and Toxic Hazards from Industrial Polluters*, February 1989 (1987 data), 50 pp. This study analyzes toxics-release reports together with demographic data to reveal the disporportionate toxic burden borne by low-income, minority citizens.
West Virginia Citizen Action Group	*Toxics in Our Midst: An Examination of Toxic Chemical Releases in West Virginia*, July 1989 (1987 data), 34 pp. This report represented the first extensive analysis of TRI data in West Virginia.
Greenpeace USA	*Mortality and Toxics Along the Mississippi River*, September 1988 (1987 data), 112 pp. This Greenpeace report (prepared in conjunction with Public Data Access, Inc.) statistically explores elevated toxics levels and excess mortality along the Mississippi River.
Greenpeace USA Great Lakes	*We All Live Downstream: The Mississippi River and the National Toxics Crisis*, December 1989 (1987 data), 186 pp. This extensive study of toxics in the Mississippi River uses TRI to supplement other data sources.
Natural Resources Defense Council	*A Who's Who of American Toxic Air Polluters: A Guide to More than* 1500 *Factories in* 46 *States Emitting Cancer-Causing Chemicals*, June 1989 (1987 data), 100 pp. NRDC identifies the largest sources of 11 unregulated cancer-causing chemicals released to air. The verified information includes useful chemical-by-chemical and state-by-state listings.

Source: Taken from Working Notes (1991).

2.2.2.2 *Measures Indexed to Output* Wastes or emissions can be indexed to the mass of products, number of products, or dollar value of products. All of these are relatively straightforward, and this is far from an all-inclusive list. As an example of the complexity of the issues surrounding output-based measures of pollution prevention, consider an electronics facility. If the facility decreases its emissions by 50% while producing 50% less memory storage assemblies, it could be said that the emissions per unit of production had remained constant. But what if each memory assembly has twice the storage capacity as previously? Should it be reported that emissions per unit of memory storage produced has been cut in half? These questions remain unresolved. Some of the advantages and disadvantages of output-based measures of pollution prevention are given in Table 2-8.

2.2.2.3 *Measures Indexed to Input* Wastes or emissions can be indexed to mass of raw materials, dollar value of raw materials, or number of employees. These indices are not in common use, but may occasionally have some advantages over other measures. The advantages and disadvantages are summarized in Table 2-9.

2.2.2.4 *Measures Indexed to Throughput* When a material is used exclusively as an intermediate, and is not a raw material or product, then the previous indices may not be appropriate. To resolve this and other problems with measurement indices, the Natural Resources Defense Council (NRDC) has proposed throughput-based indices. The NRDC defines throughput as the sum of the amount of chemical consumed on site, the amount of chemical shipped off site and the amount of the chemical released to all media of the environment. There are instances in which a production ratio for the final product would measure pollution prevention in the same way that a throughput index would. For example, if a process is changed so that the intermediate is created in the same quantities to produce the same amount of product, but so that the intermediate is consumed faster and thus has fewer opportunities to be released, both indices would provide a good measure of pollution prevention. A throughput index would fail to recognize a successful pollution prevention project in which the process was modified in order to create a greater quantity of product from the same amount of intermediate, because the index would decrease at the same rate as the releases of the intermediate.

2.2.2.5 *Measures Indexed to Activity* An activity index provides information about the number of times a specific waste-producing activity occurs. In some cases, an activity index provides a more accurate measure of pollution prevention than a production ratio. In the "Toxic Chemical Release Inventory Reporting Form R and Instructions," one example given for an activity index is for a facility that produces dyes in a batch processing mode. In between different colors of dye, the equipment is rinsed out with

Table 2-8. Advantages and disadvantages of output-based measures for normalizing the absolute amount of hazardous waste generated by production processes

Metric	Advantages	Disadvantages
Number of units of product	Widely applicable to different manufacturing processes and facilities Readily available information	May not be highly correlated with hazardous waste generation when a production process or facility produces several heterogeneous products (e.g., products that are substantially different in weight or size) May be inconsistent over time because changes in product weight and/or size are ignored May be confidential information for some production processes and/or facilities May not be very useful for non-manufacturing businesses Does not incorporate changes in output quality or product mix Difficult to aggregate from process level to higher levels if product is different for each process

Mass of product	Most applicable to primary and secondary producers Easy to aggregate from the process level to higher levels Directly comparable to the weight of hazardous waste generated Provides a consistent measure of production over time	May not be very useful for nonmanufacturing businesses May not be highly correlated with hazardous waste generation for producers using hazardous materials for cleaning their products May be confidential information for some processes and/or facilities Does not incorporate changes in output quality or product mix
Value of product	Widely applicable to different processes and facilities Easy to aggregate from the process level to higher levels U.S. Bureau of the Census collects information annually from facilities on the value or product shipments and the value added by production (although these data are not available to states) Changes in constant-dollar output prices may reflect changes in output quality	May be confidential information for some processes and/or facilities The effect of inflation and other changes in product prices having no connection with changes in production must be eliminated National price indices for deflating output prices may not accurately reflect each facility's particular output mix Prices may not be available for some intermediate products (i.e., products that are inputs to other production processes at a facility)

Source: RTI (1989).

Table 2-9. Advantages and disadvantages of input-based measures for normalizing the absolute amount of hazardous waste generated by production processes

Metric	Advantages	Disadvantages
Mass of raw materials	Readily available information from many facilities Most applicable to primary and secondary producers Not likely to be confidential information Easy to aggregate from the process level to higher levels Directly comparable to the weight of hazardous waste generated	May not be highly correlated with hazardous waste generation for producers using hazardous materials for cleaning their products May not be highly correlated with hazardous waste generation for production if hazardous materials in the raw materials changes over time May not be inconsistent over time if the proportion of a facility's production processes that generate hazardous waste changes significantly over time
$ value of raw materials	Readily available information from many facilities Most applicable to primary and secondary producers Not likely to be confidential information Easy to aggregate from the process level to higher levels	The effects of inflation and other changes in the price of raw materials having no connection with changes in production or value must be eliminated National indices of producer prices for deflating raw-material prices may not accurately reflect each facility's particular mix of raw materials May not be highly correlated with hazardous waste generation for producers using hazardous materials for cleaning their products

		May not be highly correlated with hazardous waste generation for production processes where the proportion of hazardous materials in the raw materials changes over time May be inconsistent over time if the proportion of a facility's production processes that generate hazardous waste changes significantly over time
Total number of employees	Readily available information from many facilities U.S. Bureau of the Census collects this information annually from facilities Widely applicable to different processes and facilities Not likely to be confidential information (although may be for individual processes) Easy to aggregate from the process level to higher levels	May not be highly correlated with hazardous waste generation unless part-time employees are converted to full-time equivalent employees (e.g., converting two half-time employees to one full-time equivalent employee) May be inconsistent over time, because technological changes will increase the productivity of employees over time May not be highly correlated with hazardous waste generation over time if the proportion of nonproduction workers changes significantly over time May be inconsistent over time if the proportion of a facility's production processes that generate hazardous waste changes significantly over time

Source: RTI (1989).

a solvent that is not used for anything else at the facility. In this case, the quantity of dyes produced by the facility bears no relation to the releases of the solvent. The Form R instructions suggest an activity index based on the number of times the equipment is rinsed out, so that the effectiveness of techniques for reducing solvent emissions for each rinse could be measured. It could be argued that an even better activity index for this facility might be the number of colors produced in a year, because one pollution prevention technique the facility might use would be to institute market forecasting so that each color is produced all at once each year. If the number of colors produced drops, solvent releases are expected to drop; and if solvent releases drop with the same number of colors produced, this would be due to pollution prevention efforts.

2.3 SUMMARY

The primary goal of this chapter is to instill an understanding of available waste generation and management data. To this end, the major national inventories of wastes and emissions, most of which cover hazardous wastes, were reviewed. Using a variety of data sources, a picture of industrial waste generation and management patterns was provided. The main features of this picture are

1. There are tens of thousands of hazardous waste generators in the United States producing hundreds of millions of tons of hazardous waste annually.
2. The most frequently generated hazardous wastes are those labeled corrosive.
3. The chemical manufacturing industry generates more hazardous waste than any other industry.
4. Most hazardous wastes generated are in the form of wastewaters.
5. Management of hazardous wastes is divided roughly evenly on a dry basis between wastewater treatment, land treatment and disposal, recycling operations, and incineration plus reuse as fuel.
6. Nonhazardous waste, most of which is land-managed, is generated at roughly 10 times the rate of hazardous waste.

While waste and emission data, in their present form, are clearly not completely adequate for measuring pollution prevention, they are currently being used by various advocacy and regulatory groups to grade pollution prevention performance. It is essential that process designers be aware of these data, their limitations, and what the issues are in the continuing policy debate on measuring pollution prevention.

QUESTIONS FOR DISCUSSION

1. What data sources in Table 2-1 might contain information on wastes and emissions from petroleum refineries? From petroleum extraction?
2. Give examples other than those provided of wastes that would fall into categories A to G of Figure 2-1.
3. Use Figure 2-6 to determine the amount of hazardous waste generated in the EPA region you live in and compare that to the amount generated in the regions bordering the region you live in.
4. Use Table 2-6 to list the most common type of unit for managing nonhazardous waste for each of the waste categories. What is the most common type of unit overall?
5. What are the advantages and disadvantages of using throughput-based indices to normalize waste generation data? What are the advantages and disadvantages of using activity-based indices?
6. If one waste chemical at a TRI facility is generated at the same rate regardless of production or activity and another waste chemical is generated at a rate that depends on production or activity, should the facility report the same production ratio or activity index for both of these chemicals in a year where the facility had decreased production or activity?
7. Boeing measures its environmental performance by normalizing TRI emissions by the number of production employees. Does this index make sense for Boeing? What are its advantages and disadvantages?

PROBLEM STATEMENTS

1. Draw a diagram similar to Figure 2-1 showing which inventories have data on pollutant releases to different environmental media (land, air, and water).
2. Compare the relative contribution of different manufacturing sectors towards criteria pollutant emissions, toxic chemical releases and transfers, and hazardous waste generation using Figures 2-3 and 2-4 and Table 2-2. Do any sectors dominate all the inventories?
3. Calculate the hazardous waste managed per facility for each type of management method using Table 2-4.
4. Use Tables 2-4 and 2-5 to calculate the amount of waste managed in metal recovery by the primary-metals industry. Also calculate the amounts sent to incineration, wastewater treatment, disposal and surface impoundments, and underground injection in chemical manufacturing. How do these quantities compare?

OPEN-ENDED PROBLEMS

1. Use the TRI to identify large sources of emissions in your county.
2. Obtain TRI data for your county or, if little or no reporting is required in your county, a nearby county. Narrow the entries you examine by media, industry, or chemical until you have approximately 10 chemical/facility combinations to analyze, and write a report about the data you found. Discuss the limitations in the data and how these limitations affect the conclusions you were able to draw.
3. Read a publicly issued report that was based on TRI data and provide a critique of the accuracy of the statements made by the authors of the report. Did they recognize the limitations of the data they presented?

REFERENCES

Baker, R. D. and J. L. Warren, "Generation of Hazardous Waste in the United States," *Hazardous Waste & Hazardous Materials*, **9**(1), 19–35, Winter 1992.

Baker, R. D., J. L. Warren, N. Behmanesh, and D. T. Allen, "Management of Hazardous Waste in the United States," *Hazardous Waste & Hazardous Materials*, **9**(1), 37–59, Winter 1992.

Natural Resources Defense Council (NRDC), "Testimony of Jane L. Bloom Before the Transportation, Tourism and Hazardous Material Subcommittee of the House Committee on Energy and Commerce," New York, 1988.

Research Triangle Institute (RTI), "Alternatives for Measuring Hazardous Waste Reduction," (prepared for the Illinois Hazardous Waste Research and Information Center), available through NTIS as PB91-208595, April 1989.

Science Applications International Corporation (SAIC), "Summary of Data on Industrial Nonhazardous Waste Disposal Practices," Contract 68-01-7050, U.S. Environmental Protection Agency, Washington, DC, 1985.

Seinfeld, J. H., *Atmospheric Chemistry and Physics of Air Pollution*, Wiley-Interscience, New York, 1986.

Thomas, D. L., and G. D. Miller, "Using Existing Hazardous Waste Databases: Limitations and Future Needs," *Hazardous Waste & Hazardous Materials*, **9**(1), 97–111, Winter 1992.

TRI91, *Toxic Chemical Release Inventory for 1991* (database), Bethesda, MD, National Library of Medicine, Aug. 1993a.

TRI91, *Toxic Chemical Release Inventory for 1991* (database), Bethesda, MD, National Library of Medicine, Sep. 1993b.

TRI94, *Toxic Chemical Release Inventory for 1994*, Bethesda, MD, National Library of Medicine, Jul. 1996.

U.S. Department of Energy (US DOE), "Characterization of Major Waste Data Sources," DOE/CE-40762T-H2, 1991.

U.S. Department of Energy, (US DOE), "Waste Generation in Industry" (draft), Sept. 1994.

U.S. Environmental Protection Agency (US EPA), "Report to Congress: Solid Waste Disposal in the United States, Volume 1," EPA/530-SW-88-011, 1988a.

U.S. Environmental Protection Agency (US EPA), "Report to Congress: Solid Waste Disposal in the United States, Volume 2," EPA/530-SW-88-011B, 1988b.

U.S. Environmental Protection Agency (US EPA), "1986 National Survey of Hazardous Waste Treatment, Storage, Disposal and Recycling Facilities," EPA/530-SW-88-035, 1988c.

U.S. Environmental Protection Agency (US EPA), "National Survey of Hazardous Waste Generators and Treatment, Storage, Disposal and Recycling Facilities in 1986: Hazardous Waste Management in RCRA TSDR Units," EPA/530-SW-91-060, Washington, DC, 1991.

West Virginia Discharge Reduction Scorecard, National Institute for Chemical Studies, University of Charleston, Charleston, WV 25304, 1990.

Working Notes on Community Right-to-Know, U.S. Public Interest Research Group Education Fund, Washington, DC 20003, 1991.

3

INDUSTRIAL ECOLOGY

Audits of waste generation, presented in the previous chapter, represent a first step in performing macroscale studies of pollution prevention. The next step is to integrate waste generation data with production data, recognizing that the 12 *billion tons of waste generated annually in the United States cannot be ignored as a potential source of industrial materials. The conversion of wastes from something undesirable into something of value eliminates the need for the release of these substances into the environment and reduces the need for raw material extraction. Studies in what has come to be called industrial ecology (Frosch and Gallopoulos,* 1989) *or industrial metabolism (Ayers,* 1989) *examine the material efficiency of large-scale industrial systems, searching for ways to improve that efficiency. By integrating models of the uses of selected materials with data on waste generation for the same materials, it may be possible to identify targets of opportunity for recycling between industry sectors. Studies can also be conducted to determine whether the contaminants in waste are potentially recoverable, and what factors promote and hinder the "mining" of wastes. Finally, the mix of technologies used to produce goods can be examined in order to determine the most environmentally benign combination of technologies. In this chapter, examples of each of these macroscale views of the integration of feedstock and waste data are presented.*

3.1 MATERIALS INTEGRATION ACROSS INDUSTRY SECTORS

When the waste from one facility can be used directly as feedstock in another facility, substantial environmental and economic benefits are possible. The facility producing the waste is spared the burden of treatment or disposal, the facility accepting the waste can reduce the costs of feedstock materials, and the environmental burdens associated with disposal and raw material extraction are reduced. It has been estimated that U.S. industry saves $27 million and the energy-equivalent of 100,000 barrels of oil annually by exchanging waste (National Materials Exchange Network, 1992). Examples of wastes that might potentially be exchanged include surplus chemical stock, used containers, or an alkaline slurry capable of neutralizing acidic waste. At the present time, arrangements for materials integration between facilities are made more by chance than from careful planning and a thorough understanding of what types of feedstocks and wastes lend themselves to integration. However, as data on waste generation and management improve and the field of industrial ecology matures, the direct utilization of wastes will be more systematically exploited. In this section, the integration of several industrial and municipal facilities in Kalundborg, Denmark is described. This pioneering industrial park represents a very high degree of materials integration between very different facilities under different ownership. Next, avenues for waste exchange in the United States are discussed, and some successful examples of waste exchange are given.

3.1.1 The Kalundborg Park

Facilities at Kalundborg, Denmark provide a sense of the potential for material exchanges between industries. In Figure 3-1, a conceptual flow diagram of the industrial park in Kalundborg is given. There is an oil refinery, a sulfuric acid plant, a power plant, and a gypsum-board manufacturer in the park. As shown in the diagram, the power plant and the refinery exchange steam, gas, and cooling water. Waste heat from the power plant is used in district heating and to warm greenhouses and a fish farm. Gypsum from stack gas scrubbers at the power plant is used to make gypsum plaster board (Grann, 1994). The material exchanges between these facilities are substantial. For example, the power plant, which is Denmark's largest, provides 80,000 tons of gypsum per year to the plaster board manufacturer, accounting for two-thirds of the gypsum used by the facility. The power plant accepts 5000 tons of flare gas as well as 700,000 m^3 of cooling water and 500,000 m^3 of reused wastewater from the refinery. The refinery thus provides two-thirds of the power plant's water requirements.

While the Kalundborg industrial park provides an interesting model for converting waste material from one industrial sector into the feedstock of another sector, it does not give a sense of the overall magnitude of the exchanges that could take place, or the limitations to such exchanges. In fact, the facilities at Kalundborg were not originally intended for intensive integration. When material exchanges first took place, it was due to fortuitous circumstances. Because of their success, the participants are now dedicated to finding new ways to integrate their processes and reduce waste. The

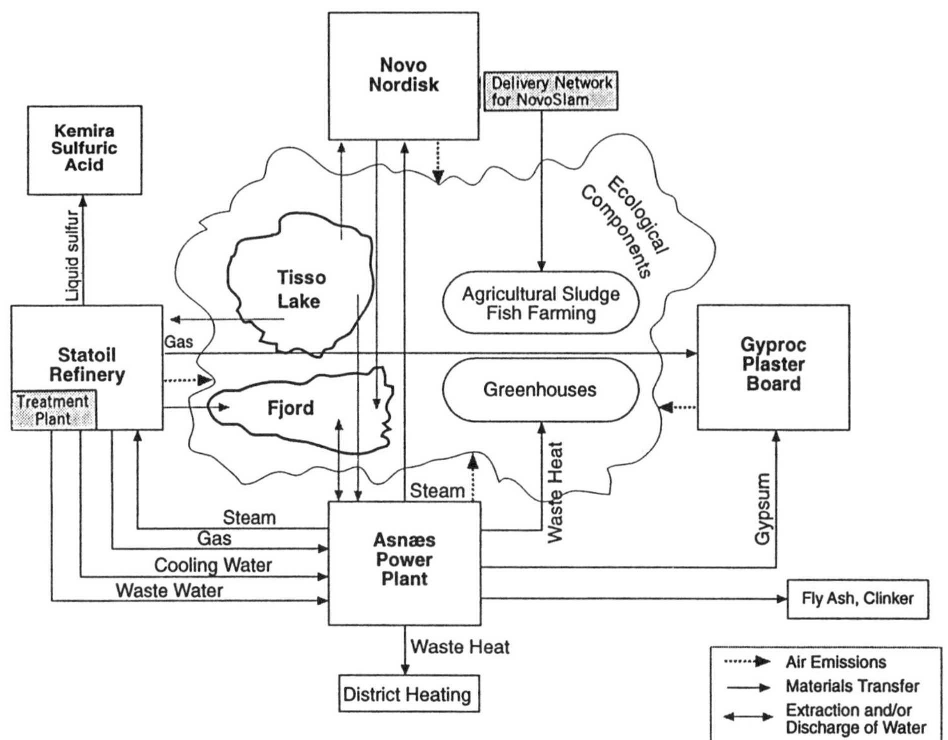

Figure 3-1. The Kalundborg industrial park (Grann, 1994).

emerging tools of industrial ecology can be used to promote a systematic improvement in materials efficiency, rather than improvement dictated by chance.

3.1.2 Waste Exchanges

In the United States, waste exchanges have been formed to act as information clearinghouses that promote the exchange of waste from generators to other businesses. Besides matching companies which must discard a material with companies whose processes require the same material, waste exchanges often provide information about commercial recycling facilities. Table 3-1 contains the addresses and phone numbers of some waste exchanges in the United States. Pacific Materials Exchange, listed in the table, supports a nationwide electronic waste exchange bulletin board called the National Materials Exchange Network. Access information for this bulletin board is given in the references to this chapter.

Table 3-1. Selected waste exchanges in the United States

Alphabetized by Name

California Integrated Waste Management Board
CALMAX
8800 Cal Center Drive
Sacramento, CA 95826
(916) 255-2369 phone
(800) 553-2962 recycling hotline

California Waste Exchange
Department of Health Services
Toxic Substances Control Division
Alternative Technology Section
714/744 P Street
Sacramento, CA 94234-7320
(916) 324-1807 phone
(916) 327-4494 fax

Great Lakes Waste Exchange
400 Ann Street, N.W. Suite 201-A
Grand Rapids, MI 49504-2054
(616) 363-3262

Indiana Waste Exchange
Purdue University School of Civil Engineering
West Lafayette, IN 47907
(317) 494-5038 phone
(800) 669-6604 phone (Indiana only)
(317) 494-6422 fax

Industrial Materials Exchange (IMEX)
Seattle-King County Environmental Health
172 20th Avenue
Seattle, WA 98122
(206) 296-4633 phone
(206) 296-0188 fax

Industrial Material Exchange Service (IMES)
2200 Churchill Road, #24
Springfield, IL 62794-0276
(217) 782-0450 phone
(217) 524-4193 fax

Industrial Waste Information Exchange
New Jersey Chamber of Commerce
5 Commerce Street
Newark, NJ 07102
(201) 623-7070 phone
(201) 623-8739 fax

Iowa State University Waste Exchange Project
205 Engineering Annex
IMSE
Ames, IA 50011
(515) 294-4056 phone
(515) 294-1682 fax

Montana Industrial Waste Exchange
Montana Chamber of Commerce
2030 11th Avenue
Helena, MT 59601
(406) 442-2405 phone
(406) 442-2409 fax

Northeast Industrial Waste Exchange (NIWE)
90 Presidential Plaza, Suite 122
Syracuse, NY 13210
(315) 422-6572 phone
(315) 422-9051 fax

Pacific Materials Exchange (PME)
S. 3707 Godfrey Boulevard.
Spokane, WA 99204
(509) 623-4244 phone
(509) 623-4276 fax

Resource Exchange Network for Eliminating Waste (RENEW)
Texas Water Commission
P.O. Box 13087
Austin, TX 78711-3087
(512) 463-7773 phone
(512) 463-8317 fax

Southeast Waste Exchange (SEWE)
Urban Institute
Department of Civil Engineering
University of North Carolina
Charlotte, NC 28223
(704) 547-2307 phone
(704) 547-2767 fax

Table 3-1. (*Continued*)

Alphabetized by Name	
Southern Waste Information Exchange (SWIX) 2035 E. Dirac Drive Suite 226 Tallahassee, FL 32310 (800) 441-7949 phone (904) 644-5516 phone (904) 574-6704 fax	Tennessee Waste Exchange Tennessee Association of Business 226 Capitol Boulevard, Suite 122 Nashville, TN 37219 (615) 256-5141 phone

Source: PIES (1992).

There are many documented instances of successful waste exchange between businesses in the United States. A few are presented here to illustrate the diversity of saleable wastes and the issues involved in marketing wastes (Hunt et al., 1986). An example illustrating the benefits for the supplier in a waste exchange is provided by the experiences of an electric power utility in North Carolina. This company was able to reduce their disposal costs and generate $2.5 million per year in revenue from the sale of fly ash and bottom ash. The ash was used as an ingredient in Portland cement, concrete, and plastics, and as an aggregate base course for unpaved roads. In another case, a fertilizer producer benefitted from being at the receiving end of a waste exchange. This facility reduced its operating costs by purchasing wastes containing the metals that were necessary nutrients in their products. Another facility, a producer of printed circuit boards, has taken active steps to make its wastes more attractive in the waste exchange market. These steps include segregating waste streams, increasing the concentration of valuable materials in waste streams, and in some cases modifying production processes so that waste has value elsewhere.

3.2 RECOVERY OF VALUABLE MATERIALS IN WASTE

In the previous section, the use of wastes from one process as feedstocks for another process was discussed. In this section, factors affecting the viability of extracting valuable materials from waste streams are described. As shown in this section, the potential for recovery of contaminants in waste streams depends on both the concentration of the contaminants and their market value. Even when these two factors indicate that a waste can be economically "mined," the existence of a recycling infrastructure and the technologies currently in place influence recycling rates.

Case studies of metals are used to illustrate the topics in this section.

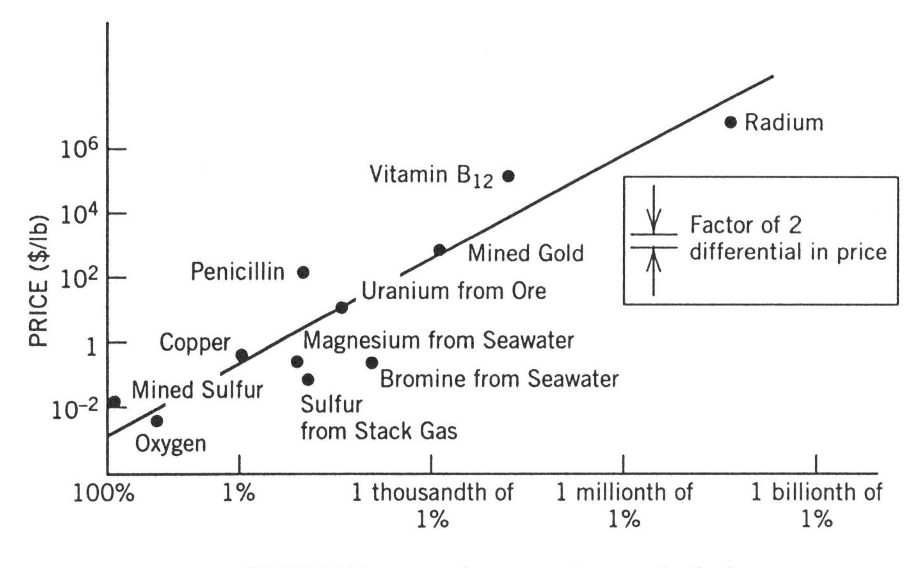

Figure 3-2. The Sherwood diagram showing the correlation between the selling price of materials and their degree of dilution in the original matrix from which they are separated (National Research Council, 1987). (Reprinted with permission from *Separation and Purification: Critical Needs and Opportunities*. Copyright © 1987 by the National Academy of Sciences. Courtesy of the National Academy Press, Washington, DC.)

Metals are attractive materials for industrial ecology studies because they can be tracked easily as they move through chemical processes.

3.2.1 Dilution and Value: The Sherwood Diagram

The extent to which industrial wastes might serve as raw materials is dependent not only on the mass of resources present but also on their concentration. Figure 3-2 shows that resources that are present at very low concentrations are recovered only if their selling price is high, while resources present at high concentrations can have a low selling price. The points in this figure, which is called the *Sherwood diagram* in honor of Professor T. K. Sherwood, largely represent the extraction of virgin resources, but a similar relationship is expected to exist for extraction of raw materials from waste.

Before the potential for recovery of valuable materials in waste streams can be assessed, however, information on waste composition and waste stream flow rates must be known. As discussed in Chapter 2, the Biennial Reporting System of hazardous wastes conducted under the Resource Conservation and Recovery Act (RCRA) includes information on waste stream flow rates and on hazardous materials present, but the quantity of hazardous material in the stream is not given. The Toxic Chemical Release Inventory

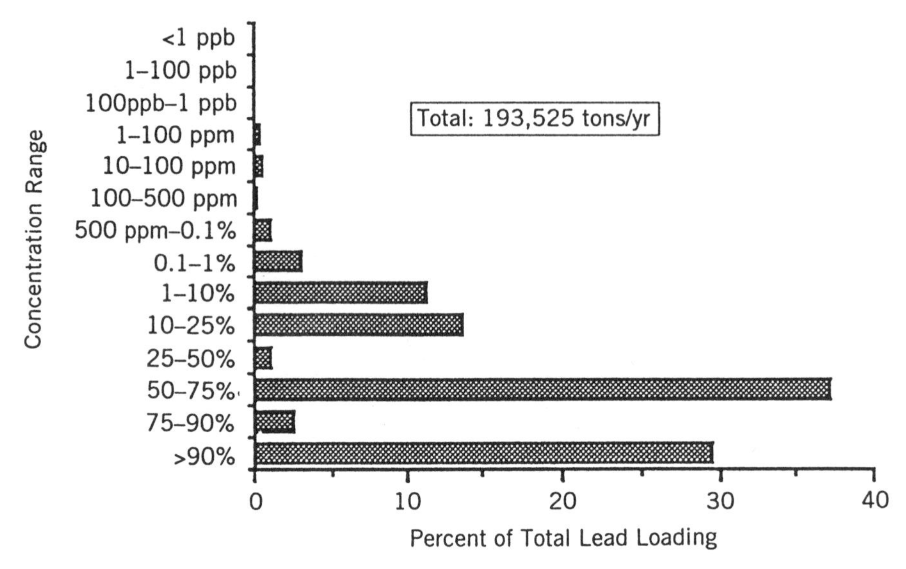

Figure 3-3. Concentration distribution of lead in industrial hazardous waste streams in 1986 (Allen and Behmanesh, 1993). (Reprinted with permission from *The Greening of Industrial Ecosystems*. Copyright © 1993 by the National Academy of Sciences. Courtesy of the National Academy Press, Washington, DC.)

(TRI) gives no information on waste stream flow rates, but does give data on the quantities released. Information about both waste composition and bulk waste stream properties for a limited set of metals is available in the National Hazardous Waste Survey. The data in this survey, which are for 1986 and include information on only a portion of the total industrial waste flow, can be used to determine concentration profiles for metals in waste streams. An example of such a profile is shown in Figure 3-3. Profiles like this can be used to estimate the degree to which waste streams can be "mined." From the Sherwood diagram and the price of the metal, the concentration at which extraction is economical can be estimated. The metals in waste streams whose metal concentrations fall above this concentration could, in principle, be economically recovered. Table 3-2 lists the minimum recoverable concentrations for 15 metals and the fraction of metals in hazardous waste streams that in principle should be economically recoverable, along with the fraction that actually is recovered. This table shows that metals in many waste streams are significantly underutilized.

There are two main points to be derived from this discussion of evaluating the potential for recovering materials from wastes: (1) information about waste stream flow rates and composition is necessary for this evaluation and (2) any practice, such as segregation of waste streams, which leads to higher concentrations of material in the waste streams, will allow for more economic recovery of materials.

Table 3-2. The potential for recovery of metals in industrial hazardous waste streams in 1986

Metal	Minimum Concentration Recoverable, from Sherwood Diagram (mass fraction)	Total Loading in Hazardous Waste Streams (tons/yr)	Recoverable Fraction of Metal in Waste	Percent of Metal Recycled from Waste Streams
Antimony	0.00405	17,000	0.74–0.87	35
Arsenic	0.00015	440	0.98–0.99	3
Barium	0.0015	59,000	0.95–0.98	1
Beryllium	0.012	5,300	0.54–0.84	11
Cadmium	0.0048	16,000	0.82–0.97	8
Chromium	0.0012	90,000	0.68–0.89	5
Copper	0.0022	110,000	0.85–0.92	10
Lead	0.074	190,000	0.84–0.95	56
Mercury	0.00012	5,400	0.99	16
Nickel	0.0066	3,600,000	1.00	0.1
Selenium	0.0002	2,000	0.93–0.95	29
Silver	0.000035	17,000	0.99–1.00	1
Thallium	0.00004	280	0.97–0.99	5
Vanadium	0.0002	4,400	0.74–0.98	1
Zinc	0.0012	270,000	0.95–0.98	12

Source: Data from Allen and Behmanesh (1993).

3.2.2 Other Factors Influencing Waste Recovery

Studies of lead, the platinum group metals, and scrap iron can be used to illustrate some of the factors other than dilution and price that can facilitate and hinder the recycling of materials in our economy.

Lead is an example of a metal that is successfully recycled. The sources of the 1.2 million tons of lead that are converted into products each year include both mining and recycling operations, with recycling supplying approximately 70% of the lead used. Most of the lead is used to make lead storage batteries. Minor uses include metal products, electronics, pigments, and additives to glass, ceramics, and plastics. Once the lead is converted into products, it is eventually either recycled or sent for disposal. The lead in batteries is extensively recycled: 900,000 tons of lead acid batteries were discarded in 1988, of which 800,000 tons were recycled (US EPA, 1989, 1990). Further, approximately 250,000 tons of lead were discarded in industrial waste in 1986, of which 150,000 tons were recycled. A simplified model of the industrial ecology of lead can be obtained by coupling this recycling data with data on industrial waste generation and on lead production and consumption. A flowchart of this model is shown in Figure 3-4. A representation of the worldwide extraction, use, and disposal of lead is shown in Figure 3-5. The flows in this figure represent annual usage rates, but since

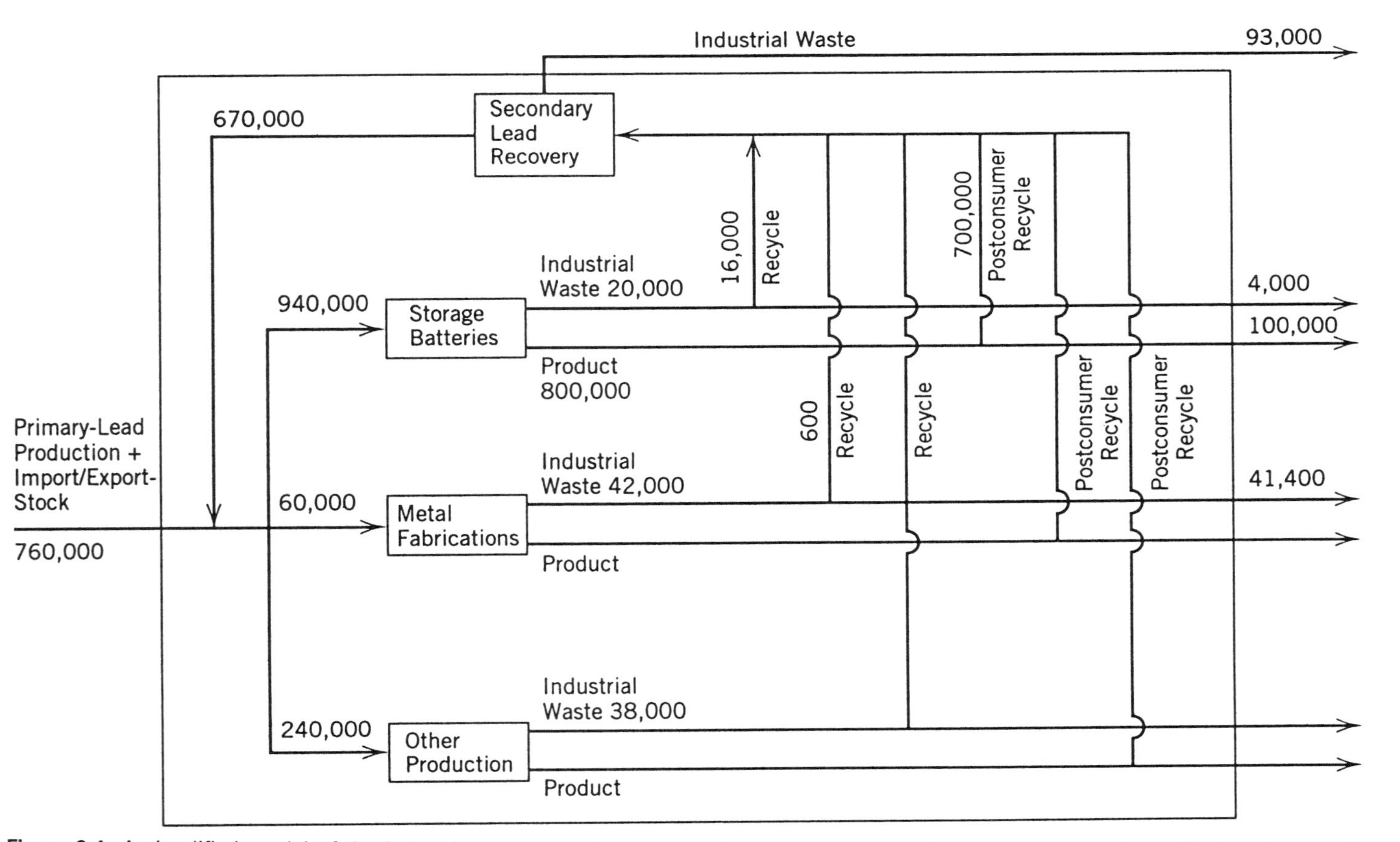

Figure 3-4. A simplified model of the industrial ecology of lead; amounts in tons per year (Allen and Behmanesh, 1993). (Reprinted with permission from *The Greening of Industrial Ecosystems*. Copyright © 1993 by the National Academy of Sciences. Courtesy of the National Academy Press, Washington, DC.)

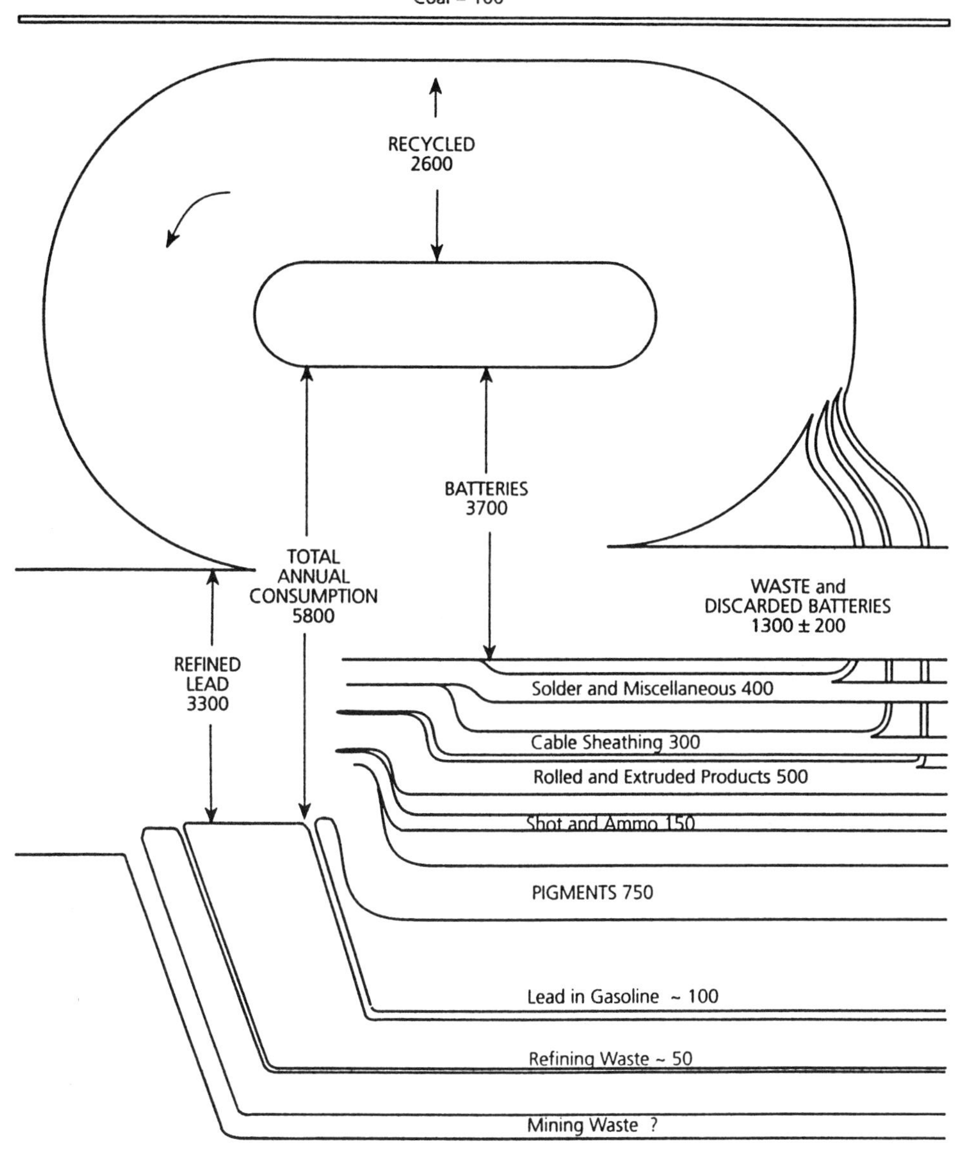

Figure 3-5. World extraction, use, and disposal of lead in 1990 in thousands of tons (Thomas and Spiro, 1994). (From *Industrial Ecology and Global Change*. Copyright © 1993 by UCAR. Reprinted with the permission of Cambridge University Press.)

some products, like batteries, have multiyear lifetimes, the amount of lead stored in products currently in use may be extensive.

The platinum group metals exhibit different use and recycling patterns than lead, as shown in Figure 3-6, which depicts a simplified model of their industrial ecology. The greatest contrast between the industrial ecology of lead and that of platinum group metals is the fate of materials used in automotive endproducts. Lead acid batteries from automobiles are recycled at a rate of over 90% (US EPA, 1989, 1990), while metals from catalytic converters are recycled at only a 12% rate (Frosch and Gallopoulos, 1989). The difference in these rates cannot be attributed to the low concentration of metals in the catalyst, because industrial catalysts of similar composition are recycled at high rates. Instead, the main reason for the low recycling rates for platinum group metals in automotive catalytic converters is the lack of an effective collection network and the presence of problematic cocontaminants. Such a network exists for lead acid automobile batteries and the batteries do not have serious cocontaminant problems.

Even when a collection network exists, technology selection can strongly influence the industrial ecology of materials, as is the case with scrap iron. The stockpile of scrap iron awaiting recycling increased from 610 million tons in 1982 to more than 750 million tons in 1987 (Frosch and Gallopoulos, 1989). One of the reasons for this increase was a technology shift from open hearth to basic oxygen furnaces in steelmaking. While open hearth furnaces could accept up to 50% scrap iron, basic oxygen furnaces accept only 20%. Thus, even when a recycling infrastructure is in place, technology selection can play a key role in encouraging or discouraging recycling.

These simple case studies of the industrial ecology of metals reveal the importance of infrastructure and technology selection in industrial metabolism. They also reveal that industrial ecology studies are still relatively elementary. More detailed analysis of potential recovery and recycling opportunities is feasible, however, as illustrated in the next section.

3.3 TECHNOLOGY AND THE INDUSTRIAL ECOSYSTEM

In previous sections, the influence that technology selection has had on the recycling of scrap iron was described and the possibility of altering existing processes so that their wastes could be used as feedstocks in other processes was discussed. In this section, the role that technology selection can play in reducing the quantity and/or toxicity of environmental releases is described.

3.3.1 Selecting Desirable Combinations of Process Technologies in Organic Chemical Manufacturing

Determining the industrial metabolism of organic compounds is much more difficult than determining the industrial metabolism of metals because organic

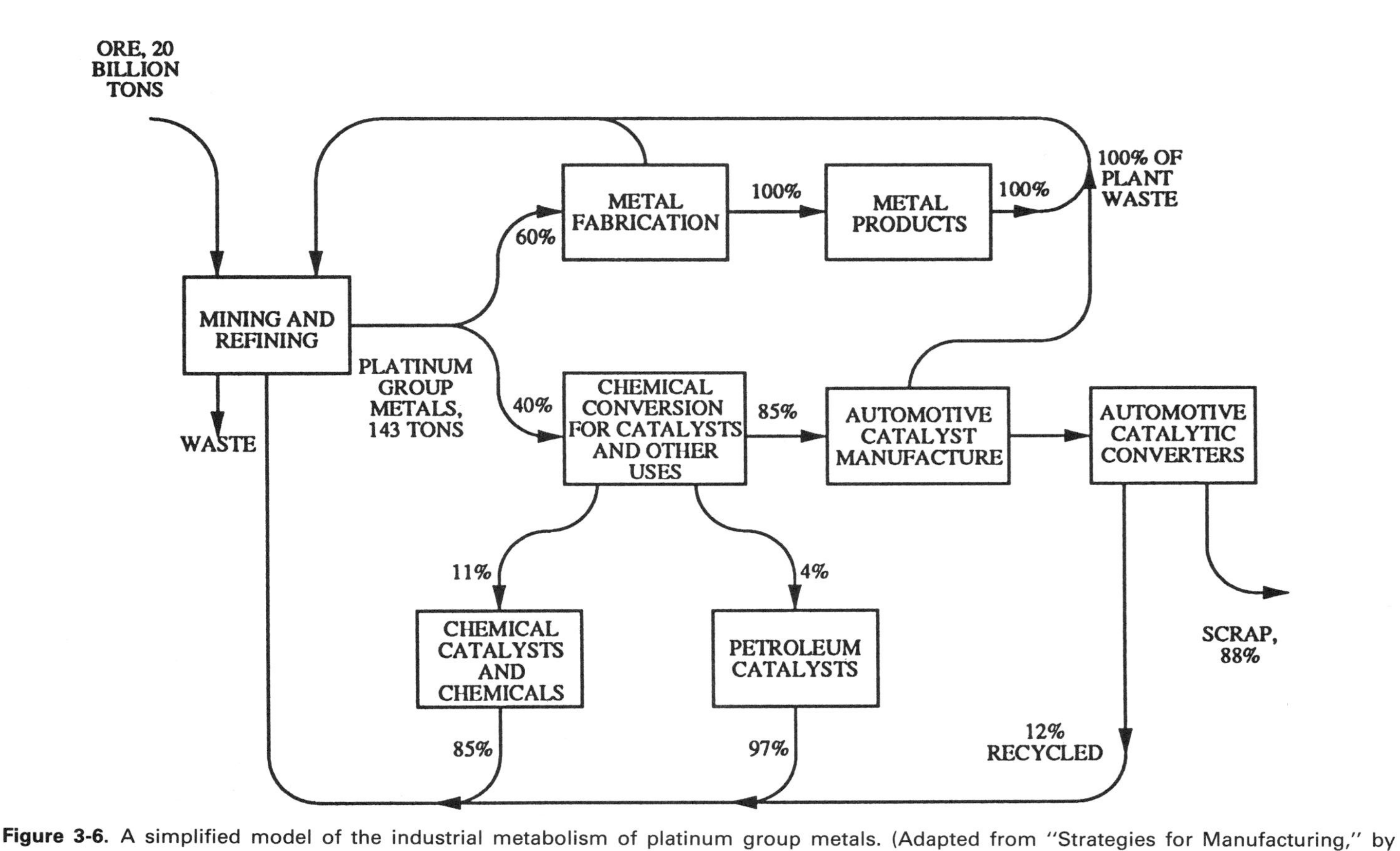

Figure 3-6. A simplified model of the industrial metabolism of platinum group metals. (Adapted from "Strategies for Manufacturing," by Robert A. Frosch and Nicholas E. Gallopoulos.)

compounds are frequently transformed into other compounds as they move through processes. One of the best frameworks available for such studies are the models of the chemical process industries developed by Rudd and coworkers (Rudd et al., 1981). In these structural models, the flows of raw materials into and products out of approximately 180 types of chemical processes were determined. These processes generate over 120 different products and form a complex web of material flows that make possible many routes to the formation of chemical products.

As a relatively simple case study of the multiple pathways available in chemical synthesis, consider methyl *tert*-butyl ether (MTBE), an oxygenated additive to gasoline that is being produced in large quantities to satisfy the provisions of the 1990 Clean Air Act Amendments. MTBE is produced by reacting methanol with isobutylene using one of two processes. In addition to having two processes to choose from in the MTBE synthesis, there are options in the synthesis of the raw materials. The methanol fed to the process can be made from methane, via carbon monoxide, or via fermentation. The isobutylene can be made by cracking petroleum, by isomerizing butenes, or by dehydrogenating butanes. These different processes provide more than a dozen routes from raw materials to MTBE. Each route has different energy requirements and rates of waste generation. Selecting the cleanest and most economical route is a difficult proposition. It is made even more difficult when the influence a particular process has on the rates of waste generation in the rest of the chemical industry is considered. For example, if methanol is produced using carbon monoxide, the carbon monoxide may be generated through partial oxidation of a material that is currently wasted. On the other hand, to convert the carbon monoxide into methanol requires hydrogen, which is an energy-intensive material.

Only a few comprehensive analyses of the technological structure of the chemical industry have been performed, and most of these studies were motivated by the energy crises of the 1970s. Their focus therefore tended to be on the role that technology selection could play in improving the energy efficiency of the chemical industry; wastes and emissions were not explicitly considered. One of the few studies available on the impact that technology selection can have on the environmental performance of the chemical industry was published in 1985 by a group at the University of California, Los Angeles (Fathi-Afshar and Yang, 1985). In this study, the mix of technologies that would result in the minimum cost of production for 124 chemical products and that would also result in the lowest gross toxicity of the chemical intermediates used in the synthesis was examined. The results are summarized in Figure 3-7. The point labeled *A* on Figure 3-7 corresponds to the cost of production of the 124 chemicals if the sole criterion for process selection is the toxicity of the intermediates required. Point *C* corresponds to the total cost of chemical production if cost is the only criterion used in selecting technologies. The line connecting points *A* and *C* is the collection of possible technology mixes. The ideal point is *B*, which represents the lowest possible

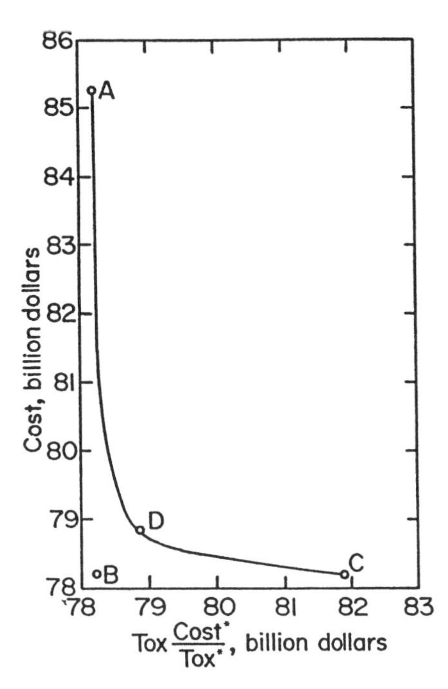

Figure 3-7. Tradeoffs between cost and the use of toxic intermediates in the chemical industry. (Reprinted from *Chemical Engineering Science*, **40**, S. Fathi-Afshar and J. C. Yang, "Design the Optimal Structure of the Petrochemical Industry for Minimum Cost and Least Gross Toxicity of Chemical Production," pp. 781–797, copyright © 1985, with kind permission from Elsevier Science Ltd, The Boulevard, Langford Lane, Kidlington OX5 1GB, UK.)

cost and the lowest possible toxicity of intermediates, but which is not obtainable using current technologies. Clearly, a mix of technologies located somewhere near point *D* represents the best compromise between economics and the environmental objective of lowest toxicity of intermediates.

In addition to selecting desirable mixes of technologies, this type of analysis can help to identify critical areas for technology improvement. If, for example, a mix of technologies was selected that resulted in the lowest total cost of chemical production, the chemical intermediates that contribute the most to toxicity for this particular mix of technologies could be identified. Such an analysis is presented in Figure 3-8 and reveals that just a few chemical intermediates dominate the toxicity of the entire hypothetical system: toluene diisocyanate (TDI), phosgene, and methylene diphenylene isocyanate (MDI). Finding technological alternatives to these intermediates could dramatically reduce overall toxicity.

The example described above demonstrates the utility of structural analyses of the chemical industry. Similar studies could be done, examining the tradeoffs between economics, energy consumption, and waste generation, if process-specific waste generation data were available. Unfortunately few such sources of data are available (Allen and Jain, 1992) and the quality and timeliness of the data are questionable. Thus, studies of the industrial ecology of organic compounds remain at a rather rudimentary level.

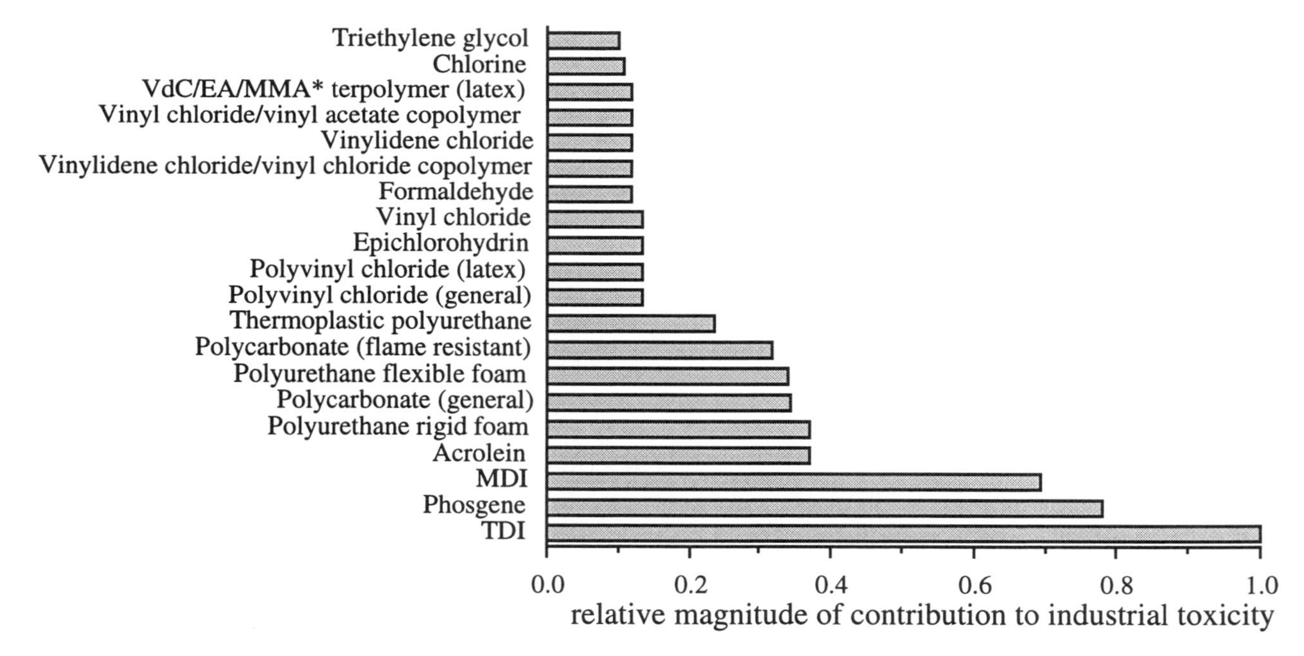

Figure 3-8. Major contributors to the total toxicity of the chemical intermediates industry. (Reprinted from *Chemical Engineering Science*, **40**, S. Fathi-Afshar and J. C. Yang, "Design the Optimal Structure of the Petrochemical Industry for Minimum Cost and Least Gross Toxicity of Chemical Production," pp. 781–797, copyright © 1985, with kind permission from Elsevier Science Ltd, The Boulevard, Langford Lane, Kidlington OX5 1GB, UK.)

3.4 SUMMARY

Industrial ecology is an emerging discipline in which the flows of materials in the economy are studied and more efficient use of the materials is sought. One way to increase material efficiency is to reduce the disposal of wastes by converting them into something of value. The utilization of wastes not only eliminates their release into the environment but also reduces the need for raw material extraction. One way to utilize wastes is to recover the valuable components in waste streams. In this chapter, it was shown that there is a relationship between the concentration of a waste in a waste stream and its potential for economic recovery. However, dilution and value are not the only factors that influence recovery rates. Another way to utilize waste is through waste exchange, in which a facility markets a waste they must otherwise dispose of or uses wastes from another process as a source of feedstock for their processes. An industrial park in the city of Kalundborg, Denmark, has successfully integrated several manufacturing facilities and a utility that exchange waste and energy. There are a number of waste exchange information clearinghouses in the United States that provide opportunities for exchanging materials. Finally, studies in industrial ecology can lead to a better understanding of the ways in which technology selection influences the environmental impacts of industrial processes.

QUESTIONS FOR DISCUSSION

1. If you were given the task of finding the best way to dispose of a waste stream at a facility, how would you find out if the waste stream could be useful as a feedstock in another industry sector? What barriers might make a plant manager in the other industry sector reluctant to accept your facility's waste as a feedstock? What might you do to "sell" such a plant manager the idea of accepting your waste as a feedstock?
2. The data in Table 3-2 indicate that some metals in waste streams are underrecovered. Describe how you might go about researching the potential for starting up a business that recovered metal from waste streams. What are some of the factors that might act as a barrier to the operation of such a business?

PROBLEM STATEMENT

1. Using the information on MTBE production given in Section 3.3.1, draw a diagram that depicts all the possible routes to MTBE production.
2. Table 3-2 shows that large quantities of metals are present at recoverable concentrations in hazardous waste streams. Estimate the value of the

recoverable metals in these wastes. Metal prices can be obtained either from the *Minerals Yearbook* published by the U.S. Department of the Interior or may be estimated from the Sherwood diagram.

3. To recover valuable materials from waste streams, the minimum amount of energy that must be invested is determined by the entropy that must be overcome in concentrating the valuable materials. The following simple example illustrates the magnitude of these energy requirements.
   a. Calculate the entropy of mixing that must be overcome in separating a metal present at 1 part per million (ppm) (mole fraction basis) in an aqueous waste stream. Assume ideal behavior so that

$$\Delta S = -R \sum x_i \ln x_i \ ,$$

      where ΔS is the molar entropy of mixing, R is the gas constant, x_i is the mole fraction of each component in the mixture, and the summation is done over all components in the mixture.
   b. Calculate the free energy required to perform the separation if

$$\Delta G \cong -T \Delta S \ ,$$

      where ΔG is the Gibbs free energy of mixing and T is the absolute temperature. Perform your calculation at room temperature and at 1000 K.
   c. If a gallon of fuel costs $1.00 and each gallon contains approximately 10^5 Btu, what are the energy costs necessary to recover one pound of metal? Assume the metal has a molecular weight of 100.
   d. Compare your estimate of the cost required to overcome the entropy of mixing to the costs reported in the Sherwood diagram. Comment on any discrepancies.

REFERENCES

Allen, D. T. and N. Behmanesh, "Wastes as Raw Materials," in *The Greening of Industrial Ecosystems*, B. Allenby, ed., National Academy Press, Washington, DC, 1993.

Allen, D. T. and R. K. Jain, eds., *Hazardous Waste & Hazardous Materials*, **9**(1) (entire issue), 1992.

Ayers, R. V., "Industrial Metabolism," in *Technology and Environment*, J. H. Ausubel and H. E. Sladovich, eds., National Academy Press, 1989.

Fathi-Afshar, S. and J. C. Yang, "Design the Optimal Structure of the Petrochemical Industry for Minimum Cost and Least Gross Toxicity of Chemical Production," *Chem. Eng. Sci.*, **40**, 781–797, 1985.

Frosch, R. A. and N. E. Gallopoulos, "Strategies for Manufacturing," *Sci. Am.* 144–152, Sep. 1989.

Grann, H., "The Industrial Symbiosis at Kalundborg, Denmark," paper presented at the National Academy of Engineering's Conference on Industrial Ecology, Irvine, CA, May 9–13, 1994.

Hunt, G. et al., *Accomplishments of North Carolina Industries—Case Summaries*, North Carolina Dept. Natural Resources and Community Development, Jan. 1986 in PIES, 1992.

National Materials Exchange Network (electronic bulletin board system), Pacific Materials Exchange [phone (800) 858-6625; modem settings: no parity, 8 data bits; voice help: 509/325-0507], Spokane, WA, Nov. 1992.

National Research Council, *Separation and Purification: Critical Needs and Opportunities*, National Academy Press, Washington, DC, 1987.

PIES, Pollution Prevention Information Exchange System (electronic bulletin board system), Science Applications International Corporation, Falls Church, VA, Sep. 1992.

Rudd, D. F., S. Fathi-Afshar, A. A. Trevino, and M. A. Stadtherr, *Petrochemical Technology Assessment*, Wiley, New York, 1981.

Thomas, V. and T. Spiro, "Emissions and Exposures to Metals: Cadmium and Lead," in *Industrial Ecology and Global Change*, R. Socolow, C. Andrews, F. Berkhout, and V. Thomas, eds., Cambridge University Press, New York, 1994.

U.S. Environmental Protection Agency (US EPA), "Characterization of Products Containing Lead and Cadmium in Municipal Solid Waste in the United States, 1970 to 2000," EPA/530-SW-89-015A, 1989.

U.S. Environmental Protection Agency (US EPA), "Characterization of Municipal Solid Waste in the United States: 1990 Update," EPA 530-SW-90-042, 1990.

4

LIFE-CYCLE ASSESSMENT

A macroscale approach to identifying opportunities for pollution prevention that is becoming increasingly popular is to consider an individual product and trace the flows of energy, raw materials, and waste streams that were required to create, use, and dispose of the product. This analysis framework is called life-cycle assessment. *Life-cycle assessments are conducted by companies wishing to improve the environmental characteristics of their products and by companies or industry groups wishing to promote their product as environmentally superior to a competitor's product. They are also used as a tool for strategic corporate planning and for government initiatives in environmental labeling, acquisition, procurement, and regulation. In this chapter, the procedures for conducting a life-cycle assessment are described, and the uncertainties inherent in the analysis are discussed. Finally, an overview of the applications of life-cycle assessments is provided.*

4.1 A FRAMEWORK FOR CONDUCTING LIFE-CYCLE ASSESSMENTS

A life-cycle assessment (LCA) consists of three components: (1) an inventory of wastes, emissions, and raw-material and energy use associated with the entire life cycle of a product from raw-material extraction to ultimate disposal; (2) an assessment of the environmental impacts associated with the wastes, emissions, and raw-material and energy use; and (3) an improvement analysis where mechanisms for reducing adverse environmental impacts are sought (SETAC, 1991). Of these three life-cycle assessment steps (inventory, impact assessment, and improvement analysis), only the inventory stage is well developed. In contrast, the framework for assessing life-cycle impacts is largely conceptual (Vigon, 1994), and while elementary improvement analyses can be made, such analyses are limited by the limitations of the tools for evaluating life-cycle impacts. Because of the relatively advanced state of life-cycle inventories, as opposed to life-cycle impact assessment and life-cycle improvement analysis, this chapter focusses primarily on the inventory stage of a life-cycle assessment. Brief descriptions of life-cycle impact assessment and even briefer descriptions of improvement analysis are provided.

4.1.1 Life-Cycle Inventories

The basic framework for compiling an inventory of wastes, emissions, and raw-material and energy use associated with the manufacture, use, and disposal of a product is shown conceptually in Figure 4-1. Examples of each of the major life-cycle stages and recycle loops are given in this figure. Note that for a complex product that is produced from many materials, life-cycle inventories become quite complex. The wastes, emissions, and raw-material and energy use of each significant material must be estimated.

Some of the data from a life-cycle inventory for a commodity chemical, ethylene, are shown in Table 4-1. The quantities reported are for the life-cycle stages up to and including the production of ethylene from petroleum naphtha or natural gas. This includes the production of petroleum or natural gas in the field, the transportation of the feedstock from the oil or gas field to the refinery, the production of naphtha from the petroleum, and the cracking of naphtha or natural gas to form ethylene. The raw materials, energy use, emissions, and wastes for these life-cycle stages are reported in the table.

Two features of Table 4-1 are particularly worthy of note. One is that the petroleum and natural-gas feedstock used to produce ethylene is reported in energy, rather than mass units. This is common practice in life-cycle inventories of carbon-based products. Often, the feedstock or "embodied" energy is not reported separately as it is in Table 4-1. Instead, the feedstock and fuel consumption energies are simply lumped together. A second feature that is implicit in Table 4-1 is the aggregated nature of the emissions data. For example, the 6 g of nitrogen oxides emitted per kilogram of ethylene produced includes emissions from combustion systems in the oil or gas fields, combustion emissions from the ships and vehicles used to transport the petroleum, and combustion emissions from refinery and ethylene production furnaces. Thus, the nitrogen oxide emissions are broadly distributed over a large geographic area. A life-cycle inventory generally does not include data on where such emissions occur.

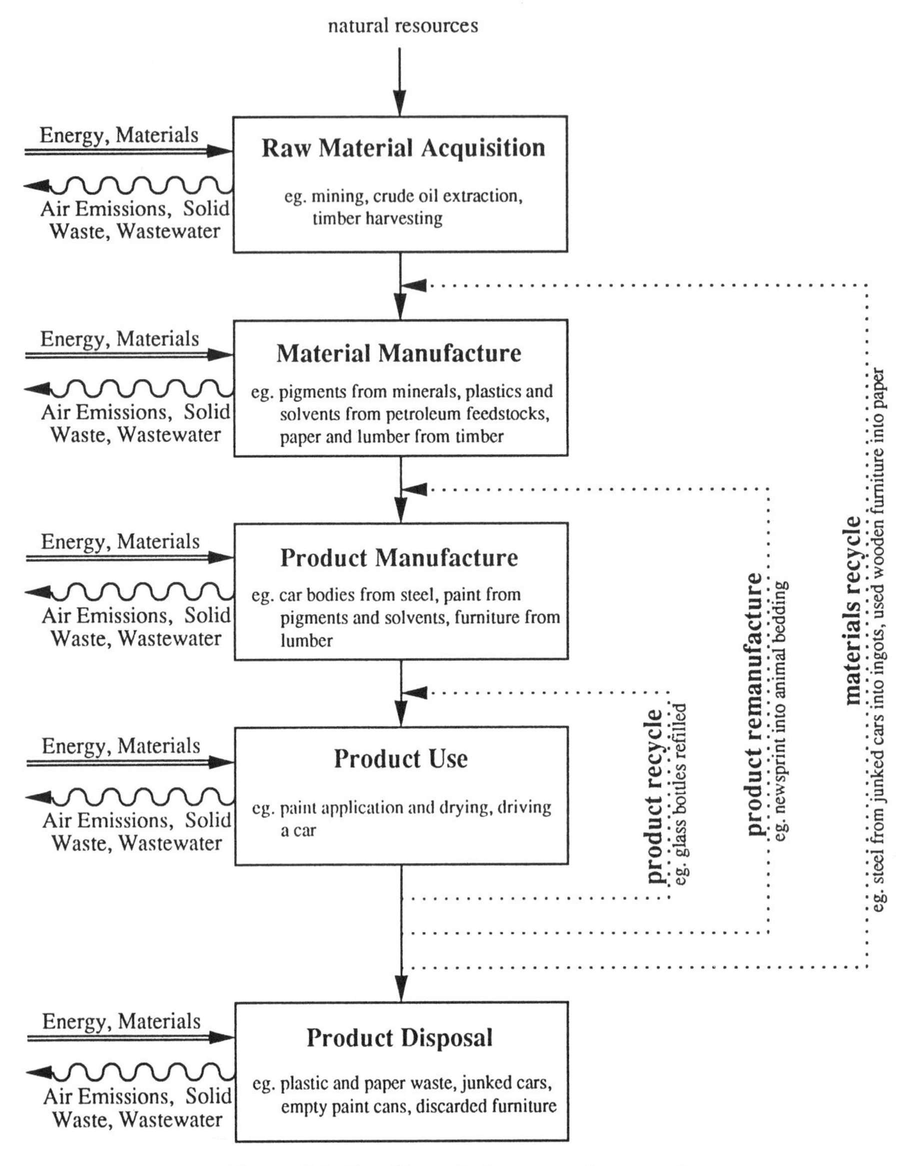

Figure 4-1. The life-cycle inventory framework.

4.1.2 Life-Cycle Impact Assessments

The data presented in Table 4-1 are a useful summary of the total raw-material use, energy use, waste generation, and emissions associated with producing a kilogram of ethylene. If the goal is to use these data to improve the environmental performance of ethylene production processes, it is not

Table 4-1. Gross inputs and outputs associated with the production of one kilogram of ethylene based on the operation of 19 European ethylene manufacturers

	Category	Unit Average
Fuels, MJ	Coal	0.94
	Oil	1.75
	Gas	6.06
	Hydro	0.12
	Nuclear	0.32
	Other	<0.01
	Total	9.19
Feedstock, MJ	Coal	<0.01
	Oil	31.45
	Gas	28.82
	Total	60.28
Total Fuel Plus Feedstock		69.47
Raw materials, mg	Iron ore	200
	Limestone	100
	Water	1,900,000
	Bauxite	300
	Sodium chloride	5,400
	Clay	20
	Ferromanganese	<1
Air emissions, mg	Dust	1,000
	Carbon monoxide	600
	Carbon dioxide	530,000
	Sulfur oxides	4,000
	Nitrogen oxides	6,000
	Hydrogen sulfide	10
	Hydrogen chloride	20
	Hydrocarbons	7,000
	Other organics	1
	Metals	1
Water emissions, mg	COD[a]	200
	BOD[b]	40
	Acid as H^+	60
	Metals	300
	Chloride ions	50
	Dissolved organics	20
	Suspended solids	200
	Oil	200
	Phenol	1
	Dissolved solids	500
	Other nitrogen	10

Table 4-1. *(Continued)*

	Category	Unit Average
Solid waste, mg	Industrial waste	1,400
	Mineral waste	8,000
	Slags and ash	3,000
	Nontoxic chemicals	400
	Toxic chemicals	1

[a]Chemical oxygen demand.
[b]Biological oxygen demand.
Source: APME (Brussels), Boustead (1993).

clear which of the entries in Table 4-1 should be of most concern. The challenges associated with using life-cycle inventory data directly are apparent in a case where a product designer must choose between polyethylene and polystyrene. As shown in Tables 4-2 and 4-3, a kilogram of polyethylene requires less energy but more water than polystyrene. Further examination of the environmental profiles of the two materials adds to the ambiguity of choosing an environmentally preferable product. For example, producing a kilogram of polyethylene requires less chlorine but releases more hydrogen chloride air emissions than producing a kilogram of polystyrene. Because the inventories are ambiguous, determining which is the preferred choice requires that the environmental impacts of the life-cycle inventory entries be evaluated. Methods for performing life-cycle impact assessment are still under development (Vigon, 1994), and a comprehensive review of the methods is beyond the scope of this chapter. Nevertheless, a sense of the state of development of these tools can be obtained by examining a system developed jointly by the Swedish Environmental Institute and Volvo. In this system, commonly referred to as the *environmental priority strategies* (EPS) system, total environmental loads are determined by assigning an environmental index to each element in the inventory. As shown in Table 4-4, these indices are expressed as environmental load units per kilogram of the inventoried item. For example, if producing a kilogram of ethylene results in the emission of 0.53 kg of carbon dioxide and 0.006 kg of nitrogen oxides, then the environmental load associated with these two inventory entries is

$$0.53 \text{ kg } CO_2 \times \frac{0.09 \text{ environmental load units}}{\text{kg } CO_2} = 0.048 \text{ environmental load units,}$$

and

$$0.006 \text{ kg } NO_x \times \frac{0.22 \text{ environmental load units}}{\text{kg } NO_x} = 0.0013 \text{ environmental load units.}$$

Table 4-2. Gross inputs and outputs associated with the production of one kilogram of low-density polyethylene

	Category	Unit Average
Fuels, MJ	Coal	3.28
	Oil	3.58
	Gas	12.38
	Hydro	0.54
	Nuclear	1.67
	Other	0.21
	Total	21.66
Feedstock, MJ	Coal	<0.01
	Oil	33.87
	Gas	33.02
	Wood	<0.01
	Total	66.89
Total Fuel Plus feedstock		88.55
Raw materials, mg	Iron ore	200
	Limestone	150
	Water	24,000,000
	Bauxite	300
	Sodium chloride	8,000
	Clay	20
	Ferromanganese	<1
Air emissions, mg	Dust	3,000
	Carbon monoxide	900
	Carbon dioxide	1,250,000
	Sulfur oxides	9,000
	Nitrogen oxides	12,000
	Hydrogen chloride	70
	Hydrogen fluoride	5
	Hydrocarbons	21,000
	Other organics	1
	Metals	5
Water emissions, mg	COD	1,500
	BOD	200
	Acid as H^+	60
	Nitrates	5
	Metals	250
	Ammonium ions	5
	Chloride ions	130
	Dissolved organics	20
	Suspended solids	500
	Oil	200

Table 4-2. *(Continued)*

	Category	Unit Average
	Hydrocarbons	100
	Dissolved solids	300
	Phosphate	5
	Other nitrogen	10
Solid waste, mg	Industrial waste	3,500
	Mineral waste	26,000
	Slags and ash	9,000
	Toxic chemicals	100
	Nontoxic chemicals	800

Source: APME (Brussels), Boustead (1993).

By summing the environmental load units over all the inventory elements, a total environmental load associated with a product or material can be estimated. Volvo has used this system to evaluate product designs and alternative materials. Figure 4-2 and Table 4-5 summarize the application of the EPS system to a car bumper material selection problem.

In critically examining the EPS method, it is clear that the choice of environmental indices can have a major impact on the analysis. The factors involved in developing the EPS system environmental indices are (Horkeby, 1993):

1. *Scope*: the general impression of the environmental impact
2. *Distribution*: the extent of the affected area
3. *Frequency or intensity*: the regularity and intensity of the problem in the affected area
4. *Durability*: the permanence of the effect
5. *Contribution*: the significance of one kilogram of the emission of the substance in relation to the total effect
6. *Remediability*: the relative cost to reduce the emission by one kilogram

The environmental index of a substance is the product of the values assigned to these six factors. Sensitivity analyses can be performed to determine the extent to which changes and uncertainties in these factors change the value of the index.

Note that the EPS system is both qualitative and subjective, which makes it representative of the current state of life-cycle impact assessments.

Table 4-3. Gross inputs and outputs associated with the production of one kilogram of general-purpose polystyrene

	Category	Unit Average
Fuels, MJ	Coal	1.30
	Oil	6.44
	Gas	20.71
	Hydro	0.23
	Nuclear	0.84
	Other	0.14
	Total	29.66
Feedstock, MJ	Coal	<0.01
	Oil	34.34
	Gas	37.38
	Total	71.72
Total Fuel Plus Feedstock		101.38
Raw materials, mg	Iron ore	300
	Limestone	200
	Water	5,000,000
	Bauxite	1,800
	Sodium chloride	14,000
	Clay	20
	Ferromanganese	<1
Air emissions, mg	Dust	3,100
	Carbon monoxide	1,400
	Carbon dioxide	1,600,000
	Sulfur oxides	34,000
	Hydrogen sulfide	2
	Nitrogen oxides	24,000
	Hydrogen chloride	40
	Hydrogen fluoride	1
	Hydrocarbons	26,000
	Metals	10
Water emissions, mg	COD	1,800
	BOD	80
	Acid as H^+	200
	Metals	1,100
	Ammonium ions	10
	Chloride ions	500
	Dissolved organics	50
	Suspended solids	1,000
	Oil	200
	Hydrocarbons	500
	Dissolved solids	500
	Other nitrogen	20

Table 4-3. (*Continued*)

	Category	Unit Average
Solid waste, mg	Industrial waste	3,000
	Mineral waste	14,000
	Slags and ash	5,000
	Toxic chemicals	<1
	Nontoxic chemicals	45,000

Source: APME (Brussels), Boustead (1993).

Table 4-4. A selection of environmental indices from the environmental priority strategies (EPS) system; units are environmental load units (ELU) per kilogram

Raw Materials		Air Emissions		Water Emissions	
Cobalt	76	Carbon dioxide	0.09	Nitrogen	0.1
Chromium	8.8	Carbon monoxide	0.27	Phosphorous	0.3
Iron	0.09	Nitrogen oxide	0.22		
Manganese	0.97	Nitrous oxide	7.0		
Molybdenum	1,500	Sulfur oxide	0.10		
Nickel	24.3	CFC-11	300		
Lead	180	Methane	1.0		
Platinum	350,000				
Rhodium	1,800,000				
Tin	1,200				
Vanadium	12				

Source: Steen and Ryding (1992).

4.2 COMPLEXITIES AND UNCERTAINTIES ENCOUNTERED IN PERFORMING LIFE-CYCLE ASSESSMENTS

The previous section outlined the basic framework for performing life-cycle inventories and life-cycle impact assessments. This section uses a series of case studies to illustrate some of the complexities and ambiguities generally inherent in LCAs.

The first level of uncertainties in life-cycle assessment are uncertainties in the life-cycle inventory. Because inventories are the basis of the remaining steps in life-cycle assessment, these uncertainties have an effect on all the results of an assessment, not just in the inventory stage. Researchers at the U.S. EPA concluded that inventory data may be unobtainable because they are proprietary, because there is no methodology for obtaining data, or because there are insufficient resources devoted to obtaining data (Freeman et al., 1992). Also, sources used to obtain data are rarely presented, except very briefly. In some cases, generic data from national databases, industrial

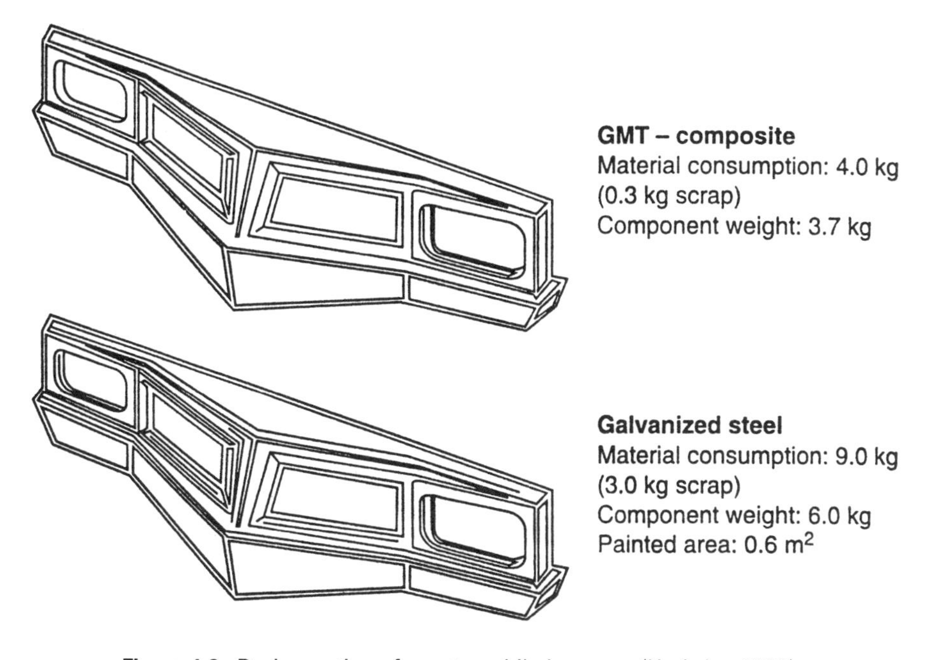

Figure 4-2. Design options for automobile bumpers (Horkeby, 1993).

averages, or professional judgments are used. When it cannot be determined which data come from detailed specific sources and which are derived from other less certain sources, it is impossible to determine whether the uncertainty of the data has a significant impact on the final results of the study.

In the following example, the results of life-cycle inventories for paper and polyethylene grocery sacks that were prepared for the Council for Solid Waste Solutions (an association of plastics manufacturers) are used to illustrate the potential effect of one type of uncertainty common in life-cycle inventories.

Example 4-1 Uncertainties and Assumptions in Life-Cycle Inventories At the supermarket checkstand, consumers are generally asked to choose between having their purchases placed in unbleached paper or polyethylene grocery sacks. Many consumers base this choice on their perception of the relative environmental impacts of these two products. Figures 4-3 and 4-4 give graphic representation to the life-cycle stages of grocery sacks. For paper sacks, there are steps involving timber harvesting, pulping, papermaking, product use, and waste disposal. For polyethylene sacks, there are steps involving petroleum extraction, ethylene manufacture, ethylene polymerization, polymer processing, product use, and waste disposal. The product use stage of the life cycle of both types of sack is less important than the other

Table 4-5. Environmental load units (ELUs) for automobile front ends

Materials and Processes	Production			Product Use[a]			Waste: Incineration			Waste: Reuse			Total
	ELU/kg	kg	ELU	ELU/kg	kg	ELU	ELU/kg	kg	ELU	ELU/kg	kg	ELU	ELU
Plastic GMT Composite													
Production													
GMT material	0.58	4.0	2.32										2.32
Reused production scrap	−0.58	0.3	−0.17										−0.17
Compression molding	0.03	4.0	0.12										0.12
Product use: petrol				0.82	29.6	24.27							24.27
Recycling: GMT material							−0.21	3.7	−0.78				−0.78
Total sum			2.27			24.27			−0.78				25.76
Galvanized Steel													
Production													
Steel material	0.98	9.0	8.82										8.82
Steel stamping	0.06	9.0	0.54										0.54
Reused production scrap	−0.92	3.0	−2.76										−2.76
Spot welding (spots)	0.004	48	0.19										0.19
Painting (m^2)	0.01	0.6	0.02										0.02
Product use: petrol				0.82	48.0	39.36							39.36
Recycling steel material										−0.92	6.0	−5.52	−5.52
Total sum			6.81			39.36						−5.52	40.65

[a]The ELU/kg figure is based on one year of use. For the automobile, an 8-year life is assumed, so the second product use entry is 8 times the actual weight.

Source: Ryding et al. (1993).

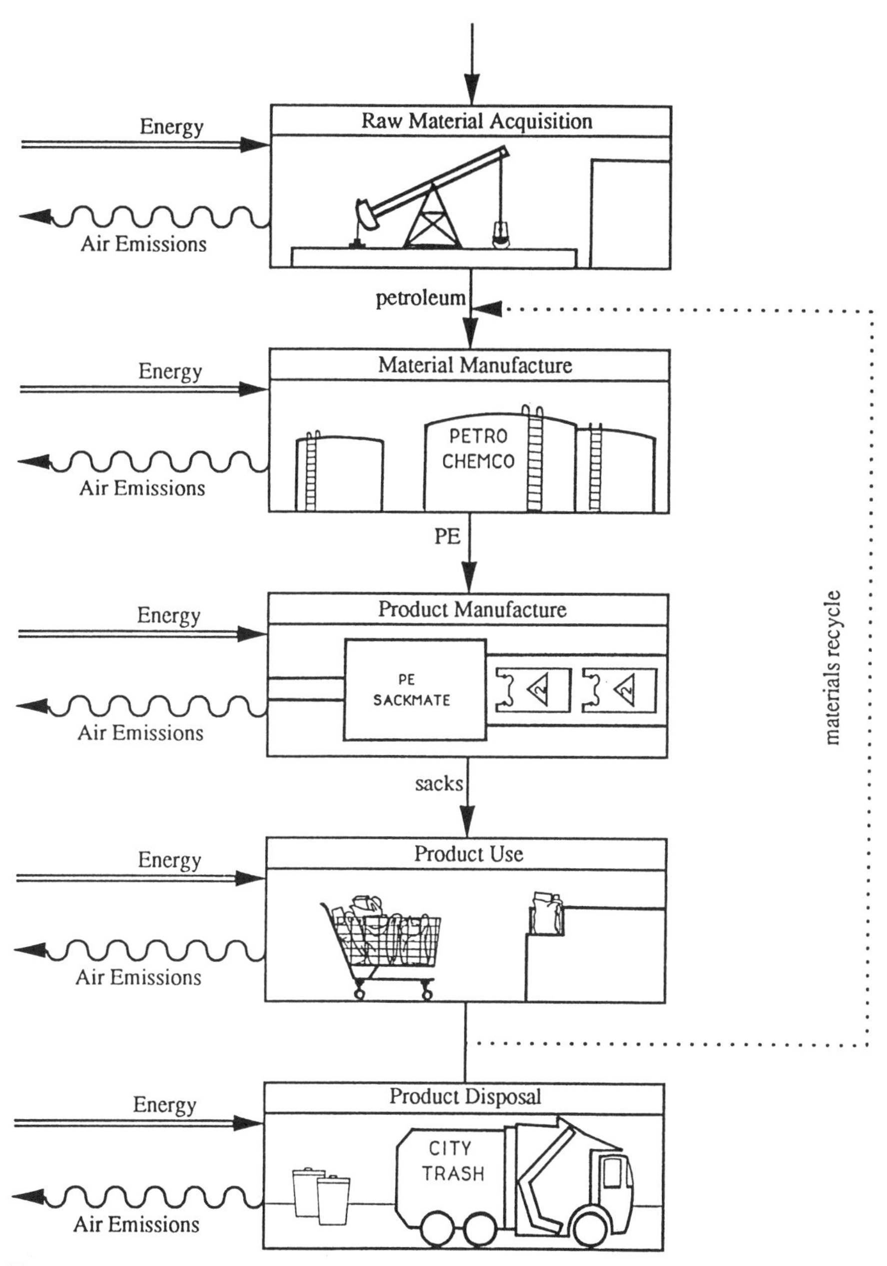

Figure 4-3. Life-cycle inventory framework for polyethylene grocery sacks (Allen et al., 1992). (Copyright © American Institute of Chemical Engineers Center for Waste Reduction Technologies. Published November 1992. ISBN 0-8169-058-1-9, Pub. Code C-1.)

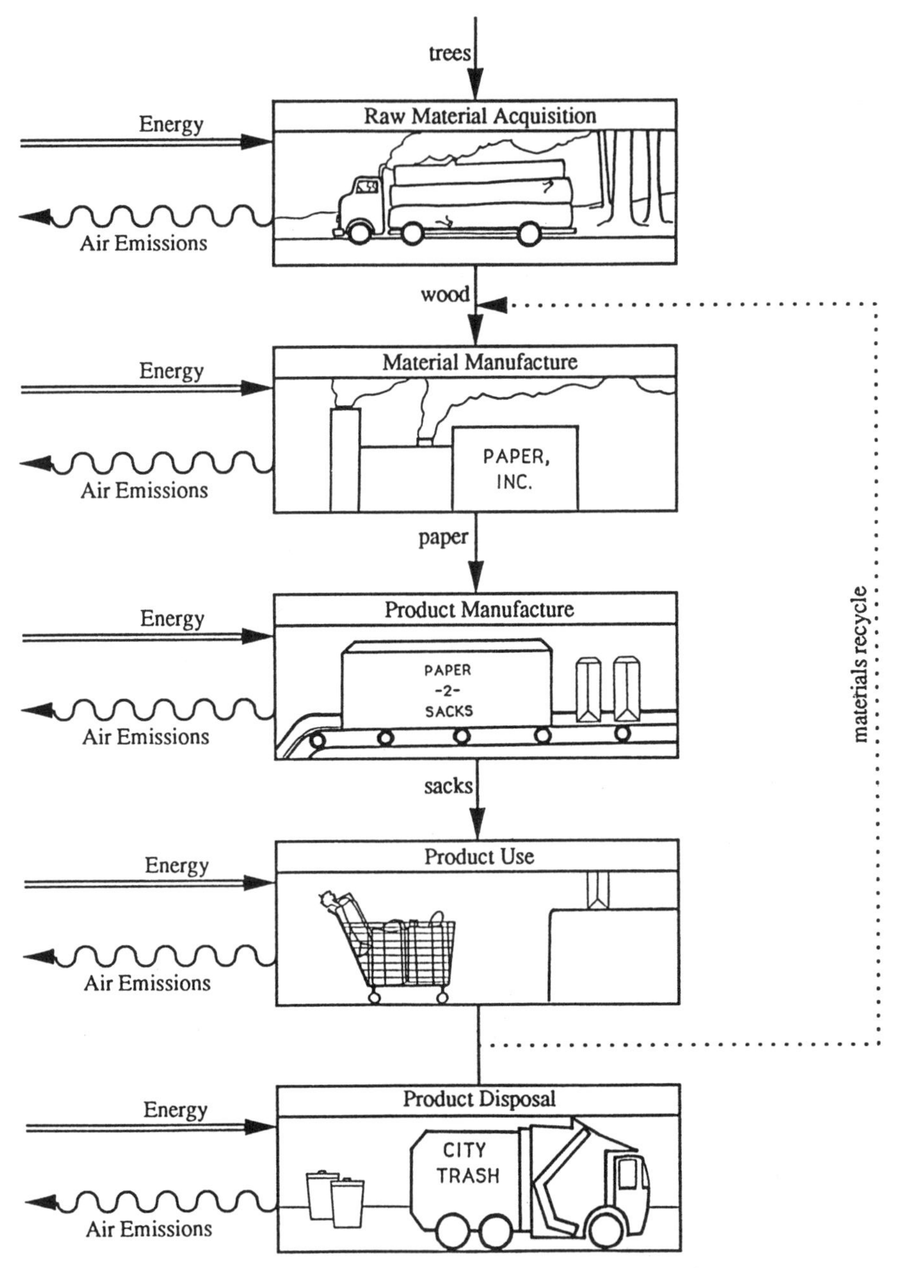

Figure 4-4. Life-cycle inventory framework for paper grocery sacks (Allen et al., 1992). (Copyright © American Institute of Chemical Engineers Center for Waste Reduction Technologies. Published November 1992. ISBN 0-8169-058-1-9, Pub. Code C-1.)

Table 4-6. Air emissions and energy requirements for polyethylene and paper grocery sacks

	Air Emissions, oz/sack		Energy Required, Btu/sack	
Life-Cycle Stages	Paper	Polyethylene	Paper	Polyethylene
Materials manufacture plus product manufacture plus product use	0.0516	0.0146	905	464
Raw-materials acquisition plus product disposal	0.0510	0.0045	724	185

Source: Franklin Associates, Ltd. (1990a). [Reprinted with permission of Franklin Associates, Ltd. and American Plastics Council (formerly Council for Solid Waste Solutions).]

stages in terms of energy requirements and waste generation. Figures 4-3 and 4-4 represent a modification of the general framework for life-cycle inventories given in Figure 4-1. One modification is that the product reuse and product remanufacture loops present in Figure 4-1 are missing in Figures 4-3 and 4-4. This is because almost all recycled grocery sacks are returned to the material manufacture stage rather than undergo product remanufacture or consumer grocery-sack reuse. Also, for the sake of simplicity, the only wastes considered in this example are air emissions, so solid waste and wastewater do not appear in Figures 4-3 and 4-4. Life-cycle energy and air emission inventories of paper and polyethylene grocery sacks are given in Table 4-6. These data might be used as a partial basis for comparing the two products. Before a quantitative comparison can be made, however, differences in product use must be considered. Although both types of sacks are designed to have a capacity of $\frac{1}{6}$ barrel, fewer groceries are generally placed in polyethylene sacks than in paper sacks, even if the practice of double-bagging paper sacks (one sack inside the other), used in some stores, is taken into account. There is no general agreement on the number of polyethylene grocery sacks needed to hold the volume of groceries usually held by a paper sack. Reported values range from 1.2 to 3. Using the data in Table 4-6, determine the amount of energy required and the quantity of air pollutants released both for a single polyethylene sack and for the number of paper sacks required to carry the same volume of groceries as a plastic sack. Perform the calculations at both the lowest reported and highest reported ratio of polyethylene to paper sacks and at 0% and 100% recycle. How much does the uncertainty in the equivalence ratio of paper to plastic sacks affect the results? Does recycling have a similar effect on both types of sacks?

Solution To calculate the air emissions over the life cycle of the number of paper sacks equivalent to one polyethylene sack at 0% recycle, obtain the

Table 4-7. Air emissions and energy requirements for paper and polyethylene sacks over the range of possible equivalencies

Recycle Rate, %	Air Emissions, oz/polyethylene sack equivalent: Paper (1.2:1)	Air Emissions: Paper (3:1)	Air Emissions: Polyethylene	Energy Requirements, Btu/polyethylene sack equivalent: Paper (1.2:1)	Energy Requirements: Paper (3:1)	Energy Requirements: Polyethylene
0	0.0855	0.0342	0.0191	1360	543	649
100	0.0430	0.0172	0.0146	754	302	464

air emissions for the entire life cycle of a paper sack and convert it to the equivalent grocery carrying capacity of a polyethylene sack, as follows:

$$\left(\frac{0.0516\text{ oz}}{\text{paper sack}} + \frac{0.0510\text{ oz}}{\text{paper sack}}\right)\left(\frac{\text{paper sack}}{1.2\text{ polyethylene-sack}}\right) = \frac{0.0855\text{ oz}}{\text{polyethlene-sack equivalent}}.$$

At 100% recycle, there are no raw-materials acquisition or product disposal stages, as follows:

$$\left(\frac{0.0516\text{ oz}}{\text{paper sack}}\right)\left(\frac{\text{paper sack}}{1.2\text{ polyethylene sack}}\right) = \frac{0.0430\text{ oz}}{\text{polyethlene-sack equivalent}}.$$

Results for both types of sack at both recycle rates and equivalence ratios (three polyethylene sacks per paper sack and 1.2 polyethylene sacks per paper sack) are given in Table 4-7. These results are compared graphically in Figure 4-5. Reported energy requirements for the life cycle of both types of sacks are very nearly equal and the choice of the paper : plastic sack ratio for determining equivalent use is crucial to the results. Also, reported energy requirements for paper sacks are more sensitive to recycle rates than are those for polyethylene sacks. Reported air emissions for paper sacks are higher than those of polyethylene sacks for the entire range of equivalence ratios. As with energy requirements, reported air emissions for paper sacks are more sensitive to recycle rates than are those for polyethylene sacks.

In the previous example, a large uncertainty existed in the equivalency factor for the two products. There can also be large uncertainties in the values obtained for emissions, energy use, and raw-material requirements. Air and water emissions, in particular, can be very difficult to determine with any certainty, as shown in the next chapter.

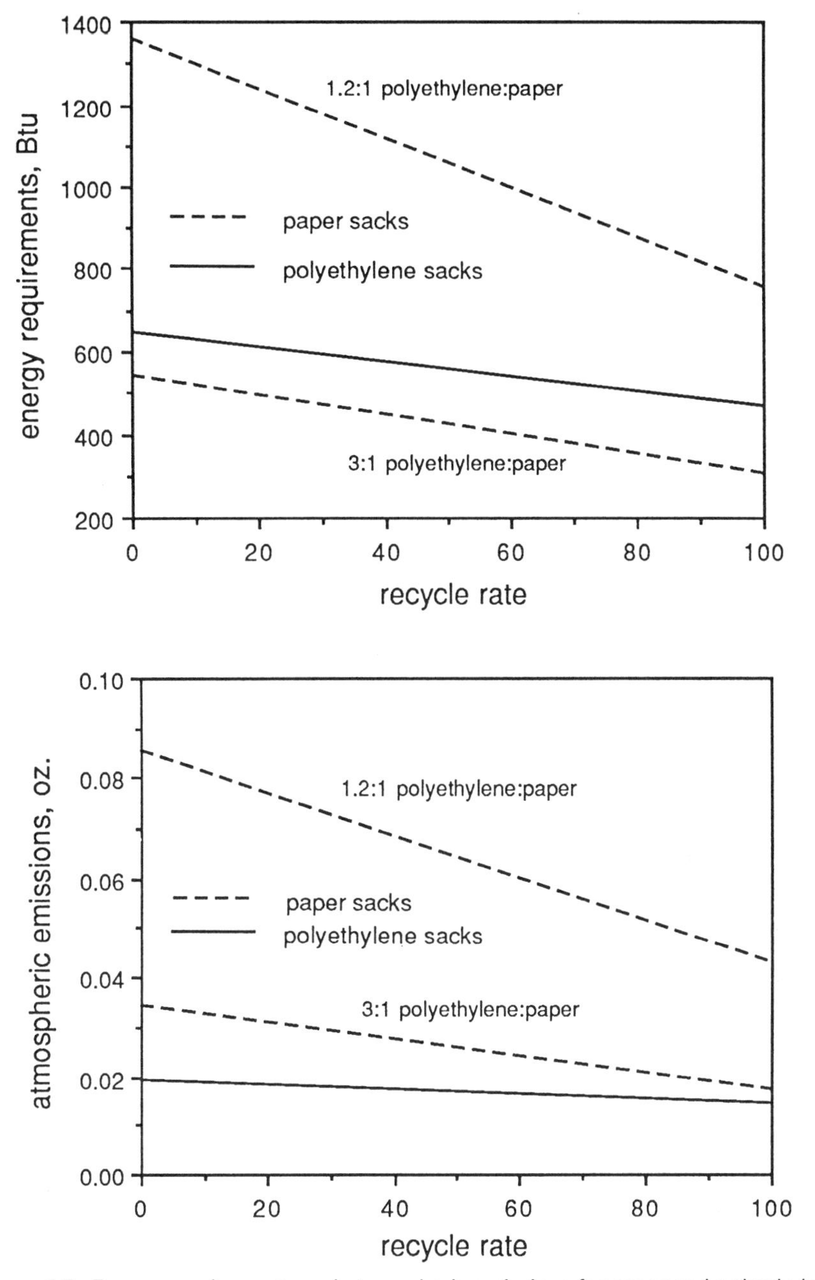

Figure 4-5. Energy requirements and atmospheric emissions for paper and polyethylene grocery sacks as a function of recycle rate.

The uncertainties in the inventory step of life-cycle inventories can also be examined by comparing different life-cycle studies of identical products. This is illustrated in Tables 4-8 and 4-9, where the results of comparisons between paper and polystyrene cups for hot drinks are presented. These data are difficult to compare partly because the findings of the two studies are categorized differently and the study boundaries are not identical. However, if it can be assumed that the categories of water effluents and air emissions that Hocking reported in his summary (Table 4-8) represent a substantial majority of emissions, then the total air emissions and waterborne wastes can be derived from Hocking's data. Also, an upper bound for industrial solid waste can be estimated from Hocking's data by subtracting the mass of the finished product from the total mass of the raw materials. These comparisons are given in Table 4-10. The first thing to note in making these comparisons is that postconsumer solid waste reported in the Hocking study for polystyrene cups is roughly one-third of the postconsumer solid waste reported in the Franklin study (Table 4-9). In contrast, postconsumer solid waste estimates for the paper cups are similar. Atmospheric emission estimates taken from Hocking's data are lower than Franklin's data for both types of hot drink container, which could be because some of the high-mass emissions are not reported in the Hocking paper. Waterborne waste estimates taken from Hocking's data are higher than Franklin's for paper cups and lower than Franklin's for polystyrene cups. Estimates of industrial solid waste taken from Hocking's data are higher than Franklin's data for both types of cup. In some cases the values reported in the two studies are different by more than 100% and even by an order of magnitude for industrial solid waste, which gives an indication of the uncertainties involved in compiling life-cycle inventories.

In addition to the uncertainties inherent in taking an inventory of the environmental burdens of a product, there are complexities associated with the relative importance of some types of impacts over others. These complexities must be addressed in the life-cycle impact assessment, which is the second stage of life-cycle assessment. In the following example, data from the earlier life-cycle inventory of paper and polyethylene grocery sacks are used. This time, the focus is on the qualitative differences in the impacts of air emissions due to the two types of sacks.

Example 4-2 Tradeoffs in Life-Cycle Impacts While polyethylene grocery sacks generate lower amounts of atmospheric emissions then paper sacks at all recycle rates (see Example 4-1), there are qualitative differences between the emissions. Table 4-11 shows the reported differences in air emissions for polyethylene and paper sacks. Is it valid to directly compare the total mass of air emissions due to the two products when determining relative environmental impacts? Also, comment on the differences in impacts between the raw materials used to make paper and the raw materials of polyethylene.

Table 4-8. Raw-material, utility, and environmental summary per 10,000 hot-drink containers

Item	Paper Cup[a]	Polystyrene Cup[b]
Raw materials		
Wood and bark (lb)	620–810	0
Petroleum fractions (lb)	62–120	70
Other chemicals (lb)	40	1.1
Finished weight (lb)	222	33
Wholesale cost	2.5×	×
Utilities		
Steam (lb)	2000–2700	~160
Power (Btu)	340,000	6100–9200
Cooling water (ft^3)	200	82
Water effluent		
Volume (ft^3)	200–680	0.3–1
Suspended solids (lb)	7.8–10	Trace
BOD (lb)	7–10	0.002
Organochlorines (lb)	1–2	0
Metal salts (lb)	0.2–4	0.7
Air emissions		
Chlorine (lb)	0.1	0
Chlorine dioxide (lb)	0.04	0
Reduced sulfides (lb)	0.4	0
Particulates (lb)	1–3.3	0.003
Pentane (lb)	0	1.2–1.6
Sulfur dioxide (lb)	~2	~0.3
	Recycle potential	
To primary user	Possible, though washing can destroy	Easy, negligible water uptake
After use	Low, hot melt adhesive or coating difficulties	High, resin reuse in other applications
	Ultimate disposal	
Proper incineration	Clean	Clean
Heat recovery (Btu)	2×10^6	0.6×10^6
Mass to landfill (lb)	222	33
Biodegradable	Yes; BOD to leachate, methane to air	No; essentially inert

[a]Made from fully bleached kraft pulp; information from British Columbia pulp mills. Note that these data have been severely criticized; see Letters to *Science*, June 7, 1991.
[b]Made from molded polystyrene foamable beads.

Source: Modified with permission from M. B. Hocking, "Paper Versus Polystyrene," *Science*, Vol. 251, pp. 504–505, Feb. 1991. Copyright © 1991 American Association for the Advancement of Science.

Table 4-9. Environmental impact data per 10,000 cups

	Atmospheric Emissions	Waterborne Wastes	Industrial Solid Waste		Postconsumer Solid Waste	
Type of Cup	(lb)	(lb)	(lb)	(ft^3)	(lb)	(ft^3)
Process pollutants						
Foam polystyrene	5.0	1.7	5.2	0.1	120.3	13.7
LDPE[a]—coated paperboard	7.4	2.0	28.1	0.6	218.3	7.3
Wax-coated paperboard	7.0	3.1	32.5	0.7	266.2	8.8
Fuel-related pollutants						
Foam polystyrene	6.8	0.5	13.4	0.3	—	—
LDPE-coated paperboard	10.7	1.0	26.2	0.5	—	—
Wax-coated paperboard	14.8	1.4	38.5	0.8	—	—
Total Pollutants						
Foam polystyrene	11.8	2.1	18.6	0.4	120.3	13.7
LDPE-coated paperboard	18.1	2.9	54.3	1.1	218.3	7.3
Wax-coated paperboard	21.8	4.5	71.0	1.4	266.2	8.8

[a]Low-density polyethylene.

Source: Franklin Associates, Ltd. (1990b). [Reprinted with permission of Franklin Associates Ltd. and American Plastics Council (formerly Council for Solid Waste Solutions).]

Table 4-10. Life-cycle inventory data for hot-drink containers from two practitioners (per 10,000 cups)

	Mass per 10,000 Hot-Drink Containers, lb			
	Foam Polystyrene		LDPE-Coated Paperboard	
Type of Impact	Franklin[a]	Hocking[b]	Franklin[a]	Hocking[b]
Atmospheric emissions	11.8	1.5–1.9[c]	18.1	3.5–5.8[c]
Waterborne wastes	2.1	0.7	2.9	9–16
Industrial solid waste	18.6	38[d]	54.3	500–750[d]
Postconsumer solid waste	120	33	220	222

[a]Franklin (1990b).
[b]From Hocking (1991).
[c]Estimate of total based on summation of individually reported emissions.
[d]Upper bound based on total mass of raw materials less mass of finished product.

Table 4-11. Profile of atmospheric emissions for paper and plastic grocery sacks

	Atmospheric Emissions Per 10,000 Sacks (lb)			
	Polyethylene Sacks		Paper Sacks	
Pollutant Category	0% Recycling	100% Recycling	0% Recycling	100% Recycling
Particulates	0.8	0.8	24.6	2.8
Nitrogen oxides	2.1	1.7	9.2	8.0
Hydrocarbons	5.8	3.2	4.9	3.9
Sulfur oxides	2.6	2.7	13.6	10.6
Carbon monoxide	0.7	0.6	7.0	6.5
Aldehydes	0.0	0.0	0.1	0.1
Other organics	0.0	0.0	0.3	0.2
Odorous sulfur	—	—	4.5	0.0
Ammonia	0.0	0.0	0.0	0.0
Hydrogen fluoride	—	—	—	—
Lead	0.0	0.0	0.0	0.0
Mercury	—	—	—	—
Chlorine	—	—	—	—

Source: Franklin Associates, Ltd. (1990a). [Reprinted with permission of Franklin Associates Ltd. and American Plastics Council (formerly Council for Solid Waste Solutions).]

Solution The emissions of particulates, nitrogen oxides, and sulfur oxides are higher for paper sacks than for polyethylene. As might be expected, higher levels of hydrocarbon emissions are assigned to polyethylene sacks. These hydrocarbons are also very likely to be qualitatively different from the hydrocarbon emissions generated by paper-sack production. It would be difficult to assess the respective environmental impacts of the hydrocarbon emissions without a much more detailed description of the emissions. Also, lack of emission data from other sources within the life cycle (i.e., incineration and emissions from landfills) makes the comparison of polyethylene and paper sacks incomplete and any comprehensive comparison difficult. The raw material used to make polyethylene sacks is petroleum, a nonrenewable resource, whereas paper sacks are based on forest resources that are generally regarded as renewable. These different raw-material requirements change the overall environmental impacts of these two products.

While in Example 4-1 the data in the inventory suggested that polyethylene sacks, over their entire life cycle, created fewer air emissions and possibly required less energy than did paper sacks, it should now be clear that in most cases inventories alone cannot be used to identify the most environmentally benign option from a set of very different alternatives.

These examples are intended to illustrate two key facets of life-cycle assessment. First, the complexity of the data needed for the inventory necessitates that a number of assumptions and estimates be accepted, and the results of the inventory depend on these assumptions. Second, even if the data produced in a life-cycle inventory were above reproach, an assessment of the impacts described in the inventory would be necessary before an environmentally preferable product or process can be identified. This assessment introduces a whole new tier of assumptions and estimates.

4.3 APPLICATIONS OF LIFE-CYCLE ASSESSMENT

Even with their limitations, as outlined above, life-cycle assessments have been used in a variety of applications in both the public and private sectors. Many of these assessments have been well publicized. Among the applications of LCA that are particularly well known are comparisons between cloth and disposable diapers (Franklin Associates, Ltd., 1990c), plastic and paper cups (Franklin Associates, Ltd., 1990b, Hocking, 1991), and polystyrene clamshells and coated paper wrappers for hamburgers (see Svoboda and Hart, 1993). These studies have done much to raise awareness about LCA as an analysis tool, but they have also generated a significant amount of confusion and skepticism about the value of LCA.

This confusion and skepticism concerning specific product comparisons has diverted attention away from many applications of life-cycle assessments and concepts that are less controversial. In this section, the spectrum of uses to which LCA has been applied is described and the various applications are illustrated with case studies. Some of the common motivations for performing an LCA, as identified by product manufacturers, are identified in Figure

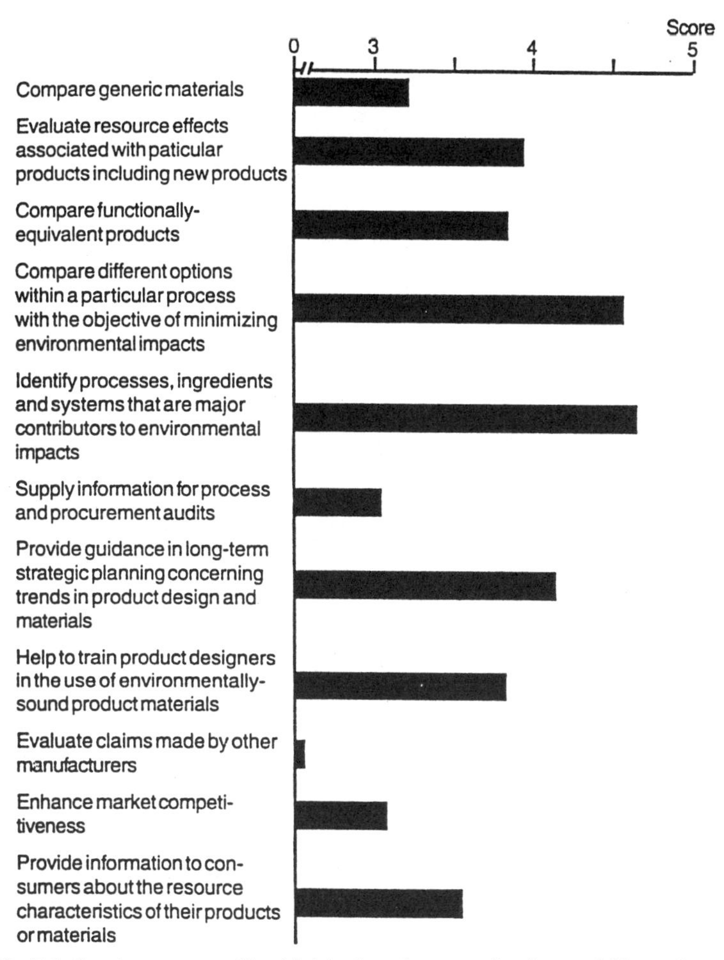

Figure 4-6. Relative importance (5 = high) of various applications of life-cycle assessment, as assigned by product manufacturers surveyed by the Swedish Waste Research Council (Ryding, 1994).

4-6. Figure 4-7 provides similar data from the viewpoint of public policymakers. These data were assembled by the Swedish Waste Research Council (Ryding, 1994) and were based on an international survey of over 40 organizations actively involved in LCA. The data indicate that the most common reasons for performing an LCA are to improve the environmental performance of products and to inform long-term policy decisions. Product comparisons are a less common motivation. Figure 4-8 expands on the information needs that drive LCAs and Figure 4-9 indicates that one of the most valued features of LCAs are their emphasis on multiattribute interdisciplinary analysis.

These data indicate that many information needs are driving the growing

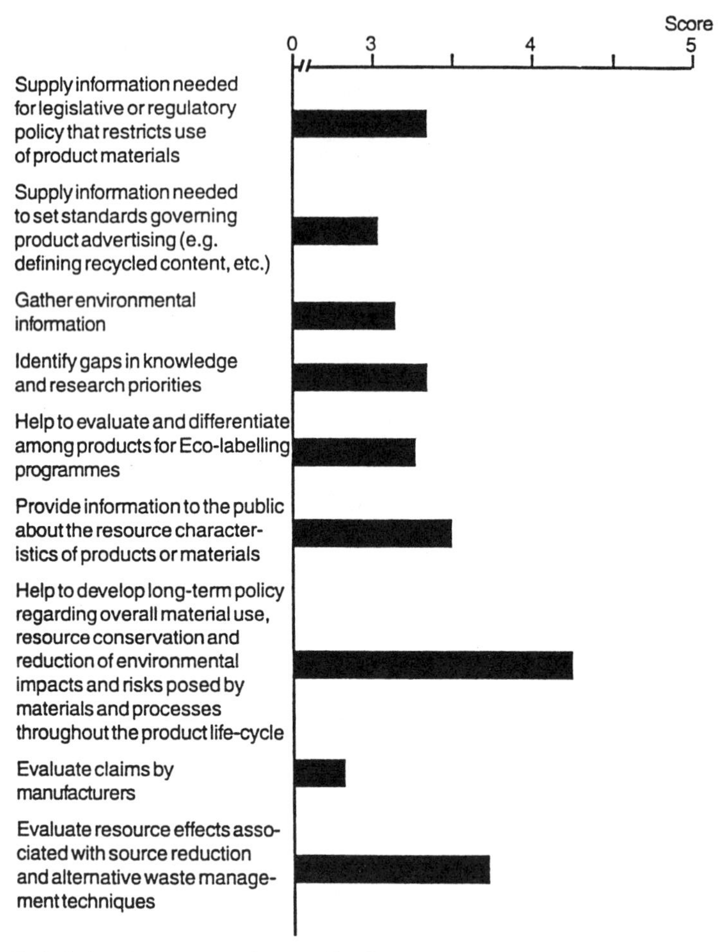

Figure 4-7. Relative importance (5 = high) of various applications of life-cycle assessment, as assigned by public policymakers surveyed by the Swedish Waste Research Council (Ryding, 1994).

use of LCA. In this section, the current uses of LCA and life-cycle concepts are organized into a number of public- and private-sector application categories. Examples of LCAs that address product improvement, product comparisons, and strategic planning in the private sector are discussed. Ecolabeling and the use of life-cycle assessments in drafting regulations are among the public-sector applications, and they are also illustrated in this section with descriptions of case studies.

4.3.1 Product Improvement

One of the most common uses of life-cycle assessments is in identifying critical areas in which the environmental performance of a product can

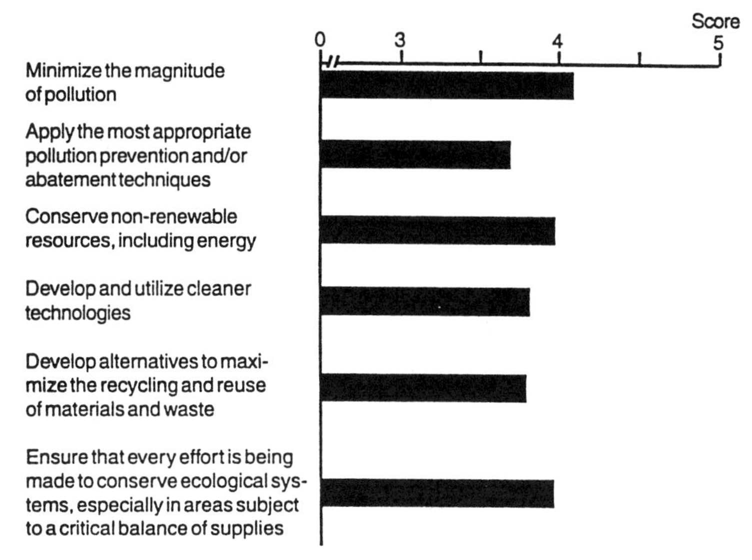

Figure 4-8. Relative importance (5 = high) of goals defined for life-cycle assessment (Ryding, 1994).

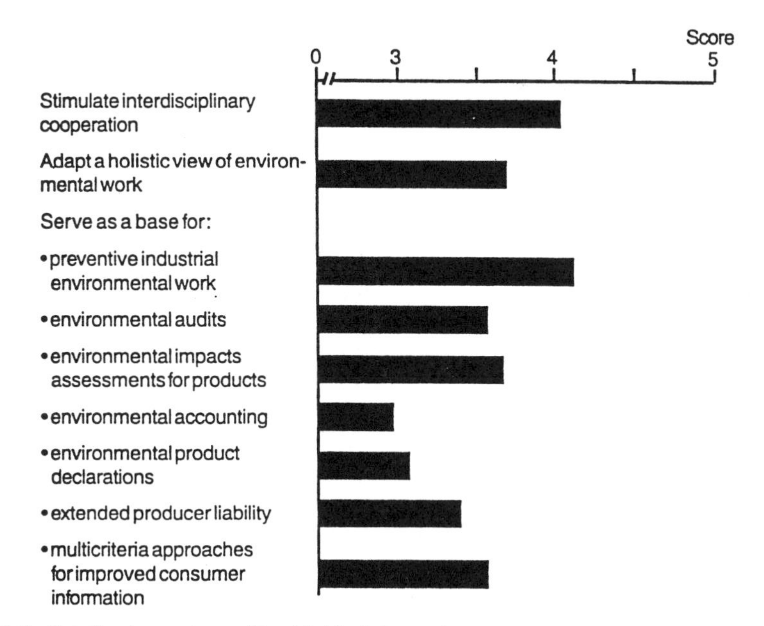

Figure 4-9. Relative importance (5 = high) of the various advantages of life-cycle assessment (Ryding, 1994).

Table 4-12. Average gross energy required to produce one kilogram of polyethylene (all grades) in Europe

Fuel Type	Fuel Production and Delivery (MJ)	Delivered Energy (MJ)	Feedstock[a] Energy (MJ)	Total Energy (MJ)
Electricity	5.31	2.58	0.00	7.89
Oil fuels	0.53	2.05	32.76	35.34
Other fuels	0.47	8.54	33.59	42.60
Totals	6.31	13.17	66.35	85.83

[a]Feedstock energy is defined as the calorific value of materials that are input into the processes required to produce polyethylene.

Source: APME (Brussels), Boustead (1993).

be improved. Often, a life-cycle assessment will only confirm conventional wisdom. For example, the data shown in Table 4-12 indicate that the vast majority of the fuel required to produce a kilogram of polyethylene is the oil, coal, or other hydrocarbon that is diverted from fuel use in order to make ethylene and polyethylene. Thus, improvements in the energy efficiency of polyethylene products are most effectively made by reducing the mass of polyethylene. This is not a surprising result to anyone familiar with chemical manufacturing processes. Some life-cycle assessments, however, do yield surprises. Consider the data presented in Figure 4-10, which reports the results of a life-cycle inventory prepared for the American Fiber Manufacturers Association. This study examined the life cycle of a woman's blouse made of 100% knitted filament polyester (Franklin Associates, 1993). The study revealed that 82% of the energy consumed during the life cycle of a blouse (50 wearings total with 2 wearings between washings) is associated with hot water cleaning and machine drying. As shown in Figure 4-10, the energy demand associated with cleaning could be reduced by 90% if the garment is washed in cold water and line- rather than machine-dried. This life-cycle assessment demonstrated that the greatest opportunities for improved energy efficiency for garments are in the development of fabrics, dyes, and detergents that are compatible with cold-water washing.

Another life-cycle inventory that yielded a surprising result was done for Patagonia, Inc. (Hopkins et al., 1994). Patagonia is a privately held company known primarily for the high-performance outdoor clothing and travel-related products that it designs and markets. Sales and distribution are handled through dealers, Patagonia owned retail outlets, and mail order. The study focussed on the transportation and distribution of one of Patagonia's garments throughout its life cycle. The rationale for focusing on environmental improvements in the transportation and distribution of a garment may not be immediately apparent. Previous studies of garment life cycles have not found transportation and distribution to be significant in their impacts.

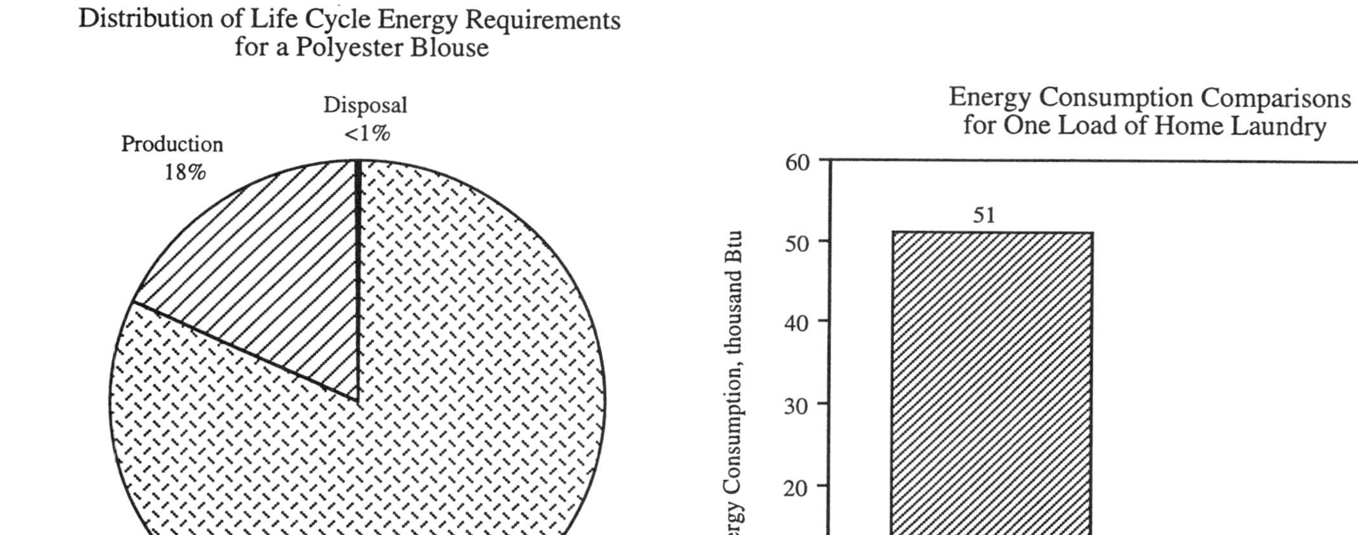

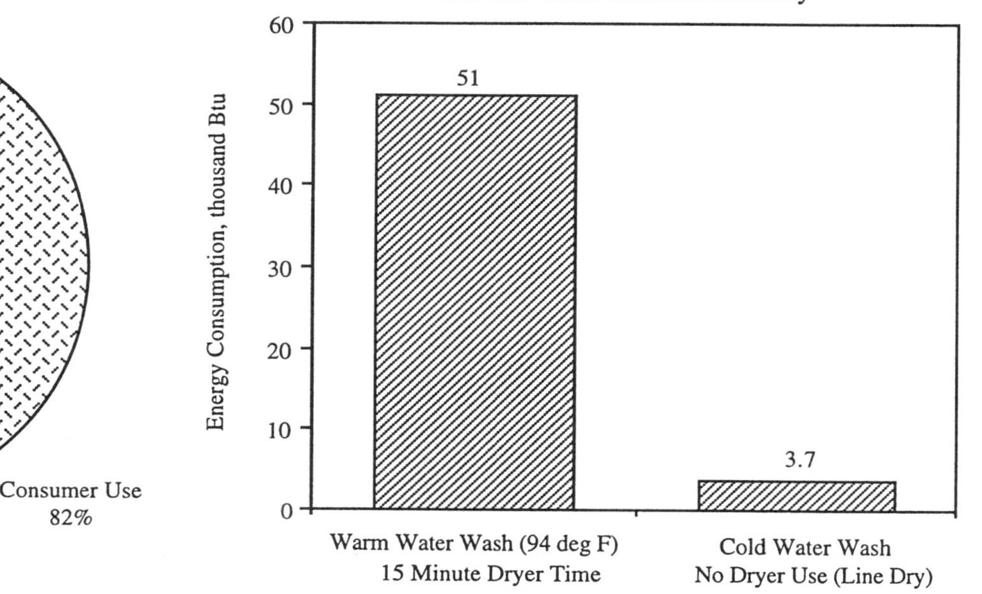

Figure 4-10. A life-cycle inventory of a woman's blouse revealed that the greatest improvements in the energy efficiency of the life cycle could be obtained by changing consumer washing and drying methods (Franklin Associates, Ltd., 1993). [Reprinted with permission of Franklin Associates, Ltd. and American Fiber Manufacturers Association, Inc.]

Such studies generally assume, however, that transport of products and material is done exclusively by truck and rail. These modes of transport are efficient, but they are seldom rapid enough for Patagonia's customer demands. Limited space in retail outlets and highly seasonal sales necessitates small but frequent product deliveries. For example, during the holiday season, the urgency of keeping store shelves stocked often means that products are shipped by air. The use of air transport dramatically changes the relative importance of transportation and distribution in the life-cycle inventory for a garment. Estimates of the energy burdens associated with air deliveries and combined truck and rail (intermodal) deliveries compared to manufacturing energy burdens are shown in Figure 4-11.

Another group of applications in which life-cycle assessments can lead to new insights are those that involve products with multiple components. Consider the life-cycle inventory of a computer workstation, performed by an industry team coordinated by the Microelectronics and Computer Technology Corporation (MCC, 1993). Workstations contain a variety of major and minor components, including the cathode-ray tube (display), plastic housings, semiconductors, and printed wiring boards. The life-cycle inventory conducted by MCC was able to identify, for a variety of environmental inventory categories, which components were of primary concern. For example, product disposal was dominated by issues related to cathode-ray tubes. Hazardous waste generation was dominated by semiconductor manufacturing. Energy usage was dominated by the consumer use stage of the life cycle. Somewhat surprisingly, semiconductor manufacturing was a significant component of material use. Results are shown in Figure 4-12. This study was used to prioritize research and technology needs for the microelectronics industry.

A final type of application in which LCA can be useful in product improvement is the evaluation of product design options. Table 4-13 illustrates this use of LCA. A variety of product reformulations and packaging alternatives for a liquid fabric conditioner are presented in this table, and the energy and solid waste burdens associated with each of these options are compared, revealing some of the tradeoffs associated with each design.

The examples cited above demonstrate that life-cycle assessments of products can be valuable from a variety of perspectives. They can help to identify areas for environmental improvement that, at times, can be surprising. They can help identify processes, components, ingredients, and systems that are major contributors to environmental impacts and they can be used to compare options for minimizing environmental impacts.

4.3.2 Product Comparisons

There is a general consensus supporting the use of life-cycle inventories and life-cycle assessments in targeting areas and identifying methods for the environmental improvement of a product. Far more controversial, however,

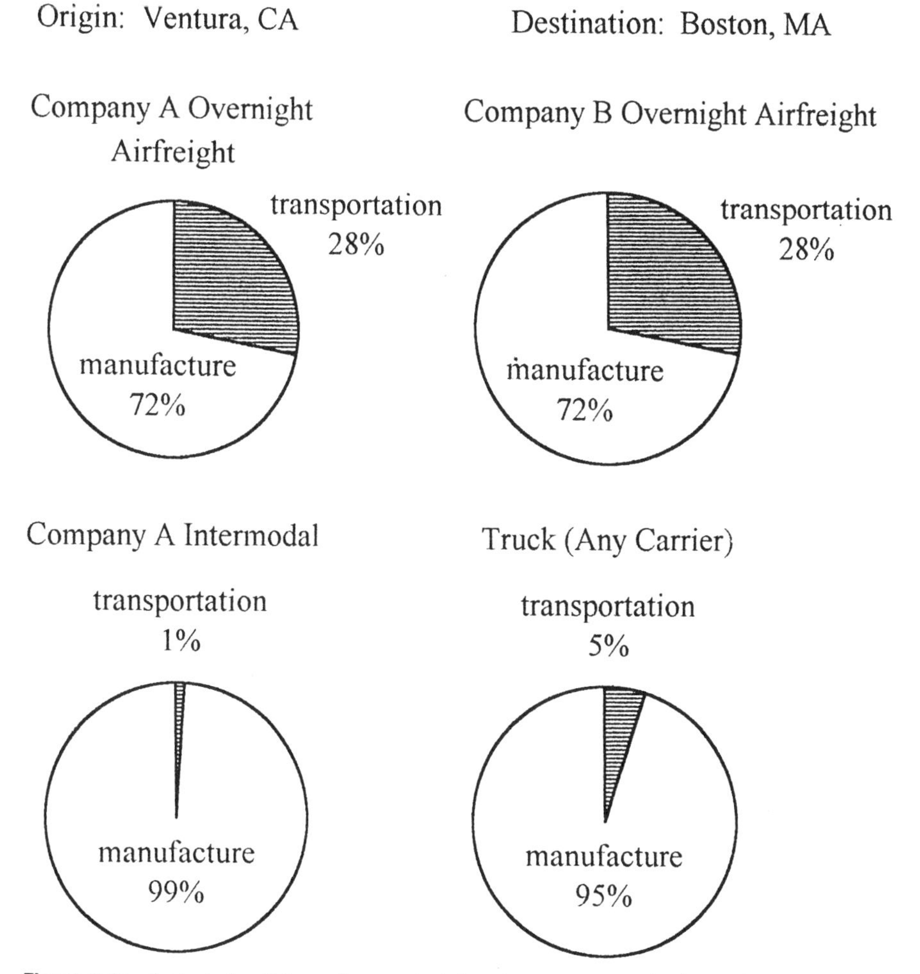

Figure 4-11. A study for Patagonia revealed that the energy required to deliver a garment to a customer or a retail outlet by overnight airfreight can make the transportation component of a garment life cycle significant. (From "Quantifying and Reducing Environmental Impacts Resulting from Transportation of a Manufactured Garment," L. Hopkins, D. T. Allen, and M. Brown, *Pollution Prevention Review*, copyright © 1994 Executive Enterprises, Inc. Reprinted by permission of John Wiley & Sons, Inc.)

is the use of life-cycle inventories and life-cycle assessments to compare products that serve similar functions. Three case studies are presented in this section to help illustrate some of the controversies: a comparison of cloth and disposable diapers, a comparison of beverage packaging systems, and a comparison of sandwich packaging products.

A comparison between cloth and disposable diapering systems illustrates several critical features of product comparisons. First, it highlights the ambi-

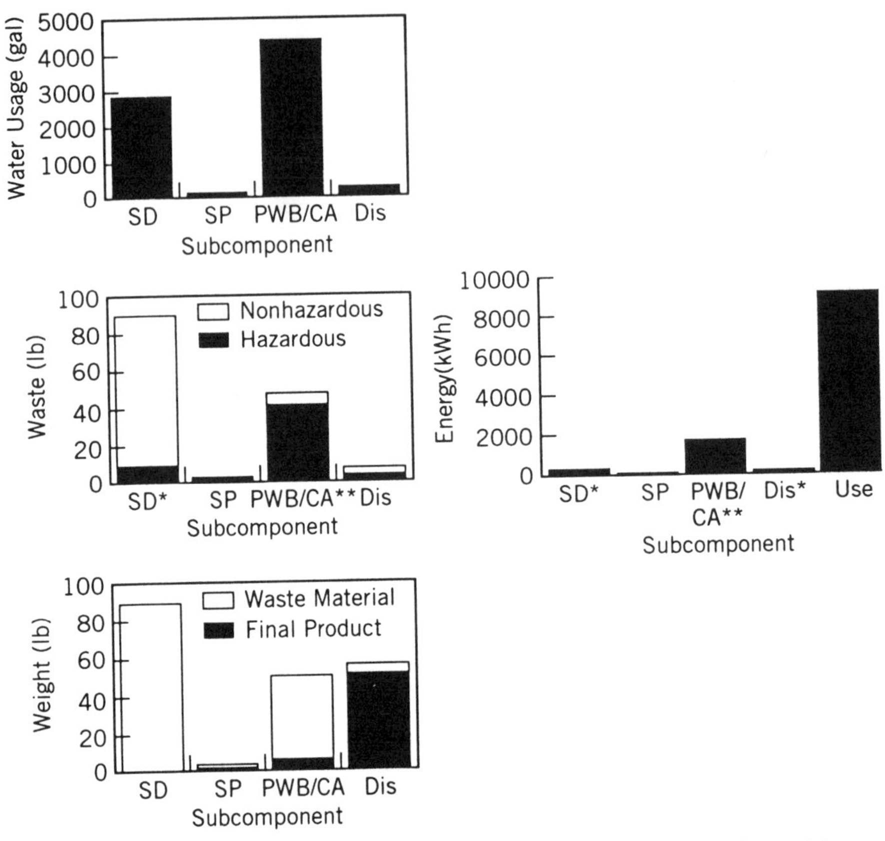

Figure 4-12. Energy consumption during production and consumer use and material use, water use, and waste generation during manufacture for a computer workstation (MCC, 1993). Note that different subcomponents of the workstation (SD = semiconductor, SP = semiconductor packaging, PWB/CA = printed wiring board and computer assemblies, DIS = display) dominate different inventory categories.

guity associated with determining the environmental superiority of one product over another. Data reported in Table 4-14 indicate that the use of cloth diapers lowers the generation of solid waste but increases water use relative to disposable diapers. Energy requirements may or may not be significantly different between the two systems, depending on whether the cloth diapers are home- or commercially laundered. Thus, there is no clear environmentally superior product unless the relative importance of energy demand, water demand, and solid waste generation can be explicitly evaluated. A second point illustrated by the comparison of diapering systems is related to product reuse. Cloth that has been retired from diapering finds many applications, including use as rags where the cloth diaper may replace the

Table 4-13. Summary of a life-cycle inventory of packaging alternatives for a liquid fabric conditioner

Strategies for Packaging Improvement	% Decrease in Energy Needs			% Decrease in Emissions		
	Process	Transport	Feedstock	Solid	Aqueous	Airborne
Incorporate 25% recycled plastic	3	0	9	9	(+4)	4
Encourage 25% consumer recycling	3	2	11	11	(+4)	5
Triple-concentrate (3×) product	55	53	56	55	54	55
Market product soft pouch	3	18	67	85	(+12)	24
Market 3× product in soft pouch	68	73	89	95	63	75
Market 3× product in paper carton	53	58	94	91	40	62
Encourage 25% composting for 6	53	58	94	92	40	62

Source: Procter & Gamble (1993).

Table 4-14. Relative environmental burdens for cloth and disposable diapers

Impact	Disposable	Cloth	
		Commercially Laundered	Home-Laundered
Energy requirements	0.50	0.55	1.0
Solid waste	4.1	1.0	1.0
Waterborne wastes	0.14	0.95	1.0
Water volume requirements	0.27	1.3	1.0

Source: Based on data from Franklin Associates, Ltd. (1990c).

use of a paper towel. The methods used for assigning material, energy, and waste credits for these end-of-life uses influences the overall burdens associated with the cloth diaper, but not the disposable diaper. Thus, it is virtually impossible to treat the cloth and disposable products in a consistent manner in the comparison. Finally, as with paper and plastic grocery sacks, comparison of diapering systems illustrates the importance of determining product equivalency. In order to effectively compare any two product systems, the material use, energy use, and waste generation must be defined on an equivalent basis. In the case of comparing diapering systems, the functional unit for determining product equivalency might be the diapering of a baby for a week. With this as the functional unit, it is, in principle, possible to determine how many cloth and disposable diapers would be required by an average child. In practice, the exercise is not simple since cloth diapers generally require more frequent changes, some consumers use multiple cloth diapers for a single change, and because diapering is not the only use of cloth diapers in a home with small children.

The results of a life-cycle assessment performed to compare different soft-drink delivery systems reinforces the importance of the choice of functional unit. Data given in Table 4-15 are from an LCA of 12-oz aluminum cans, 16-oz nonrefillable glass bottles, and 64-oz polyethyleneterephthalate (PET) plastic bottles. Two choices of functional unit are presented: 1000 gal of beverage delivered and a container delivered to the customer. Clearly, the choice of functional unit matters. The large PET bottle is efficient when the functional unit is defined as a fluid volume delivered to the customer but is inefficient if it is assumed that the customer will consume a serving, whether it is 12, 16, or 64 oz. In reality, it is probably inappropriate to compare the 64 oz bottle with 12- and 16-oz containers since the larger product is typically used for multiple servings and the smaller containers are generally used for single servings. However, it is not at all clear what the appropriate functional unit should be in comparing 12-oz aluminum cans and 16-oz PET and glass bottles. This example, while somewhat artificial, dramatizes the importance and the difficulty associated with choosing a functional unit when comparing products with similar uses.

Despite all the challenges associated with fairly conducting product comparisons, life-cycle inventories are used to discriminate between products. The decision by McDonald's to convert from polystyrene to paper-based quilt-wrap containers for their food products is one of the best known cases. In announcing their decision on November 2, 1990, McDonald's cited a 90% reduction in the volume of their packaging and reductions in energy consumption, air emissions, and water pollution as key factors in the decision. Data supporting these claims were drawn from a life-cycle inventory and are summarized in Table 4-16. In making this decision, McDonald's was effectively withdrawing as a major contributor to the effort to develop a recycling infrastructure for polystyrene, and it is unclear what the results of the product comparison would have been if aggressively recycled polystyrene were compared to the quilt wrap products that are difficult to recycle. As a result, the response to the decision was mixed. The Environmental Defense Fund's January 1991 newsletter called it a "major victory for environmentalists." The *New York Times* described the "Greening of the Golden Arch" by stating that "McDonald's is at last showing some McSense on the environment." In contrast, the *Los Angeles Times* concluded that McDonald's "found itself doing the wrong thing for the wrong reason." A comprehensive examination of this case study has been performed by the National Pollution Prevention Center at the University of Michigan (Svoboda and Hart, 1993), highlighting the uncertainties in this complex decision.

Table 4-15. Energy and environmental impacts for three soft drink container groups using two different functional units

Container Group	Recycling Rates 0%	50%	100%	Container Group	Recycling Rates 0%	50%	100%
	Energy Required/1000 gal(MM Btu)				Energy Required/Container (Btu)		
PET (64-fl-oz bottles)	21.2	17.9	14.6	PET (64-fl-oz bottles)	10,600	8,950	7,300
Aluminum (12-fl-oz can)	50.0	32.9	15.9	Aluminum (12-fl-oz can)	4,687	3,084	1,491
Glass (16-fl-oz bottles)	49.0	35.0	20.9	Glass (16-fl-oz bottles)	6,125	4,375	2,612
	Atmospheric Emissions/1000 gal (lb)				Atmospheric Emissions/Container (oz)		
PET (64-fl-oz bottles)	62.0	53.4	44.8	PET (64-fl-oz bottles)	0.50	0.43	0.36
Aluminum (12-fl-oz can)	137.0	91.7	48.3	Aluminum (12-fl-oz can)	0.21	0.14	0.07
Glass (16-fl-oz bottles)	217.4	145.4	73.5	Glass (16-fl-oz bottles)	0.43	0.29	0.15
	Solid Waste/1000 gal (lb)				Solid Waste/Container (oz)		
PET (64-fl-oz bottles)	513.1	351.3	189.5	PET (64-fl-oz bottles)	4.1	2.8	1.5
Aluminum (12-fl-oz can)	1,938	1,068	198.2	Aluminum (12-fl-oz can)	2.9	1.6	0.3
Glass (16-fl-oz bottles)	7,000	3,881	762.3	Glass (16-fl-oz bottles)	14.0	7.8	1.5

Source: Based on data from Franklin Associates, Ltd. (1989). (Reprinted with permission of Franklin Associates, Ltd.; NAPCOR.)

Table 4-16. Energy requirements and environmental emissions for the production, delivery, and disposal of 10,000 sandwich packaging products

Type of Packaging	Total Energy Requirement (MM Btu)	Energy Credit from Incineration (MM Btu)	Net Energy Requirement (MM Btu)	Atmospheric Emissions (lb)	Waterborne Wastes (lb)	Total Solid Waste (lb)	Total Solid Waste (ft^3)
Standard paper wrap	1.5	0.1	1.4	4.5	0.8	63.7	2.0
Layered paper wrap	3.5	0.2	3.3	9.7	1.4	129.5	4.1
Polystyrene foam container	6.5	0.4	6.1	13.8	2.5	159.8	16.5
Paperboard container	9.2	0.5	8.8	25.7	4.3	382.4	11.7
Paperboard collar (optional for use with either wrap)	2.7	0.1	2.5	8.3	1.4	117.1	3.5

Source: Svoboda and Hart (1993).

4.3.3 Strategic Planning

The McDonald's case study described above illustrates how life-cycle inventories have been used at a corporate level to select material suppliers. LCAs can play other roles in corporate decisionmaking if the definition of life-cycle assessment is broadened to include qualitative and semiquantitative life-cycle concepts. "The life cycle concept is based on the recognition that a 'cradle to grave' perspective is critical to any evaluation" and that "an inherently integrated concept . . . is the best way to allow for the evaluation of economic, environmental and energy dimensions of a problem at the same time" (SETAC, 1994).

AT&T has used life-cycle concepts in a variety of settings. Using the qualitative matrices shown in Figure 4-13, AT&T evaluated a variety of substitutes for lead solder. As shown in this figure, qualitative evaluations of the environmental, manufacturing, toxicity, and political factors that impact the choice of a solder throughout the life cycle of a set of alternative materials are depicted in matrices. Thus, the approach taken by AT&T builds on the traditional methods of life-cycle assessment, which consider only environmental impacts, by adding in manufacturing and other factors. These other factors are evaluated throughout the product life cycle, and the information is then condensed into a summary matrix that helps the designer evaluate alternatives. In this case the semiquantitative analysis using life-cycle concepts led AT&T to conclude that bismuth, indium, and epoxy solders were not preferable to the existing blend containing lead (Allenby, 1992). AT&T is also developing a semiquantitative life-cycle-based approach to evaluating the environmental concerns associated with their facilities. For each facility, the ecological impacts, energy usage, solid wastes, liquid wastes, and gaseous wastes are evaluated for the life cycle of the facility, including siting, principal business activities, facility operations, and closure (Allenby, 1994).

Scott Paper Company, as well as Dow Chemical Company and AT&T, has employed analysis frameworks that utilize life-cycle concepts to strategically evaluate product lines or core businesses. The framework used by Scott evaluates resource and environmental issues at each stage of the life cycle, including natural resources, raw materials, manufacture, product/packaging development, and product use and disposal. At each life-cycle stage, ecological and human health factors are evaluated. Examples of this scoring system for human health factors in the manufacturing stage are given in Table 4-17. A similar framework, used by Dow, is given in Figure 4-14. Each framework addresses a range of impacts and a variety of stages in the life cycles of their products.

Rohm and Haas, a pharmaceuticals manufacturer, has used a somewhat simpler scheme that is very similar to Scott's. The framework for their methodology is shown in Figure 4-15. This framework is used to semiquantitatively evaluate human health and environmental effects over a portion of a

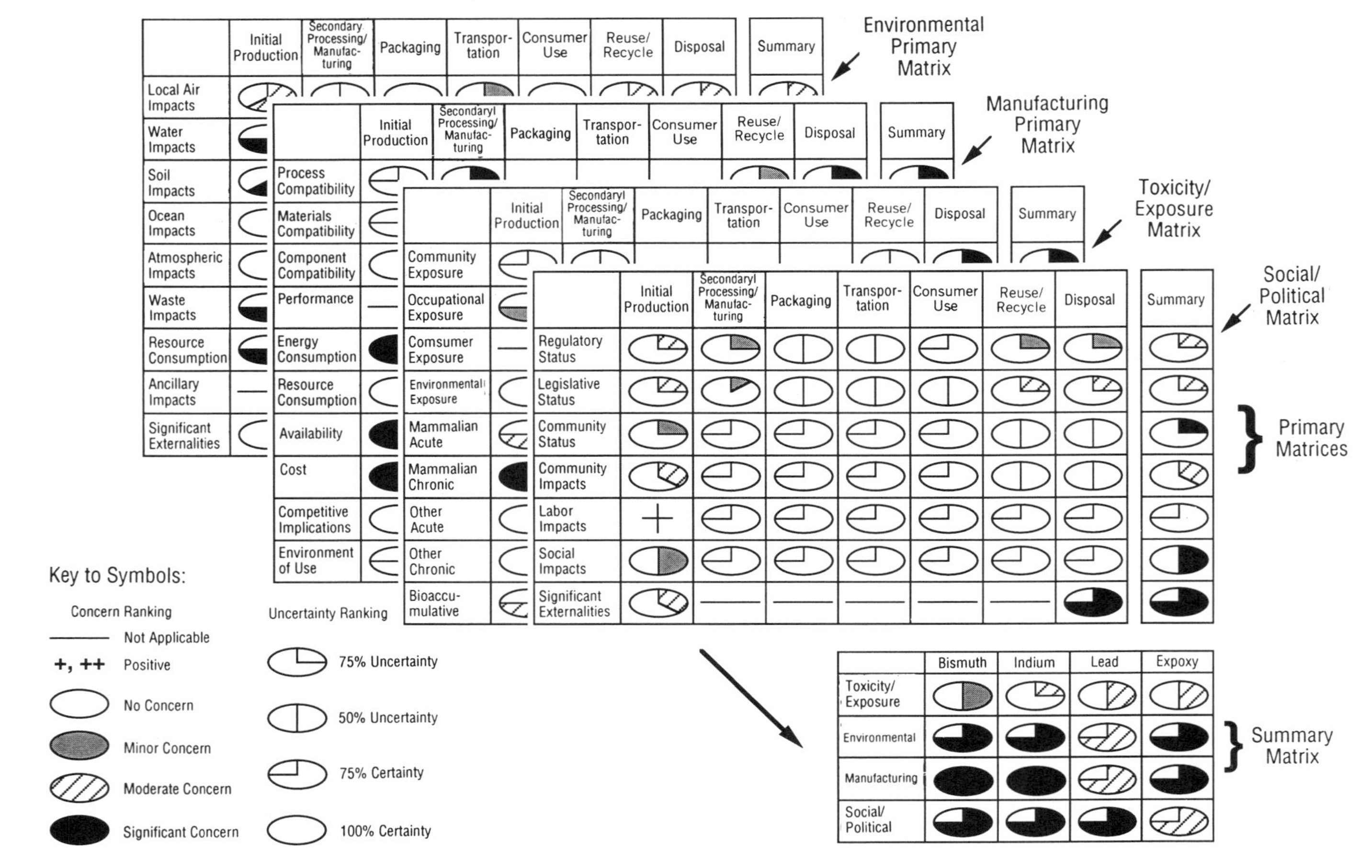

Figure 4-13. Qualitative matrix analysis used by AT&T in the analysis of substitutes for lead solder (Allenby, 1992; Richards and Frosch, 1994). (Reprinted from D. J. Richards and R. A. Frosch, eds., *Corporate Environmental Practices*: *Climbing the Learning Curve*. Copyright © 1994 by

Table 4-17. Human health evaluation criteria for the subsystem manufacturing stage; these criteria are one component of a life-cycle product evaluation framework used by Scott Paper Company

Human Health Criteria	Possible Scores		
Number of manufacturing sites	Lower	Intermediate	Higher
Population surrounding manufacturing site locations	Sparse	Moderate	Dense
Status of manufacturing process			
Degree of automation	Higher	Intermediate	Lower
Complexity (number of process steps)	Lower	Intermediate	Higher
Age, upgrade status	New or upgrade	Moderate old and upgraded	Old and not upgraded
Process stability (susceptibility to upset)	Stable (low upset)	Moderate	More subject to excursions
Hazardous chemicals and/or carcinogens present	No	Yes (either)	Yes (both)
Production volume	Lower	Intermediate	Higher
Routine air emissions	Lower	Intermediate	Higher
Hazardous waste volume	Lower	Intermediate	Higher
Wastewater releases	Lower	Intermediate	Higher

Source: Consoli (1993).

product's life cycle. After the environmental analysis is complete, it is integrated with information on the product's competitive advantage using the product strategy matrix shown in Figure 4-16. Rohm and Haas has used this framework to incorporate environmental concepts into decisions concerning product growth and development.

These comparisons between products and core businesses are quite different from the product comparisons described in the previous section. In this case, rather than comparing products with similar functions to determine which is environmentally preferable, the immediate goal is to compare the relative magnitudes of the environmental burdens of products with different functions and markets. The overall goal is to incorporate life-cycle environmental thinking into corporate decisions in environmental strategic planning, research and development, product/process design, manufacturing, decommissioning, and closure/restoration.

4.3.4 Public-Sector Uses

The previous sections have focused on private sector uses of life-cycle concepts and life-cycle assessments. There are, however, a growing number of

Environmental Dimensions	RME	RMP	MFG	DST	CNV	NDU	DSP	RCY
Safety Fire Explosion								
Human Health								
Residual Substances								
Statospheric Ozone Depletion								
Air Quality								
Climate Change								
Natural Resource Depletion								
Soil Contamination								
Waste Accumulation								
Water Contamination								
Public Perception Gap								
Competition								

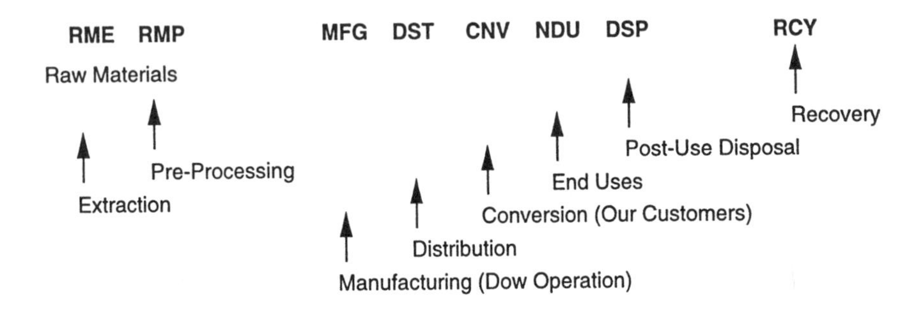

RATING SYSTEM

Hazard or Effect	Exposure (Volume, Frequency) HIGH	MEDIUM	LOW
HIGH	-9	-9	-9
MEDIUM	-9	-3	-1
LOW	-3	-1	

Vulnerabilities

9	Proven Implemented Solution
3	Project Initiated Resources Allocated
1	Project Identified

Opportunities

Figure 4-14. Life-cycle product evaluation framework used by Dow Chemical Company (Noesen, 1993).

public-sector applications of life-cycle concepts and life-cycle assessments, and some of these are summarized in Table 4-18. The range of applications include environmental labeling, acquisition, procurement, and regulation. Examples of each of these public-sector uses are given below.

Among the most visible governmental uses of life-cycle concepts and

HEALTH & ENVIRONMENTAL RANKING (HER) INDEX SCREENING

PRODUCT NO. 1

HUMAN HEALTH	**SCORE**		
MANUFACTURING			
1. Number of manufacturing sites	Lower	Intermediate	Higher
2. Population surrounding manufacturing site locations	Sparse	Moderate	Dense
3. Status of manufacturing process			
a. Degree of automation	Higher	Intermediate	Lower
b. Complexity (number of process steps)	Lower	Intermediate	Higher
c. Age, upgrade status (equipment)	New or upgrade	Moderate old and upgraded	Old and not upgraded
d. Process stability (susceptibility to upset)	Stable (low upset)	Moderate	More subject to excursions
4. Hazardous chemicals -- carcinogens present	No	Yes (either)	Yes (both)
5. Production volume	Lower	Intermediate	Higher
6. Routine air emissions	Lower	Intermediate	Higher
7. Hazardous waste volume	Lower	Intermediate	Higher
8. Wastewater releases	Lower	Intermediate	Higher
		MANUFACTURING SCORE	
DISTRIBUTION			
9. Number of distribution sites	Lower	Intermediate	Higher
10. Distribution volume	Lower	Intermediate	Higher
11. Package size	Bulk	Intermediate	Small
12. Population surrounding distribution sites	Sparse	Moderate	Dense
13. Product hazard (MSDS)	Lower	Intermediate	Higher
		DISTRIBUTION SCORE	
CUSTOMER (PRIMARY)			
14. Number of customer sites	Lower	Intermediate	Higher
15. Population surrounding customer sites	Sparse	Moderate	Dense
16. Customer sophistication	Lower	Intermediate	Higher
17. Susceptibility of upset during customer use	Lower	Intermediate	Higher
18. Consequence of upset	Lower	Intermediate	Higher
19. Potential for misuse	Lower	Intermediate	Higher
20. Customer disposal risk	Lower	Intermediate	Higher
21. Regulatory/customer pressure to alter product environemntal compatibility (e.g., solvents, packaging)	Lower	Intermediate	Higher
22. Relative environmental compatibility of RH component to others available	More	Equal	Less
		CUSTOMER SCORE	
		HER INDEX SCREENING SCORE	

Figure 4-15. The evaluation framework for life-cycle assessments used by Rohm and Haas (Fava et al., 1993).

assessments are environmental labels. Environmental labels are issued in a variety of countries, as shown in Figure 4-17. In many countries (Austria, Canada, France, Germany, the Nordic countries, the Netherlands, and the European Community), life-cycle concepts are used to determine the criteria that are considered in awarding the labels, and multiple criteria are considered. This represents a significant shift away from single-attribute labels, focused on a single stage of the life cycle (e.g., recycled content labels). Some of the principles and practices underlying environmental labeling can be illustrated by examining the system employed in the Netherlands. In the Netherlands, the Dutch ecolabel (Stichting Milieukeur) is awarded on the

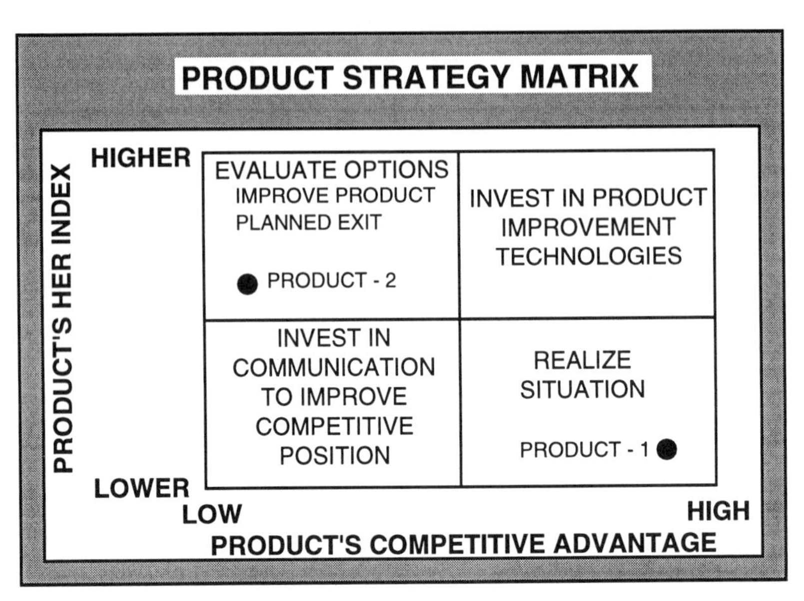

Figure 4-16. The product strategy matrix used by Rohm and Haas to incorporate environment concerns into strategic business planning (Fava et al., 1993).

basis of a qualitative matrix of environmental criteria. Label criteria are developed by a board of experts for product groups and once the criteria have been set, any manufacturer, importer, or licensee can submit an application for their product to be approved by a certification authority. The first such label was awarded in September 1993 to a writing paper product. The Dutch ecolabel is thus typical of many labeling programs: (1) multiple life-cycle-based criteria are set and (2) products are evaluated and labels are awarded, but the consumer seeing the label generally has no immediate access to the full range of data on environmental burdens and must accept the tradeoffs between burdens used by the "experts" in setting the criteria. The Netherlands is now defining a substantial expansion to this system that will provide more information to consumers and retailers. In the next generation of labeling, a simplified description of five environmental indicators will be provided on the labels of all products. These indicators are energy, waste, resources, emissions, and nuisance. Background data for the labeling will be publicly available. The information will take into account "all the environmental considerations related to a product in each phase of its life cycle ('from the cradle to the grave')" (National Environmental Policy Plan Plus, 1994). Thus, ecolabels are evolving.

Many governments are themselves users of ecolabeling data. For example, in the United States, Executive Order 12873, issued by President Clinton, mandates the procurement of environmentally preferable goods and services by federal agencies. The Environmental Protection Agency is using the life-

Table 4-18. Environmental policies relating to products

Economic Commission for Europe (United Nations)
A task force is developing guidelines for "environmental product profiles," a qualitative description of the environmental impacts of a product for use by commercial and institutional buyers
European Community
Draft law requiring specific percentages of recovery (recycling, incineration, and composting) for product packaging
European Community ecolabel
Canada
The National Packaging Protocol is a voluntary program with packaging reduction targets and dates
Environmental choice ecolabel
Denmark
Ban on domestically produced nonrefillable bottles and aluminum cans
Fee imposed on waste delivered to landfills and incinerators as an incentive to recycling and to support clean technology
Clean Technology Action Plan
Germany
Packaging Waste Law, passed in 1991, gives manufacturers responsibility for collecting and recycling various kinds of packaging at specified rates by certain dates
Manufacturer take-back-and-recycle laws have been proposed by the government for automobiles, electronic goods, and other durables
Mandatory deposit refund on plastic beverage containers (except milk)
Blue Angel product ecolabel
Japan
Recycling law, passed in 1991, sets target recycling rates at ~60% for most discarded materials by the mid-1990s; includes product redesign strategies for packaging and durable goods
Ecomark product ecolabel
Netherlands
National Environmental Policy Plan sets national targets and timetables for implementing clean technology, including redesign of products
Voluntary agreements reached with industry targeting 29 priority waste streams and reduction of packaging waste
Norway
Tax on nonreturnable beverage containers
Deposit-refund on old car bodies
Sweden
Ban "in principle" on the use of cadmium
Voluntary deposit-refunds for glass and aluminum beverage containers

Source: US Congress OTA, (1992).

Netherlands

Figure 4-17. Environmental labels from around the world.

cycle concept to establish guidelines for the procurement process. Governments can also use life-cycle concepts in setting research and development policy, stimulating markets, and setting regulations. Examples include the following:

- The U.S. EPA is considering using life-cycle concepts in rulemaking activities under the Clean Air Act Amendments. Life-cycle concepts would be used to assess materials cleaning options, including the use of chlorinated solvents, aqueous degreasers, and semiaqueous degreasers.
- In developing effluent guidelines for industrial laundries under the Clean Water Act, the U.S. EPA became concerned that regulations on the laundries might cause a shift from the use of cloth wiping towels to disposable towels, resulting in a transfer of pollutant loadings from water to landfills. The EPA will use a life-cycle framework to examine the tradeoffs between laundry effluent levels, the use of disposable wipes, and other alternatives.
- The U.S. Department of Defense now incorporates life-cycle environmental costs into procurement decisions. DoD 5000.2-M (Part 4, Section F) provides the following guidance: "During each phase of the acquisition process, identify and analyze the potential environmental consequences of each alternative being considered. This analysis includes environmental impacts of each alternative throughout the system's life cycle."
- Life-cycle assessments were performed by the U.S. Department of Energy to examine the effects of mandating the use of electric (electricity-powered) vehicles and to assess the environmental impacts of a variety of energy options, including renewables.

These are just a few of the many emerging public policy applications of life-cycle concepts and life-cycle assessments.

4.4 SUMMARY

Life-cycle assessment can be a useful tool in identifying the environmental impacts associated with a product or a material. It integrates environmental impacts over the entire life cycle, from "cradle to grave," and has applications in product design, strategic environmental planning, and public policymaking. Only a few of the case studies demonstrating the value of these tools could be presented in this chapter, but the diverse applications presented here provide a glimpse of the range and power of this emerging tool. It is clear that while life-cycle assessments are in their infancy, they are developing rapidly. Interested readers should follow closely the publications of the Life Cycle Assessment Advisory Group of the Society for Environmental Toxicology and Chemistry (e.g., SETAC, 1991), and the Life Cycle Assessment group at the EPA's Risk Reduction Engineering Laboratory (e.g., US EPA, 1991). These publications will outline accepted procedures for performing life-cycle assessments as they develop.

QUESTIONS FOR DISCUSSION

1. The life-cycle impact assessment frameworks used by AT&T, Dow, and Rohm and Haas in Section 4.3.3 address a range of impacts and a variety of life-cycle stages. Compare and contrast: (a) the life-cycle stages considered in the three systems, (b) the environmental impact categories considered in the three systems, and (c) the methods used to evaluate and integrate life-cycle data. Note the level of detail surrounding the life-cycle stages that involves each company's operations. Estimate the amount of effort required to do the assessments for each system. If the resources for a complete life-cycle assessment are unavailable, are there ways to use life-cycle concepts to improve corporate environmental performance?
2. In the garment life-cycle studies of Figures 4-10 and 4-11, energy was used as a surrogate for the total environmental burdens associated with manufacturing and distributing a garment. Treating energy consumption as a surrogate is a gross simplification, but a more comprehensive assessment would require a detailed and expensive data collection effort. What specific environmental issues would you consider important if you were undertaking a more complete assessment of the environmental impacts associated with different modes of transportation during manufacture, distribution, and disposal of the garment? Can the overall conclusions drawn from the Patagonia study (that air freight is the most energy intensive and intermodal freight is the least) be applied to other firms? What other firms? Are uncertainties in routes, driver competence, vehicle loads, and vehicle maintenance likely to affect the conclusions of this study? Also, packaging required for shipment was not included in this study. What effect might the inclusion of this factor have on the results of the study?

PROBLEM STATEMENTS

1. Perform a mass balance for ethylene production using the data of Table 4-1. Establishing a framework similar to that shown in Figure 4-18 may help you in performing the mass balance. Use the mass balance data to estimate concentrations of pollutants in water emissions and to estimate the amount of air required by the ethylene manufacturing processes.

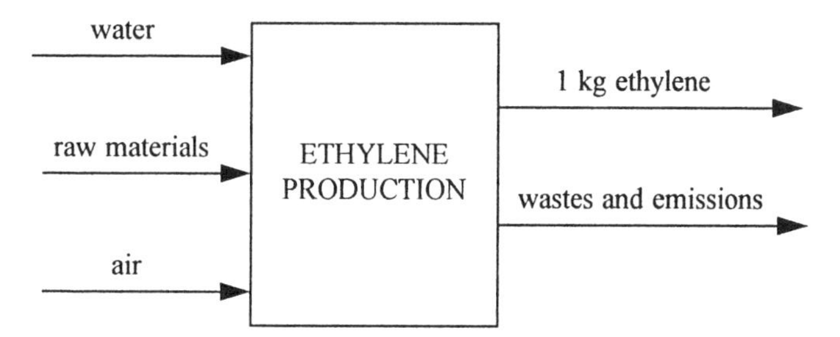

Figure 4-18. Mass-balance diagram for ethylene production.

Finally, comment on whether mass balances are an adequate tool for testing the reliability of life-cycle inventory data.

2. Repeat Problem 1 for the data of Tables 4-2 and 4-3.
3. In this problem you will need to refer to the automobile bumper systems of Figure 4-2 whose assessments are summarized in Table 4-5 (adapted from Graedl and Allenby, 1995).
   a. Suppose that global warming is thought less likely to occur than had previously been assumed, and that as a result the ELU/kg for product use is lowered to 0.6. What effect does this have on the comparative ratings of the two front ends?
   b. A new high-strength honeycomb steel has been developed and is being considered for use in automobile front ends. Rather than the steel front end weighing 6.0 kg, a satisfactory front end weighing only 4.0 kg can be formed from 6.0 kg of the new steel. The new steel's improved properties, which are due to added trace alloying elements, have negligible effects on processing or recycling, so the same ELU/kg assessments apply to those stages. Compute the ELU values for the new front end and compare them with the two options in the table.
   c. What messages to designers are implied in the analyses in parts a and b of this problem?

OPEN-ENDED PROBLEM

1. In this problem you are asked to compare two studies (A and B) that were conducted to determine the benefits of electric vehicles (EVs) over gasoline-powered vehicles.
   A. One study, conducted by the Electric Power Research Institute (EPRI, 1994), concluded that emissions of criteria air pollutants within the heavily urbanized areas of California would be 72–99% lower in the year 2011 if EVs replaced ultra-low-emission (gasoline-powered) vehicles (ULEVs). The fuel cycle from production (power plants and refineries) to tailpipe emissions was considered in this study. Results

Table 4-19. California ULEV passenger car emissions in 2011 in average grams per miles (g/mi) over the first 10 years of vehicle use

	Fuel		Vehicle[c]		
Emission Type	Production[a]	Marketing[b]	Tailpipe	Evaporative	Total Emissions
Reactive organic gases	0.017	0.06	0.049	0.065	0.191
Nitrogen oxide	0.035	0.00	0.284	0.000	0.319
Carbon monoxide	0.011	0.00	1.078	0.000	1.089
Particulate matter[d]	0.008	0.00	0.010	0.000	0.018

[a]Fuel production emissions are approximated based on calculations incorporating petroleum refinery operations in 2011. Note that there are no refineries in the SDG&E (San Diego Gas and Electric) district. Therefore, the ULEV emission rates for that area would be somewhat lower than the values shown in this table.
[b]California Air Resources Board (CARB) estimate, April 1994.
[c]Computed using CARB EMFAC and BURDEN models.
[d]Particulate matter levels exclude tire wear.

Source: Copyright © 1994 Electric Power Research Institute. EPRI TB-104068. *EV Emissions Benefits in California*. Reprinted with permission.

of the study are given in Tables 4-19 and 4-20. The authors of the study indicated that their results tend to underestimate the benefits of EVs over ULEVs because factors that do not affect EV emissions, such as equipment tampering and improper vehicle maintenance, can increase ULEV emissions. They also noted that ULEV emissions are released at street level, where they pose a greater risk than the emissions from power plants. Furthermore, EVs produce no cold start, evaporative, or refueling emissions, and would reduce engine oil disposal as well as pollution to oceans, rivers, and groundwater from petroleum, gasoline, and motor oil spills.

B. In another study (Graedl and Allenby, 1995), the life-cycle impacts of fuel to provide power for gasoline and electric automobiles were considered. The life cycles of the two energy sources are shown graphically in Figures 4-19 and 4-20. The findings of this study were mixed, as shown in Figure 4-21. Carbon monoxide and hydrocarbon emissions were found to be higher for gasoline-powered vehicles, while sulfur dioxide and particulate emissions were found to be higher for EVs. Nitrogen oxide emissions depend on the particular gasoline-powered vehicle or power plant technology in use. Total energy use and greenhouse gas emissions over the lives of the vehicles are roughly equal for the two vehicle systems, as shown in Figure 4-22 and 4-23. Again, it is noted that gasoline-powered vehicle emissions occur where the cars are operated and power plant emissions generally occur away from populated areas. Moreover, it is easier to control emissions at a few point sources rather than at a multitude of small emission sources.

Table 4-20. EV versus ULEV passenger car emissions in 2011 for electricity generation within air quality districts in g/mi

Emission Type	ULEV Emissions in California[a]	EV Emissions[b] (Power Plants within Air-Quality Districts[d])			
		LADWP	PG&E	SDG&E	SCE
Reactive organic gases	0.191	0.0024	0.0026	0.0017	0.0055
Nitrogen oxide	0.319	0.0138	0.0231	0.0040	0.0117
Carbon monoxide	1.089	0.0239	0.0398	0.0047	0.0272
Particulate matter[c]	0.018	0.0046	0.0038	0.0027	0.0049

[a]ULEV emission rates are 10-year averages and include vehicle exhaust and evaporative emissions from EMFAC and BURDEN model runs and emissions associated with fuel production, storage, and distribution.
[b]EV emission rates reflect marginal power plant emissions from ELFIN model runs. They assume an average 10-year passenger car energy efficiency of 0.26 kWh/m and an average transmission and distribution system loss of 8%.
[c]Particulate matter levels exclude tire wear.
[d]The energy for EVs would come from power plants in the following areas:

LADWP: 71% in four-county South Coast Air Quality Management District (AQMD)
SCE: 81% in South Coast AQMD and Ventura County Air Pollution Control District (APCD)
PG&E: 83% in nine-county Bay Area AQMD, three-county Monterey Bay AQMD, and San Luis Obispo APCD
SDG&E: 38% in San Diego APCD

Source: Copyright © 1994 Electric Power Research Institute. EPRI TB-104068. *EV Emissions Benefits in California*. Reprinted with permission.

Identify areas of difference and consensus between the two studies. Examine not only the reported emissions but also methodological issues such as the choice of functional unit, the use of national or site specific data, the life-cycle stages considered, and the impact categories chosen. Suggest reasons for the differences between the two assessments in their estimates of atmospheric emission patterns for the vehicles. Identify and prioritize data collection efforts that would be needed to reconcile these two studies. Be as specific as possible and rationalize your prioritizations. Can you identify any significant considerations that were neglected in the two studies?

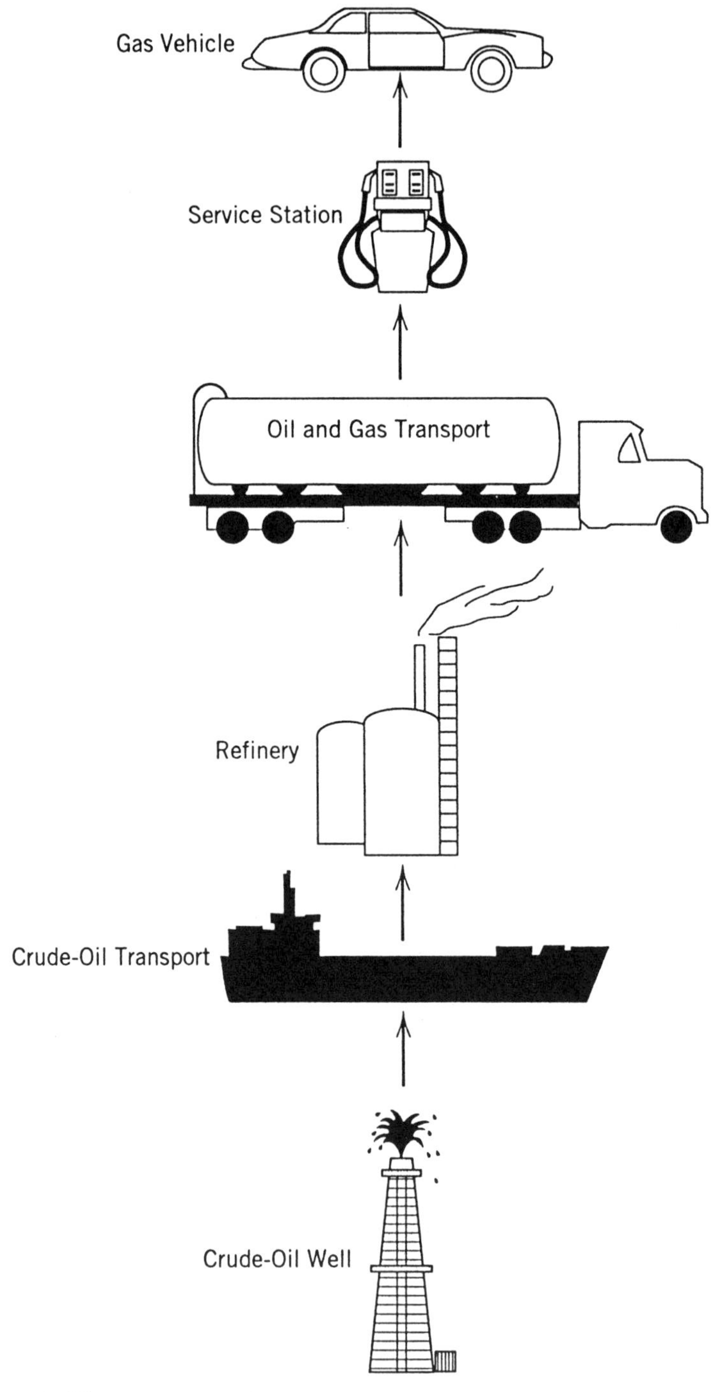

Figure 4-19. Life-cycle stages of gasoline motive power for motor vehicles (Graedl and Allenby, 1995).

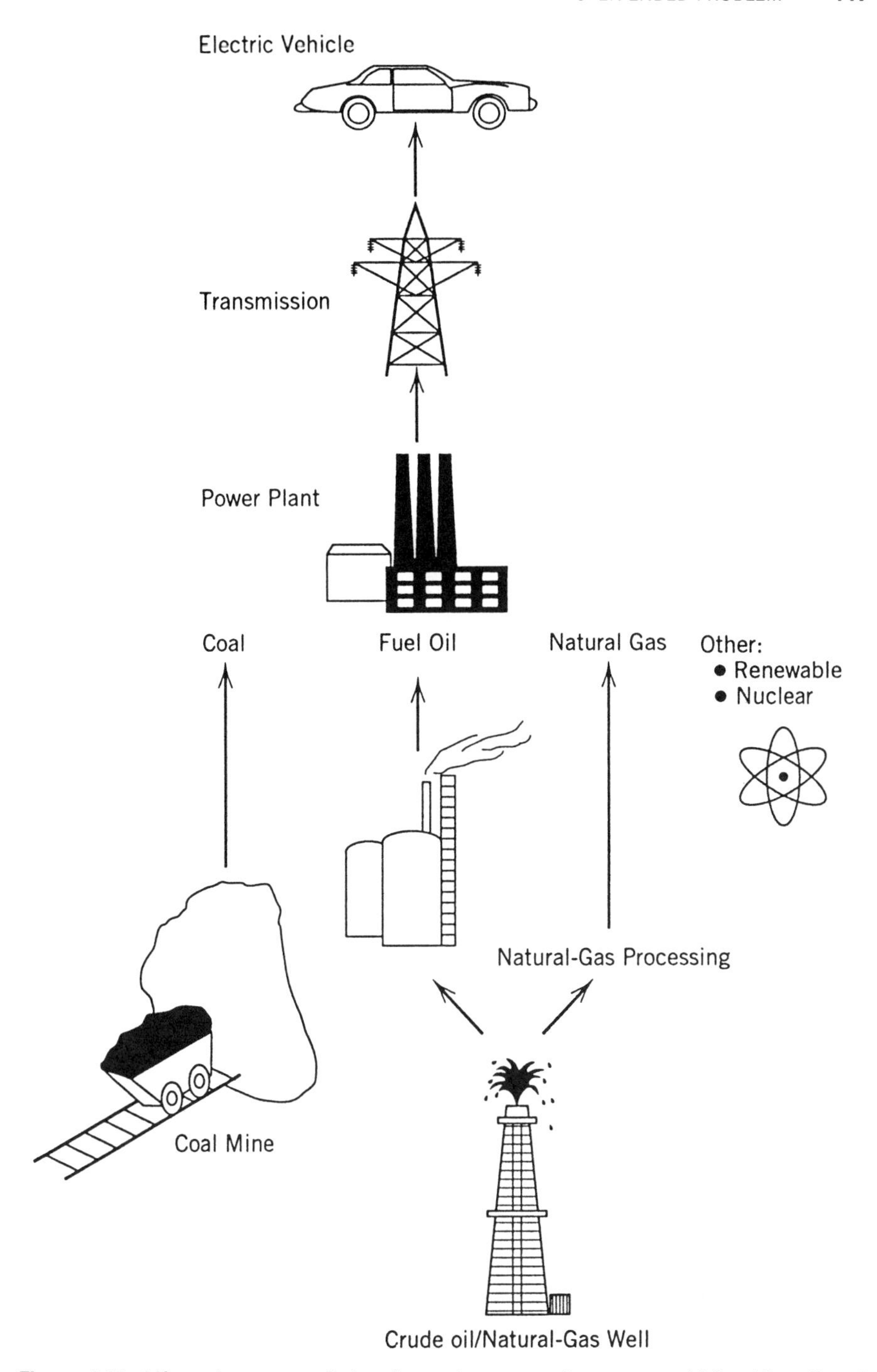

Figure 4-20. Life-cycle stages of electric motive power for motor vehicles (Graedl and Allenby, 1995).

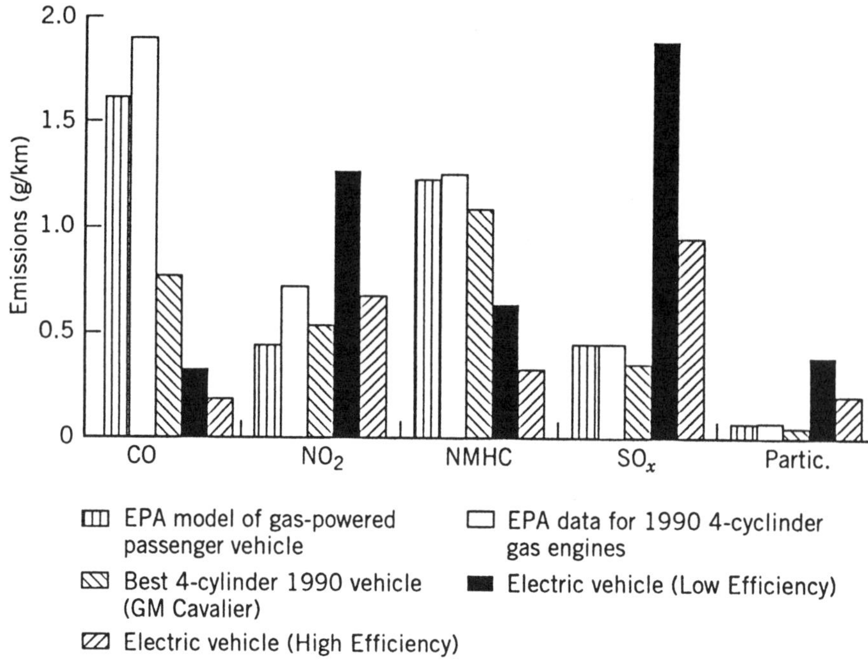

Figure 4-21. Comparison of criteria air pollutant emissions for gasoline- and electricity-powered motor vehicles (Graedl and Allenby, 1995).

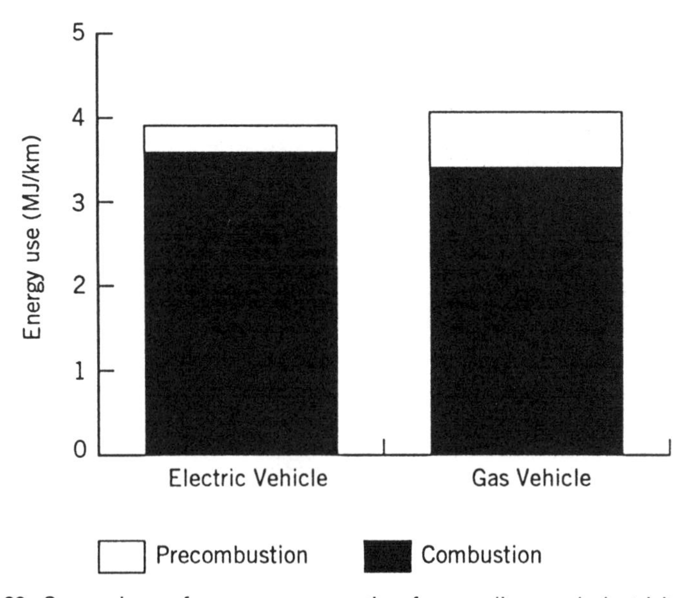

Figure 4-22. Comparison of energy consumption for gasoline- and electricity-powered motor vehicles (Graedl and Allenby, 1995).

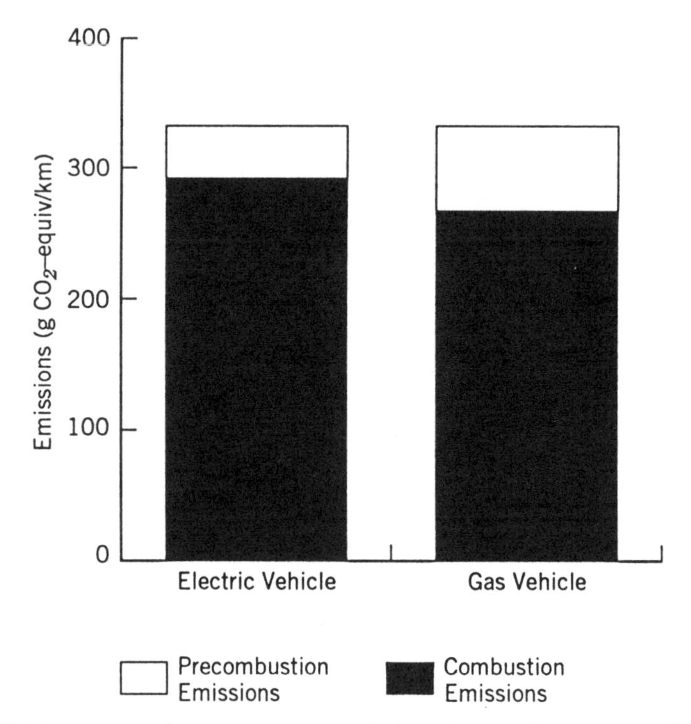

Figure 4-23. Comparison of greenhouse gas emissions for gasoline- and electricity-powered motor vehicles (Graedl and Allenby, 1995).

REFERENCES

Allen, D. T., N. Bakshani, and K. S. Rosselot, *Pollution Prevention: Homework and Design Problems for Engineering Curricula*, American Institute of Chemical Engineers Center for Waste Reduction Technologies, New York, 1992.

Allenby, B. R., "Design for Environment: Implementing Industrial Ecology," Ph.D. Dissertation, Rutgers University, 1992.

Allenby, B. R., personal communication, 1994.

Boustead, I., "Ecoprofiles of the European Plastics Industry, Reports 1–4," PWMI, European Centre for Plastics in the Environment, Brussels, May 1993.

Consoli, F., Scott Paper Company, personal communication, 1993.

Electric Power Research Institute, Inc. (EPRI), "EV Emissions Benefits in California," Technical Brief, Customer Systems Division, RB2882, Palo Alto, CA, May 1994.

Fava, J. A., R. Denison, B. Jones, M. A. Curran, B. Vigon, S. Selke, and J. Barnum, "A Technical Framework for Life Cycle Assessments," Society of Environmental Toxicology and Chemistry Workshop, Smuggler's Notch, VT, Aug. 18–23, 1993.

Franklin Associates, Ltd., "Comparative Energy and Environmental Impacts for Soft Drink Delivery Systems" (report prepared for the National Association for Plastic Container Recovery), Prairie Village, KS, March 1989.

Franklin Associates, Ltd., "Energy and Environmental Profile Analysis of Polyethylene and Unbleached Paper Grocery Sacks" (report prepared for the Council for Solid Waste Solutions), Prairie Village, KS, June 1990a.

Franklin Associates, Ltd., "Resource and Environmental Profile Analysis of Foam Polystyrene and Bleached Paperboard Containers," (report prepared for the Council for Solid Waste Solutions), Prairie Village, KS, June 1990b.

Franklin Associates, Ltd., "Energy and Environmental Profile Analysis of Childrens' Disposable and Cloth Diapers" (report prepared for the American Paper Institute's Diaper Manufacturers' Group), Prairie Village, KS, July 1990c.

Franklin Associates, Ltd., "Resource and Environmental Profile Analysis of a Manufactured Apparel Product" (report prepared for the American Fiber Manufacturers' Association, Washington, DC), June 1993.

Freeman, H., T. Haryen, J. Springer, P. Randall, M. A. Curran, and K. Stone, "Industrial Pollution Prevention: A Critical Review," *J. Air Waste Manage. Assoc.*, **42**, 618–656, 1992.

Graedl, T. E. and B. R. Allenby, *Industrial Ecology*, Prentice Hall, Englewood Cliffs, NJ, 1995.

Hocking, M. B., "Paper versus Polystyrene," *Science*, **251**, 504–505, Feb. 1991 (see also letters commenting on this paper, *Science*, **25**, 1361–1363, June 1991). (Further details are given in the full study by M. B. Hocking, "Relative Merits of Polystyrene Foam and Paper in Hot Drink Cups: Implications for Packaging," *Environmental Management*, **16**(6), 731–747, Dec. 1991.)

Hopkins, L., D. T. Allen, and M. Brown, "Quantifying and Reducing Environmental Impacts Resulting from Transportation of a Manufactured Garment," *Pollution Prevention Review*, **4**(4), 491–500, 1994.

Horkeby, I., AB Volvo, Göteburg, Sweden, personal communication.

Microelectronics and Computer Technology Corporation (MCC), "Environmental Consciousness: A Strategic Competitiveness Issue for the Electronics and Computer Industry," Austin, TX, March 1993.

National Environmental Policy Plan Plus (Netherlands), "Policy Document on Products and the Environment," Ministry of Housing, Spatial Planning and the Environment, The Hague, 1994.

Noesen, S., Dow Chemical Company, personal communication, 1993.

Procter and Gamble, personal communication, 1993.

Richards, D. J. and R. A. Frosch, eds., *Corporate Environmental Practices: Climbing the Learning Curve*, National Academy Press, Washington, DC, 1994.

Ryding, S., B. Steen, A. Wenblad, and R. Karlson, "The EPS system: A Life Cycle Assessment Concept for Cleaner Technology and Product Development Strategies, and Design for the Environment," paper presented at EPA Workshop on Identifying a Framework for Human Health and Environmental Risk Ranking, Washington, DC, June 30, July 1, 1993.

Ryding, S., "International Experiences of Environmentally Sound Product Development Based on Life Cycle Assessment," Swedish Waste Research Council, AFR-Report 36, Stockholm, May 1994.

Society of Environmental Toxicology and Chemistry (SETAC), "A Technical Framework for Life-Cycle Assessments," SETAC Foundation for Environmental Education, Inc., Washington, DC, 1991.

Society for Environmental Toxicology and Chemistry (SETAC), "Executive Summary of Workshop on Public Policy Applications of Life Cycle Assessment," Pensacola, FL, 1994.

Steen, B. and S. Ryding, "The EPS Enviro-Accounting Method: An Application of Environmental Accounting Principles for Evaluation and Valuation of Environmental Impact in Product Design," Stockholm: Swedish Environmental Research Institute (IVL), 1992.

Svoboda, S. and S. Hart, "McDonald's Environmental Strategy," National Pollution Prevention Center for Higher Education, Document #93–3, Dec. 1993.

United States Congress Office of Technology Assessment (US Congress OTA), "Green Products by Design: Choices for a Cleaner Environment," OTA-E-541, U.S. Govt. Printing Office, Washington, DC, Oct. 1992.

United States Environmental Protection Agency (US EPA), "Product Life Cycle Assessments: Inventory Guidelines and Principles," Cincinnati, OH, 1991.

Vigon, B., "Life Cycle Assessment," in *Pollution Prevention for Industrial Processes*, H. Freeman, ed., McGraw-Hill, New York, 1994.

MESOSCALE POLLUTION PREVENTION

Macroscale studies of pollution prevention, outlined in Chapters 2–4, are useful in identifying targets of opportunity for pollution prevention. Once the targets have been identified, the design of cleaner chemical processes and products can begin.

In this section, generic methods and specific examples for reducing wastes are examined. The focus is on the chemical process industries: chemical manufacturing and petroleum refining. Focusing on a particular industrial sector allows a fairly comprehensive treatment of waste reduction methods. In addition, an emphasis on chemical manufacturing and petroleum refining is relevant because these industries are responsible for over half of all hazardous wastes generated in the United States; they are also the source of approximately half of the releases of chemicals reported in the Toxic Chemical Release Inventory, and they are a significant source of wastes legally classified as nonhazardous (see Chapter 2).

The presentation of pollution prevention methods in this section is divided into six chapters. In Chapter 5, methods for estimating waste and emission generation rates for chemical processes are described. In Chapters 6 and 7, the building blocks of chemical processes—unit operations such as reactors, separators, heat exchangers, valves, and tanks—are examined for designs and practices that minimize wastes and emissions. Systematic tools for optimizing process designs for pollution prevention are presented in Chapter 8. Chapter 9 describes methods for evaluating the economic performance of pollution prevention projects. Finally, Chapter 10 presents three case studies that demonstrate the use of the tools developed in Chapters 5–9.

5

WASTE AUDITS AND EMISSION INVENTORIES

In order to form a plan for pollution prevention or to track the progress of a reduction program, pollutants must be identified and their release quantities estimated. Therefore, the first step toward reducing wastes and emissions is to perform thorough waste audits and accurate emission inventories. A waste audit is conducted to characterize the waste streams generated by a facility, while an emission inventory is performed to quantify direct releases of pollutants to the environment. The functions of waste audits and emission inventories are complementary; both participate in the development of pollution prevention strategies. In this chapter, the features of a pollution prevention waste audit are summarized and guidelines for the types of data that must be gathered are provided. The focus then shifts to methods for quantifying emissions that are difficult to measure. Finally, a method is presented for evaluating the waste streams and emissions at a facility in order to identify the most important streams to focus on in developing pollution prevention options.

5.1 WASTE AUDITS

The most common approach to performing a pollution prevention waste audit is to simply identify each waste stream and characterize its point of origin. Critics of this approach complain that such a procedure fails to address the interconnectedness of waste streams. They advocate a more technically challenging approach that focuses on the process generating the waste and makes use of process flowsheeting and materials accounting (Pojasek and Cali, 1991). Whether a simple audit or a materials accounting approach is used, data on waste stream compositions and flow rates are a starting point

Table 5-1. Waste auditing guides prepared by California's Department of Toxic Substances Control[a]

Industry Sector	Year of Report	Report Number
Automotive paint shops	1987	WA001
Automotive repairs	1987, 1988	WA002, WA003, WA004
Building construction	1990	WA019
Commercial printing	1989	WA005
Drug manufacturing and processing	1989	WA012
Fabricated metal products	1989	WA017
Fiberglass-reinforced and composite plastic products	1989	WA013
General medical and surgical hospitals	1988	WA006
Gold, silver, platinum, and other precious metals production and reclamation	1990	WA018
Marineyards for maintenance and repair	1989	WA016
Mechanical equipment repair shops	1990	WA020
Metal finishing industry	1988	WA007, WA015
Paint manufacturing industry	1989	WA008
Pesticide formulating industry	1987, 1990	WA009, WA022
Photoprocessing industry	1989	WA014
Printed-circuit board manufacturers	1987	WA010
Research and educational institutions	1988	WA011
Thermal metalworking industry	1990	WA023

[a]For copies of these reports, contact:
California Department of Health Services
Toxic Substances Control Program
Alternative Technology Division
Source Reduction Unit
P.O. Box 942732
Sacramento, CA 94234-7320

for pollution prevention. There are a number of practical guides for performing waste audits, including the U.S. Environmental Protection Agency's "Waste Minimization Opportunity Assessment Manual" (US EPA, 1988b), their more recent "Facility Pollution Prevention Guide" (US EPA, 1992), and a series of sample audits published by the State of California, which are listed in Table 5-1. A comprehensive list of such documents is available in the Hazardous Waste Minimization Bibliography from the State of California Department of Toxic Substances Control (CA DTSC, 1991). This chapter does not attempt to duplicate the information presented in the vast array of available waste auditing guides. Instead, the key features of a pollution prevention waste audit are discussed and examples of the types of data that must be gathered in such an audit are given.

Table 5-2. Refining residual streams

Oily sludges and other organic residuals	Spent catalysts
API separator sludge[a]	Fluid cracking catalyst or equivalent
Dissolved-air-flotation unit float (DAF float[a])	Hydroprocessing catalyst
Slop oil emulsion solids[a]	Other spent catalyst NOS[b]
Leaded tank bottoms[a]	
Pond sediments	Aqueous residuals
Primary sludge (FO37)[a]	Biomass
Primary sludge (FO38)[a]	Oil-contaminated waters (not wastewaters)[c]
Nonleaded tank bottoms	High-pH/low-pH waters
Waste oils/spent solvents	Spent sulfide solution
Other oily sludges/organic residuals NOS[b]	Spent stretford solution
	Treatment, storage, and disposal (TSD) leachate (FO39)[a]
	Other aqueous residuals NOS[b]
Contaminated soils and solids	Chemicals and inorganic residuals
Heat exchanger bundle cleaning sludge[a]	Spent caustics
Contaminated soils/solids	Spent acids
Residual coke/carbon/charcoal	Residual amines
Residual/waste sulfur	Other inorganic residuals NOS[b]
Other contaminated solids NOS[b]	
Other residuals	
Other wastes NOS[b]	

[a]RCRA-listed hazardous wastes for petroleum refining.
[b]NOS: not otherwise specified.
[c]Does not include NPDES or POTW wastewaters.

Source: From *Generation and Management of Residual Materials*: *Petroleum Refining Performance, 1991 Survey*, Publication 329, 1994. Reprinted courtesy of the American Petroleum Institute.

5.1.1 Features of a Waste Audit

Waste audits are performed at many levels of detail. In this section, waste generation data for petroleum refineries are presented in order to illustrate the rigor with which a waste audit for pollution prevention must be performed. A simple audit might result in a list of the liquid and solid wastes generated by a facility. Such a list for typical refining wastes, as identified in an American Petroleum Institute survey of U.S. refineries, is given in Table 5-2. More detailed information is given in Table 5-3, where the total quantities of waste generated and the number of refineries generating each of the wastes are listed. While these lists are informative, they do not give insight into effective waste reduction activities. As noted in the EPA's Waste Minimization Opportunity Assessment Manual (US EPA, 1988b), the questions that a pollution prevention waste audit should be designed to answer include

Table 5-3. Estimates of residual materials generated by the U.S. refining industry in 1991 (thousands of wet tons)

Residual Stream	Generated Quantity	Number of Refineries
Aqueous residuals NOS	9,036	14
Spent caustics	909	69
Biomass	855	45
Contaminated soil and solids	809	90
DAF float	406	44
Other inorganic residuals NOS	397	62
Pond sediments	372	25
Other residuals NOS	339	83
API separator sludge	210	76
FCC catalyst or equivalent	204	76
Primary sludge (F038)	177	14
Slop oil emulsion solids	165	32
Residual coke/carbon/charcoal	138	55
Residual amines	136	32
Primary sludge (F037)	130	59
Nonleaded tank bottoms	109	69
Spent acids	88	20
Oil-contaminated waters (not wastewater)	67	10
High-pH/low-pH waters	54	4
Other oily sludges/organic residuals NOS	54	43
Other contaminated soils NOS	37	59
Hydroprocessing catalysts	32	57
Spent stretford solution	25	6
Other spent catalysts NOS	23	63
Residual oils/spent solvents	21	65
TSD leachate (F039)	20	3
Residual sulfur	19	43
Spent sulfite solution	9	3
Heat exchanger bundle cleaning solids	3	43
Leaded tank bottoms	1	20
Total	14,800	113

Source: From *Generation and Management of Residual Materials: Petroleum Refining Performance, 1991 Survey*, Publication 329, 1994. Reprinted courtesy of the American Petroleum Institute.

- What waste streams are generated by the facility?
- Are the waste streams generated on a regular basis, or are they one-time events?
- Which processes or operations generate the waste?
- What is the regulatory status of the waste? Is it a hazardous waste, as defined by the Resource Conservation and Recovery Act (RCRA)? Is

it a hazardous waste under state regulations? What characteristics of the waste are responsible for its regulatory status?

- What are the inputs to the process generating the waste?
- How efficient is the process?
- Are wastes from different sources mixed?

This list of audit objectives is not intended to be comprehensive. Rather, it is meant to serve as a starting point for assembling information. The types of data that may need to be assembled in answering these basic auditing questions are listed in Table 5-4.

It is clear from the audit objectives noted above and the extensive information requirements in Table 5-4 that a pollution prevention waste audit is far more than a listing of the waste streams generated by a facility. The benefits of performing a detailed waste audit are illustrated by the following account of waste reduction from the dissolved-air flotation (DAF) unit at a petroleum refinery. DAF units are used to remove fine suspended oily solids from wastewaters by injecting an aqueous stream containing dissolved air into the wastewater, as shown in Figure 5-1. The dissolved air forms bubbles when it comes out of solution and carries suspended particles, which tend to concentrate at the bubble–wastewater interface, to the surface, where they form an emulsion. This emulsion is a RCRA-listed hazardous waste and is generated at a rate of more than 400 thousand tons per year at domestic refineries. The composition of the emulsion, on average, is 8.7% oil, 81.9% water, and 9.4% solids (API, 1983). This means that for every pound of oil and solids removed (the goal of the unit), five pounds of waste are formed. In fact, the oil in the stream entering the DAF unit is present only because it has been stabilized in suspended form in the wastewater by the presence of the solids, and reducing solids loading reduces oil loading. Therefore, if the particle loading in the wastewater stream can be reduced before the stream enters the unit, the amount of hazardous waste generated can be significantly reduced.

As part of a waste audit, one large west coast refinery characterized the particles entering their DAF unit by getting signatures of the trace metals in all the sources for this unit. They found that hardness precipitation from cooling-tower blowdown was a major source of the particulates that generated the waste emulsion. By simply rerouting the cooling-tower blowdown stream, as shown in Figure 5-2, DAF unit waste was reduced by roughly 40% (Heirigs, 1991). It is obvious from this example that a pollution prevention audit must include information concerning the mechanism of waste generation as well as waste stream data. Without knowing that solids generated waste in the DAF unit and without finding out where the solids originated, it would have never been apparent that rerouting the cooling-tower blowdown would result in significant waste reduction.

Table 5-4. Typical data needed in a waste audit

- Design Information
  - Process flow diagrams
  - Material and heat balances (for design balances and actual balances) for
    - Production processes
    - Pollution control processes
  - Operating manuals and process descriptions
  - Equipment lists
  - Equipment specifications and data sheets
  - Piping and instrument diagrams
  - Plot and elevation plans
  - Equipment layouts and work flow diagrams
- Environment information
  - Hazardous waste manifests
  - Emission inventories
  - Biennial hazardous waste reports
  - Waste analyses
  - Environmental audit reports
  - Permits and/or permit applications
- Raw-material and production information
  - Product composition and batch sheets
  - Material application diagrams
  - Material safety data sheets
  - Product and raw-material inventory records
  - Operator data logs
  - Operating procedures
  - Production schedules
- Economic information
  - Waste treatment and disposal costs
  - Product, utility, and raw-material costs
  - Operating and maintenance costs
  - Departmental cost accounting reports
- Other information
  - Company environmental policy statements
  - Standard procedures
  - Organization charts

Source: US EPA (1988b).

5.2 EMISSION INVENTORIES

An emission inventory is performed to assess the quantity of pollutants released directly to the environment by a facility. Besides playing a part in the planning of pollution prevention activities, emission inventories are sometimes required by governmental reporting regulations. As mentioned in

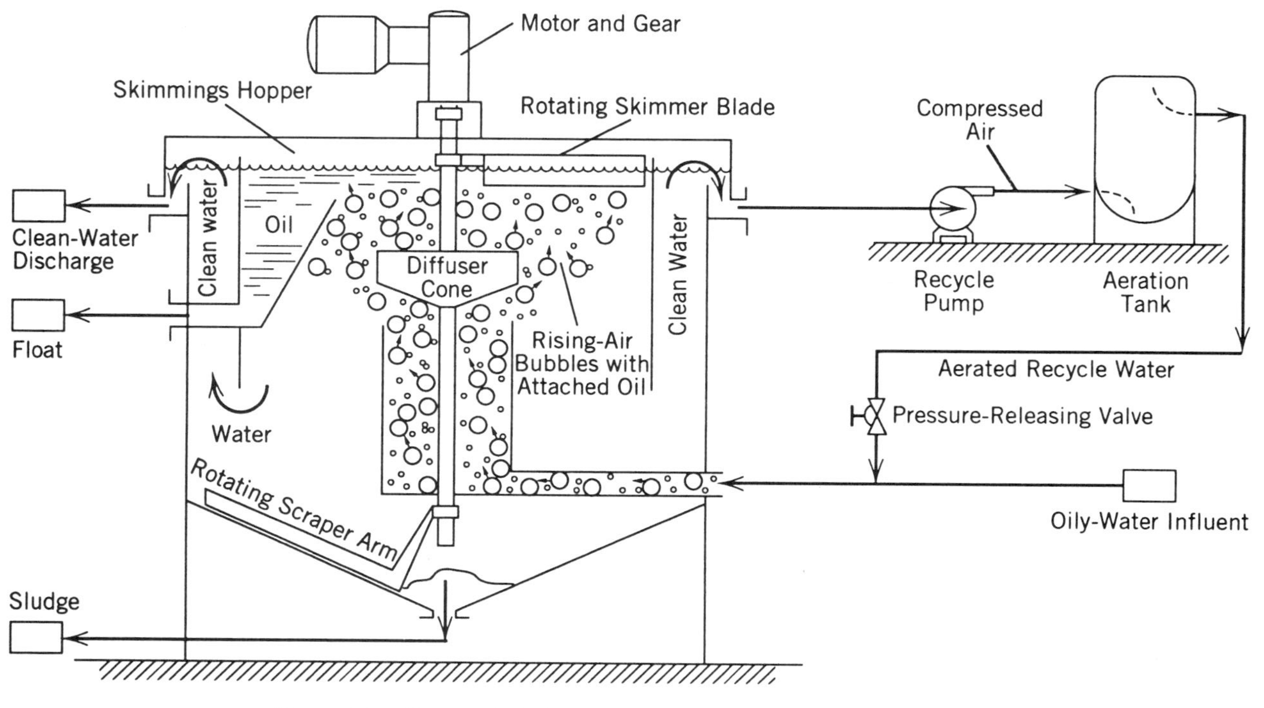

Figure 5-1. Dissolved-air flotation unit (CMA, 1990).

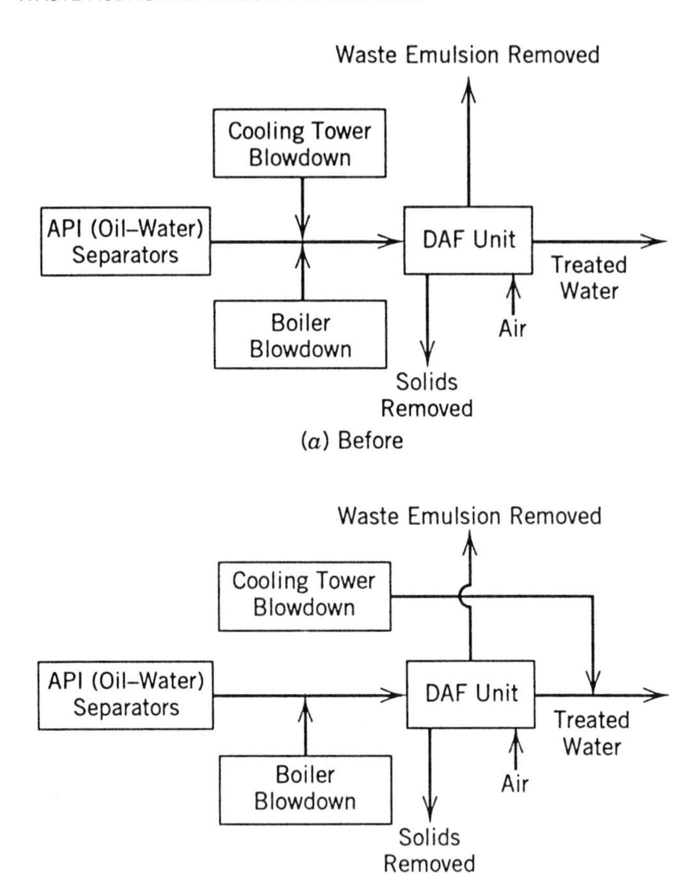

Figure 5-2. Process steam rerouting that resulted in a 40% reduction of waste (*a*) before rerouting of cooling-tower blowdown and (*b*) after rerouting.

Chapter 2, direct releases from a facility, especially air emissions, can pose significant risk to public, employee, and environmental health. Because of this risk, these releases occupy a position of importance in any waste reduction program and knowledge of their quantities is crucial. However, they can be difficult or impossible to directly measure and estimates of these releases result in significant uncertainty. Several references for information on preparing facilitywide emission inventories are listed in Table 5-5. These references include techniques for measuring or estimating emissions from sources of process, storage, transport, fugitive, and secondary emissions. Fugitive and secondary emissions, both of which make up a significant portion of the total air emissions from many facilities, are among the direct releases that are difficult or impossible to directly measure. Methods for estimating secondary and fugitive emissions are described and compared in this section.

Table 5-5. Resources for information on performing facility-wide emission inventories

Reference	Content	Availability
"Compiling Air Toxics Emissions Inventories," prepared for the United States Environmental Protection Agency by T. F. Lahre, Publication No. EPA/450/4-86-010, July 1986	Preparation of inventory	Through NTIS as PB86238086
"How to Develop Your Toxic Emissions Inventory: Approaches, Problems and Solutions," by E. G. Walther et al. in *Proceedings of the National Research and Development Conference on the Control of Hazardous Materials*, Anaheim, CA, Feb. 1991	Preparation of inventory	Through the Hazardous Materials Control Research Institute, Greenbelt, MD, (301) 982-9500
"Prepare Now for the Operating Permit Program," by M. B. Van Wormer and R. M. Iwamchuck	Usefulness of emission inventory, compendium of guides for various estimation and measurement techniques	In the April 1992 issue of *Chemical Engineering Progress*

Although these are not the only methods that an engineer is likely to need in estimating emissions from a facility, they do demonstrate that the choice of methods can have a significant impact on the estimated emission quantities. In addition, fugitive and secondary emissions are important in pollution prevention not only because of their environmental impact but also because, as seen in Chapter 7, their reduction may be among the most cost effective strategies for reducing emissions.

5.2.1 Fugitive Emissions

Fugitive emissions are unintentional releases of process fluid from equipment. Any equipment that can potentially leak (e.g., pumps, valves, and flanges) is a source of fugitive emissions. An oil refinery or large petrochemical facility might have a quarter of a million such pieces of equipment, making it impractical to measure the emissions from every source.

5.2.1.1 *Methods for Estimating Fugitive Emissions* The U.S. Environmental Protection Agency has established a number of methods for estimating fugitive emissions (US EPA, 1988a). The methods vary in the time and expense they require, and as would be expected, the more time-consuming and costly methods result in a more accurate estimate of emissions. The least accurate and least costly method for estimating fugitive emissions is to count all the potential sources of releases and apply an average "emission factor," according to the formula

$$E = m_{\mathrm{VOC}} f_{\mathrm{av}} ,$$

where E is the emission rate of volatile organic compounds (VOCs) from a component, m_{VOC} is the mass fraction of VOC in the stream serviced by the component, and f_{av} is the average emission factor. The average emission factors for fugitive emissions from organic chemical manufacturing, petroleum refining, and natural-gas plants are given in Table 5-6. These factors are intended to provide an estimate of the total emissions of volatile organic compounds from a source. For all the emission estimation methods, it is assumed that each chemical's mass fraction in the emissions is equal to its mass fraction in the process fluid's volatile organic compounds. As shown in Table 5-6, emission factors for some types of equipment vary according to the service the equipment is in: gas, light liquid, or heavy liquid. Liquids are classified based on the most volatile compound present at ≥ 20 wt%. If this compound has a vapor pressure greater than or equal to 0.04 psi (lb/in.^{2}), the equipment is in light liquid service. The following example shows how average emission factors are applied.

Example 5-1 Estimating Fugitive Emissions Using Average Emission Factors
A component count of equipment in an acrolein manufacturing facility found

Table 5-6. Average emission factors for estimating fugitive emissions from SOCMI[a] facilities, refineries, and natural-gas plants

Equipment	Service	Emission Factor, kg/hr/source		
		SOCMI[b]	Refinery[c]	Gas Plant[b]
Valves	Hydrocarbon gas	0.00597	0.027	—
	Light liquid	0.00403	0.011	—
	Heavy liquid	0.00023	0.0002	—
	Hydrogen gas	—	0.0083	—
	All	—	—	0.020
Pump seals	Light liquid[d]	0.0199	0.11	—
	Heavy liquid	0.00862	0.021	—
	Liquid[d]	—	—	0.063
Compressor seals	Hydrocarbon gas	0.228	0.63	—
	Hydrogen gas	—	0.050	—
	All	—	—	0.204
Pressure-relief valves	Hydrocarbon gas	0.104	0.16	—
	Liquid	0.0070[e]	0.0070[e]	—
	All	—	—	0.188
Flanges and other connectors	All	0.00183	0.00025	0.0011
Open-ended lines	All	0.0017	0.002	0.022
Sampling connections	All	0.015	—	—

[a]SOCMI: Synthetic Organic Chemical Manufacturing Industry.
[b]US EPA (1993) except as noted.
[c]US EPA (1985a) except as noted.
[d]This factor can be used to estimate the leak rate from agitator seals.
[e]US EPA (1985b).

that there were 1400 valves, 3048 flanges and other connectors, 27 pumps, 20 pressure-relief valves, 21 open-ended lines, and 20 sampling connections (Berglund et al., 1989). Determine the fugitive acrolein emissions in lb/yr using the average Synthetic Organic Chemical Manufacturing Industry (SOCMI) emission factors from Table 5-6. Assume that the process fluid is composed almost entirely of VOCs and that the equipment contains 87% by mass acrolein. Of the valves, 168 are in gas service; all other valves and pumps are in light liquid service. Tabulate your results by equipment type and give the percentage of total emissions due to each equipment type.

Solution For valves in gas service, estimated acrolein emissions are

Table 5-7. Acrolein emissions estimated using average emission factors

Equipment Type	Emissions, lb/yr	% of Emissions by Equipment Type
Valves	100,000	41
Flanges	94,000	38
Pumps	9,000	3.7
Pressure-relief valves	35,000	14
Open-ended lines	600	0.25
Sampling connections	5,000	2.1
Total	240,000	100

$$\left(\frac{0.87 \text{ lb acrolein}}{\text{lb VOC}}\right)(168 \text{ valves})\left(0.00597 \frac{\text{kg VOC}}{\text{hr valve}}\right)\left(2.2 \frac{\text{lb}}{\text{kg}}\right)\left(24 \frac{\text{hr}}{\text{day}}\right)\left(365 \frac{\text{day}}{\text{yr}}\right) = 17{,}000 \frac{\text{lb acrolein}}{\text{yr}}.$$

For valves in light liquid service, estimated acrolein emission are

$$\left(\frac{0.87 \text{ lb acrolein}}{\text{lb VOC}}\right)(1400 - 168 \text{ valves})\left(0.00403 \frac{\text{kg VOC}}{\text{hr valve}}\right)\left(24 \frac{\text{hr}}{\text{day}}\right)\left(365 \frac{\text{day}}{\text{yr}}\right) = 83{,}000 \frac{\text{lb acrolein}}{\text{yr}}.$$

Summing these values gives the total estimated acrolein emissions from valves, which is 100,000 lb/yr. Results for all component types are given in Table 5-7.

A second method for estimating fugitive emissions is to apply leak/no leak emission factors. Leak/no leak emission factors are given in Table 5-8. The emission factors are applied according to the equation

$$E = f_{L/NL},$$

where E is defined as before and $f_{L/NL}$ is the appropriate leak/no leak emission factor. A piece of equipment is considered to be leaking if an organic vapor analyzer (OVA) detects VOC concentrations greater than 10,000 ppm at the process fluid/air interface. This value for leaking components is intended to indicate the leak level at which it becomes cost-effective to repair a component (Dimmick and Hustvedt, 1984). The concentration

Table 5-8. Leak/no-leak emission factors for estimating SOCMI, refinery, and natural-gas plant fugitive emissions

		Emission Factor, kg/hr/source					
		SOCMI		Refinery		Gas Plant	
Equipment	Service	Leak[a]	No-Leak[b]	Leak[a]	No-Leak[b]	Leak[a]	No-Leak[b]
Valves	Gas	0.0782	0.000131	0.2626	0.0006	—	—
	Light liquid	0.0892	0.000165	0.0852	0.0017	—	—
	Heavy liquid	0.00023	0.00023	0.00023	0.00023	—	—
	All	—	—	—	—	0.098	0.0029
Pump seals	Light liquid[c]	0.243	0.00187	0.437	0.0120	—	—
	Heavy liquid	0.216	0.00210	0.3885	0.0135	—	—
	All[c]	—	—	—	—	0.150	0.020
Compressor seals	Gas	1.608	0.0894	1.608	0.0894	—	—
	All	—	—	—	—	0.442	0.025
Pressure-relief valves	Gas	1.691	0.0447	1.691	0.0447	0.863	0.0447
Flanges and other connectors	All	0.113	0.0000810	0.0375	0.00006	0.0336	0.00006
Open-ended lines	All	0.01195	0.00150	0.01195	0.00150	0.174	0.0015

[a]Screening value > 10,000 ppm.
[b]Screening value < 10,000 ppm.
[c]This factor can be used to estimate the leak rate from agitator seals.

Source: US EPA (1993).

detected by the OVA is called the screening concentration. For facilities with a lower-than-average number of leaking components, the leak/no-leak method gives a lower estimate of emissions than the average emission factor method. As an example of the application of these factors, consider once again the acrolein manufacturing facility of Example 5-1.

Example 5-2 Estimating Fugitive Emissions for an Acrolein Manufacturing Facility Using Leak/No Leak Emission Factors Recall from Example 5-1 that a component count of plant equipment found that there were 1400 valves, 3048 flanges and other connectors, 27 pumps, 20 pressure-relief valves, 21 open-ended lines, and 20 sampling connections at an acrolein manufacturing facility. The VOC streams average 87% acrolein by mass. When each component was screened with an organic vapor analyzer (OVA), it was discovered that only one pump and one valve in liquid service gave readings corresponding to a hydrocarbon concentration of >10,000 ppm (Berglund et al., 1989). Use the leak/no leak emission factors given in Table 5-8 to determine acrolein emissions in lb/yr, making the same assumptions as in Example 5-1 and using the average emission factor for sampling connections. Report the results as in Example 5-1. How do the estimates differ from those produced from average emission factors? How does the distribution of emissions differ?

Solution For nonleaking valves in gas service, the estimated emissions are

$$\left(\frac{0.87\ \text{lb acrolein}}{\text{lb VOC}}\right)(168\ \text{valves})$$
$$\times\left(0.000131\,\frac{\text{kg VOC}}{\text{hr valve}}\right)\left(2.2\,\frac{\text{lb}}{\text{kg}}\right)\left(24\,\frac{\text{hr}}{\text{day}}\right)\left(365\,\frac{\text{day}}{\text{yr}}\right)=370\,\frac{\text{lb acrolein}}{\text{yr}}.$$

For nonleaking valves in light liquid service, estimated emissions are

$$\left(\frac{0.87\ \text{lb acrolein}}{\text{lb VOC}}\right)(1400-168\ \text{valves})$$
$$\times\left(0.000165\,\frac{\text{kg VOC}}{\text{hr valve}}\right)\left(2.2\,\frac{\text{lb}}{\text{kg}}\right)\left(24\,\frac{\text{hr}}{\text{day}}\right)\left(365\,\frac{\text{day}}{\text{yr}}\right)=3400\,\frac{\text{lb acrolein}}{\text{yr}}.$$

For the leaking valve in light liquid service, estimated emissions are

Table 5-9. Acrolein emissions estimated using leak/no-leak emission factors

Equipment Type	Emissions, lb/yr	% of Emissions by Equipment Type
Valves	5,300	15
Flanges	4,100	12
Pumps	4,900	14
Pressure-relief valves	15,000	43
Open-ended lines	530	1.5
Sampling connections	5,000	14
Total	35,000	100

$$\left(\frac{0.87 \text{ lb acrolein}}{\text{lb VOC}}\right)(1 \text{ valve})\left(0.0892 \frac{\text{kg VOC}}{\text{hr valve}}\right)$$
$$\times \left(2.2 \frac{\text{lb}}{\text{kg}}\right)\left(24 \frac{\text{hr}}{\text{day}}\right)\left(365 \frac{\text{day}}{\text{yr}}\right) = 1500 \frac{\text{lb acrolein}}{\text{yr}}.$$

Summing these three yields total estimated emissions from valves, which is 5300 lb/yr. Results for all component types are given in Table 5-9. Emission estimates are almost 7 times lower using this procedure than the procedure of Example 5-1. The distribution of sources is also different. In Example 5-1, component types had emissions in the following order: valves > flanges- > pressure-relief valves > pumps > sampling connections > open-ended lines. Using the leak/no-leak factors, the order was as follows: pressure-relief valves > valves > pumps/sampling connections > flanges > open-ended lines.

An even more refined method of estimating emissions is to use EPA-provided correlations of emission rate as a continuous function of VOC screening concentration. These correlations are given for refineries and SOCMI facilities in Table 5-10. The default-zero values in this table are the emissions assigned to components whose leak rate is too low to register on an OVA. This method involves only slightly more effort than the leak/no-leak approach, with the bulk of the effort put into obtaining screening concentrations of each piece of equipment.

The most difficult and expensive way to estimate fugitive emissions is to enclose a statistically sound number of components in bags and take samples of air flowing through the bags to directly measure emissions. The definition of a statistically sound number of components, along with other details, is provided in the protocol for estimating emissions using this method (US EPA, 1993). Material used in bagging must be impermeable to hydrocarbons,

Table 5-10. Correlations for estimating fugitive emissions and their default-zero values

Equipment	Service	Leak Rate from Correlation, kg/hr[a]		Default-Zero Emission Rates,[b] kg/hr/source
		SOCMI	Refinery	
Valves	Gas	$1.87 \times 10^{-6} C^{0.873}$	$2.18 \times 10^{-7} C^{1.23}$	6.56×10^{-7}
	Light liquid	$6.41 \times 10^{-6} C^{0.797}$	$1.44 \times 10^{-5} C^{0.80}$	4.85×10^{-7}
Pump seals	Light liquid	$1.90 \times 10^{-5} C^{0.824}$[c]	$8.27 \times 10^{-5} C^{0.83}$[d]	7.49×10^{-6}[c]
	Heavy liquid		$8.79 \times 10^{-6} C^{1.04}$	
Compressor seals	Gas		$8.27 \times 10^{-5} C^{0.83}$	
Pressure-relief valves	Gas		$8.27 \times 10^{-5} C^{0.83}$	
Flanges and other connectors	All	$3.05 \times 10^{-6} C^{0.885}$	$5.78 \times 10^{-6} C^{0.88}$	6.12×10^{-7}

[a] C: screening value in ppmv.
[b] These values are applicable to all source categories.
[c] This correlation/default-zero value can be applied to compressor seals, pressure-relief valves, agitator seals, and heavy liquid pumps.
[d] This correlation can be applied to agitator seals.

Source: US EPA (1993).

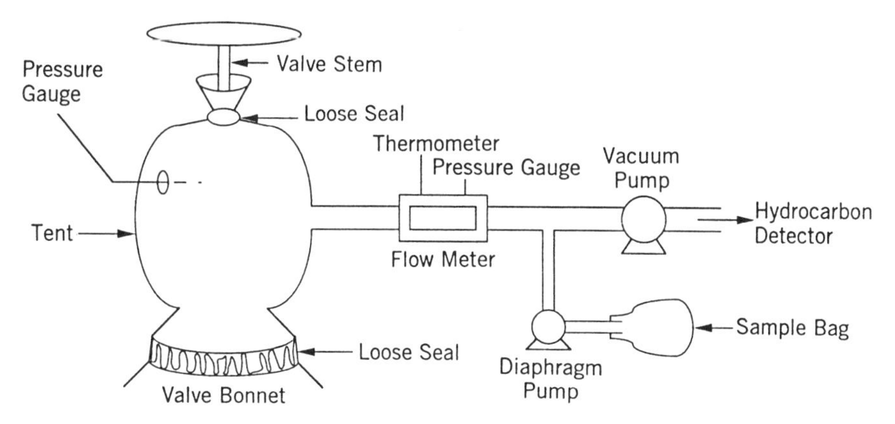

Figure 5-3. Bagging appartus for measuring fugitive-emission rates.

such as metal foil or Mylar. A typical bagging apparatus is pictured in Figure 5-3. The screening concentration obtained for the equipment that was bagged is used to develop a curve relating screening concentration to actual emissions. As in the estimation methods using leak/no-leak emission factors and EPA-provided curves, each piece of equipment, except for possibly some flanges, must be screened using an appropriate instrument.

5.2.1.2. ***The Validity of Fugitive-Emission Factors*** The EPA's fugitive-emission factors and correlation curves were originally developed from two studies published in 1980 (US EPA, 1986). In one study, sampling of more than 6000 emission sources at 13 refineries was conducted (US EPA, 1980a). Data were gathered for both screening values and mass emission rates, and it was determined from the results of this study that in the case of valves and pumps, fugitive-emission rates correlated with the volatility of the stream serviced by the equipment. This gave rise to the heavy liquid, light liquid, and gas/vapor categories for emission factors. The other study was of 24 SOCMI facilities selected to represent a cross section of the industry population (US EPA, 1980b). This study revealed that the leak frequencies in SOCMI facilities are not the same as those in refineries. The SOCMI emission factors were determined by combining the refinery mass emission and screening data with leak rates from the SOCMI study. In doing this, the assumption was made that for each equipment type (e.g., flanges), leaking pieces of equipment in refineries have on average the same emissions as leaking pieces of equipment in SOCMI facilities, and nonleaking pieces of equipment in refineries have on average the same emissions as nonleaking pieces of equipment in SOCMI facilities. The SOCMI fugitive-emission factors were revised in 1993 by combining the results of the 1980 studies with 1987/88 bagging data for connectors, light liquid pumps, gas valves, and light liquid valves in the ethylene oxide and butadiene industries.

Table 5-11. Acrolein production fugitive emissions

Estimation Technique	Fugitive Acrolein Emissions, lb/yr
Average emission factors	268,600
Leak/no-leak emission factors	63,500
EPA-provided correlation curve	19,700
Acrolein-specific correlation curve	952

Source: Berglund et al. (1989).

The methods used to generate the SOCMI emission factors have caused a great deal of controversy. SOCMI facility operators feel that especially for units handling chemicals that are toxic, explosive, or otherwise hazardous, the data used to develop the factors are not appropriate. There is controversy over not only the leak rate assumed for generating the SOCMI average emission factors, but also over emission estimates even when corrections are made for leak rate. Acrolein industry fugitive emission estimates using the four estimation methods discussed in this chapter are given in Table 5-11. The emission factors used to produce these estimates have since been revised, but the revised factors would give estimates within a factor of 2 of the values in this table. The data in Table 5-11 show that although use of the leak/no-leak emission factors results in an estimate that is 4 times lower than the average emission factor estimate, the most accurate estimate of emissions is 70 times less than the leak/no-leak estimate. This indicates that for the acrolein industry the average emission factors are inappropriate because neither the emission rate data nor the leak rate data used to generate them are representative of the acrolein industry. Use of the simpler estimation methods for these units might result in unnecessary public alarm over exposure to a highly reactive chemical that is both toxic and irritating.

A number of industry trade organizations are currently attempting to update and improve fugitive-emission estimates. Because of this, new information regarding fugitive emissions is constantly being made available. Table 5-12 provides a brief list of sources for information on fugitive emissions.

5.2.2 Secondary Emissions

Secondary emissions are releases that occur as a result of the construction or operation of a major stationary source but that do not come from the major stationary source itself (CMA, 1990). It is difficult to measure secondary emissions because the emissions from each source are dependent on location within the source, seasonal variations, current process conditions, and other factors. In addition, there are no universally applicable measurement techniques for measuring emissions from secondary sources. Instead,

Table 5-12. Sources of information on fugitive emissions

Source of Information	Content	Availability
"Protocols for Generating Unit-Specific Emission Estimates for Equipment Leaks of VOC and VHAP," from the United States Environmental Protection Agency, Publication EPA-450/3-88-010, Oct. 1988	Details procedures for estimating fugitive emissions in the synthetic organic chemical manufacturing industry; includes acceptable bagging procedures	Through NTIS as PB89-138689
"Equipment Leaks of VOC: Emissions and Their Control," W. F. Dimmick and K. C. Hustvedt, paper for presentation at the 77th Annual Meeting of the Air Pollution Control Association, San Francisco, June 24–29, 1984	Information on source counts, leak rates, reduction of fugitive emissions, reduction cost factors	Proceedings of the meeting
"A Model for Evaluation of Refinery and Synfuels VOC Emission Data, Volume I," prepared for the United States Environmental Protection Agency by R. G. Wetherold, G. E. Harris, F. D. Skinner, and L. P. Provost, Publication EPA-600/7-85-022a, May 1985	Information on source counts, leak rates, and control efficiencies	Through NTIS as PB85-215713
"Development of Refinery Plot Plans," prepared for the United States Environmental Protection Agency by D. Powell, P. Peterson, K. Luedtke, and L. Levanas, Publication EPA-450/3-78-025, June 1978	Emission factors, source counts, and phase distribution	Through NTIS as PB80-159288
"Compilation of Air Pollutant Emission Factors, Volume I with Supplement A," from the United States Environmental Protection Agency, Publication AP-42, 5th Ed.	Emission factors, reduction strategies	Through US GPO as Stock Numbers 055-000-00500-1 and 055-000-00551-6

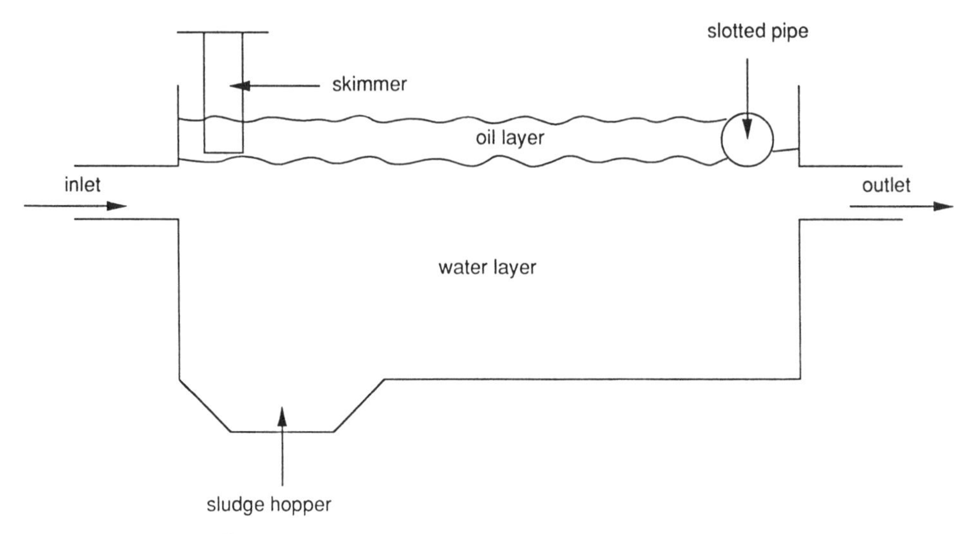

Figure 5-4. An API oil–water separator (not to scale).

techniques must be chosen carefully according to the conditions at each source. The largest class of secondary emissions in the chemical process industries are those from waste handling, treatment, and disposal. Wastewater treatment facilities in particular are considered to be a source of significant atmospheric hydrocarbon emissions (CMA, 1990), and the following discussion will focus on these treatment facilities.

Wastewater is treated before release in order to make it less toxic and, in some cases, to recover any useful material it contains. In general, wastewater treatment can be divided into three stages: (1) primary treatment that uses physical operations to remove free oil and/or suspended solids, (2) secondary treatment for the removal of dissolved contaminants through chemical or biological action, and (3) tertiary treatment for the removal of residual contaminants. Individual wastewater treatment plants are tailored to meet the needs of the waste they treat and the level of compliance they must achieve, resulting in a great deal of variation between plants.

Refining processes such as coking, sulfur recovery, steam cracking, hydrocracking, and crude desalting create large volumes of oily wastewater. An example of a primary treatment unit found in petroleum refineries is the oil–water separator. Most separators are large, open rectangular tanks through which wastewater slowly moves, allowing free oil to rise to the surface and settlable solids to sink to the bottom. As shown in Figure 5-4, a skimmer is used to push accumulated oil into a slotted pipe or drum. Volatile organics in the oil layer escape to the atmosphere and can account for as much as 15% of a refinery's air emissions. An oil–water separator is often followed by a dissolved-air flotation (DAF) unit, which was described in Section 5.1.1.

When the bubbles in a DAF unit reach the surface, they release volatile organics they acquired from the oil with which they were in contact.

Secondary treatment, for the removal of dissolved contaminants, usually relies on biological degradation of chemicals by bacteria and other microorganisms. Aerobic organisms, which require oxygen, are most commonly used. They are more effective when there is a high concentration of dissolved oxygen in the wastewater, which is usually achieved by thorough contact of the wastewater with air through mixing and agitation. This contact results in volatilization and release of organics. An alternative to aggressively agitating the wastewater to promote oxygen transfer is to allow long reaction times. When this approach is used, wastewater may be retained in a secondary treatment lagoon for days to allow time for degradation. Because of the surface area and the long retention times of these lagoons, volatile emissions can be significant.

Tertiary treatment is considered a "polishing" step. Even though tertiary treatment units often allow a great deal of contact between the wastewater and air, there are few contaminants remaining in the wastewater at this stage of treatment and the potential for hydrocarbon releases is low.

The segments of a wastewater treatment facility require support from many diverse components and processes that can themselves be a source of secondary emissions. Retention ponds, which dampen variations in wastewater quality and flow, can emit significant quantities of hydrocarbons when they are used prior to primary or secondary treatment units. Drains, manholes, and trenches are support equipment that often handle wastewaters with high organic content, elevated temperature, and high turbulence. All these factors are favorable for the volatilization of organics, and these wastewater treatment components are often open to the atmosphere to avoid creation of explosive organic concentrations. The treatment of sludge from primary wastewater treatment units is another support process. It often consists of landfilling, landfarming, or incineration, all of which result in secondary emissions to air.

5.2.2.1 *Methods for Measuring and Estimating Secondary Emissions* Methods for estimating secondary emissions generally fall into three categories: application of emission factors, calculations based on mass transfer theory, and measurement techniques. Use of emission factors is by far the simplest method, but can yield emission estimates that have high levels of uncertainty. Also, emission factors are not available for all sources of secondary emissions. As with fugitive-emission factors, when estimating releases of a particular compound, the compound's mass fraction in the VOC emissions is assumed to be the same as its mass fraction in the VOC portion of the wastewater stream. Therefore, in order to use emission factors to report emissions of a single compound, the wastewater stream must be analyzed. Application of an emission factor for estimating secondary emissions is illustrated in the following example.

Example 5-3 Estimating Secondary Emissions Using an Emission Factor For uncovered oil–water separators such as the one pictured in Figure 5-4, the volatile organic compound emission factor is 0.6 kg per 10^3 L wastewater (US EPA, 1985a). Use the emission factor to estimate the volatile organic compound emissions in kg/hr from an oil–water separator with a wastewater throughput of 7600 m^3/day. Assuming that the only volatile organic compounds in the wastewater are benzene, toluene, ethylbenzene, and xylene, and their mass fractions in the volatile organic compound portion of the wastewater are 0.00926, 0.167, 0.130, and 0.694, respectively, estimate the emissions of the individual compounds in kg/hr.

Solution Total VOC emissions are

$$
\begin{aligned}
E_{\mathrm{TOT}} &= (0.6\ \mathrm{kg}/10^3\ \mathrm{L\ wastewater})(7600\ \mathrm{m}^3/\mathrm{day})(1000\ \mathrm{L}/\mathrm{m}^3)(\mathrm{day}/24\ \mathrm{hr}) \\
&= 200\ \mathrm{kg/hr}.
\end{aligned}
$$

(Note that it would be inappropriate to report this estimate with more than one significant digit.) If R_i is the mass fraction of compound i in the VOC portion of the wastewater, then

$$E_i = R_i E_{\mathrm{TOT}}\,.$$

For benzene, toluene, ethylbenzene, and xylene, emissions are 2, 30, 20, and 100 kg/hr, respectively.

When estimating emissions using mass-transfer theory, many characteristics of the compounds and the treatment device must be found. These characteristics can be either measured, calculated from estimation methods, or taken from default values in the literature. The accuracy and difficulty of estimating emissions using mass-transfer theory depends on the amount of site-specific information gathered. The first step is to classify the equipment for which emissions are being estimated according to turbulence, biological activity, and other parameters. A typical approach for characterizing equipment is shown in the flow charts of Figures 5-5 and 5-6. Next, appropriate equations (indicated in these figures) are used to obtain estimates of gas and liquid-phase mass-transfer coefficients, which, in turn, are used to develop the overall mass-transfer coefficients that give the emission estimate. These equations are located in Table 5-13, and their parameters are defined in Table 5-14. In the next example, you will use these equations to estimate the emissions from the oil–water separator of Example 5-3.

start → Is system aerated? (yes: Diffused air?; no: Oil film layer?) → Biologically active? / Oil film thickness > 1 cm? (yes / no) → Flowthrough / Disposal

Is system aerated?	Diffused air? / Oil film layer?	Biologically active? / Oil film thickness > 1 cm?	Unit	k_l	k_g	K_{oil}	K	E
				Equations used to obtain:*				
yes	Diffused air? yes	Biologically active? yes	Flowthrough	1	2		7	20
yes	Diffused air? yes	Biologically active? yes	Disposal	1	2		7	19
yes	Diffused air? yes	Biologically active? no	Flowthrough	1	2		7	14
yes	Diffused air? yes	Biologically active? no	Disposal	1	2		7	13
yes	Diffused air? no	Biologically active? yes	Flowthrough	1,3	2,4		7	16
yes	Diffused air? no	Biologically active? yes	Disposal	1,3	2,4		7	15
yes	Diffused air? no	Biologically active? no	Flowthrough	1,3	2,4		7	12
yes	Diffused air? no	Biologically active? no	Disposal	1,3	2,4		7	11
no	Oil film layer? no	Biologically active? yes	Flowthrough	1	2		7	16
no	Oil film layer? no	Biologically active? yes	Disposal	1	2		7	15
no	Oil film layer? no	Biologically active? no	Flowthrough	1	2		7	12
no	Oil film layer? no	Biologically active? no	Disposal	1	2		7	11
no	Oil film layer? yes	Oil film thickness > 1 cm? yes	Flowthrough		2	9		18
no	Oil film layer? yes	Oil film thickness > 1 cm? yes	Disposal		2	9		17
no	Oil film layer? yes	Oil film thickness > 1 cm? no	Flowthrough		2	9		22
no	Oil film layer? yes	Oil film thickness > 1 cm? no	Disposal		2	9		23

*Numbers correspond to equations in Table 5-13 and parameters are defined in Table 5-14.

Figure 5-5. Flowchart for determining appropriate mass-transfer equations for wastewater treatment and storage units (US EPA, 1985a).

Example 5-4 Estimating Secondary Emissions Using Calculations Based on Mass-Transfer Theory The oil–water separator of Example 5-3 has an oil film layer that varies in thickness, depending on the position of the skimmer and the location with respect to the inlet of the separator. However, on average, the film is >1 cm thick. The surface area of the separator is 84 m^2. To review, wastewater throughput is 7600 m^3 per day, and the only VOCs in the wastewater are benzene, toluene, ethylbenzene, and xylene, with mass fractions in the VOC content of the wastewater of 0.00926, 0.167, 0.130, and 0.694, respectively. Use a wind speed of 4.47 m/s. The density of air at

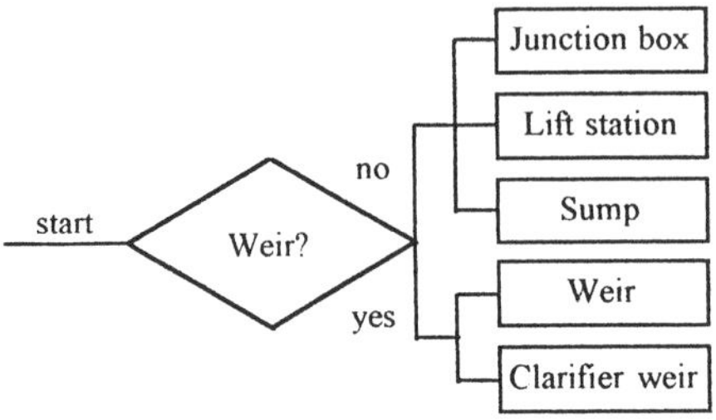

start

Weir?

no / yes

Equations used to obtain:*

Component	k_l	k_g	K_D	K	E
Junction box	3	2		7	12
Lift station	3	2		7	12
Sump	1	2		7	12
Weir			10		21
Clarifier weir	5	6		8	24

*Numbers correspond to equations in Table 5-13 and parameters are defined in Table 5-14.

Figure 5-6. Flowchart for determining appropriate mass-transfer equations for wastewater transport components (US EPA, 1985a).

the site is 1.2×10^{-3} g/cm^3, while the molecular weight and density of the oil layer are 282 g/mol and 0.92 g/cm^3, respectively. The volume fraction of oil in the wastewater is 0.1, and 10% by volume of the oil layer is comprised of VOC. The density of the volatile organic portion of the oil layer is 0.87 g/cm^3. Compound-specific data necessary to perform the calculations are given in Table 5-15. Use Figure 5-5 and Tables 5-13 and 5-14 to estimate the emissions of benzene, toluene, ethylbenzene, and xylene from this oil–water separator.

Solution From Figure 5-5, the equations to use for a nonaerated wastewater treatment component with an oil layer >1 cm thick are 2, 9, and 18 for k_g, K_{oil}, and E, respectively. Therefore, the gas-phase mass-transfer coefficient is found from

$$k_g(\mathrm{m/s}) = 4.82 \times 10^{-3}(U_{10})^{0.78}(\mathrm{Sc_G})^{-0.67}(d_e)^{-0.11},$$

where the gas-side Schmidt number is

$$\mathrm{Sc_G} = \frac{\mu_a}{\rho_a D_a}$$

and the effective diameter of the oil–water separator is

$$d_e(\mathrm{m}) = 2\left(\frac{A}{\pi}\right)^{0.5}.$$

Table 5-13. Mass-transfer equations for estimating emissions from wastewater treatment units[a]

Equation Number	Equation
	Individual Liquid (k_l) and Gas (k_g)-Phase Mass-Transfer Coefficients
1	$k_l(\text{m/s}) = 2.78 \times 10^{-6}(D_w/D_{ether})^{2/3}$ for $0 < U_{10} < 3.25$ m/s and all F/D ratios
	$k_l(\text{m/s}) = [2.605 \times 10^{-9}(F/D) + 1.277 \times 10^{-7}](U_{10})^2(D_w/D_{ether})^{2/3}$ for $U_{10} > 3.25$ m/s and $14 < F/D < 51.2$
	$k_l(\text{m/s}) = 2.61 \times 10^{-7}(U_{10})^2(D_w/D_{ether})^{2/3}$ for $U_{10} > 3.25$ m/s and $F/D > 51.2$
	$k_l(\text{m/s}) = 1.0 \times 10^{-6} + 144 \times 10^{-4}(U^*)^{2.2}(Sc_L)^{-0.5}$; $U^* < 0.3$ $k_l(\text{m/s}) = 1.0 \times 10^{-6} + 34.1 \times 10^{-4}U^*(Sc_L)^{-0.5}$; $U^* > 0.3$ for $U_{10} > 3.25$ m/s and $F/D < 14$, where $U^*(\text{m/s}) = 0.01U_{10}(6.1 + 0.63U_{10})^{0.5}$ $Sc_L = \mu_L/(\rho_L D_w)$ where $F/D = 2(A/\pi)^{0.5}$
2	$k_g(\text{m/s}) = 4.82 \times 10^{-3}(U_{10})^{0.78}(Sc_G)^{-0.67}(d_e)^{-0.11}$, where $Sc_G = \mu_a/(\rho_a D_a)$ $d_e(\text{m}) = 2(A/\pi)^{0.5}$
3	$k_l(\text{m/s}) = [8.22 \times 10^{-9}\text{ J POWR } 1.024^{(T-20)}O_t 10^6 \text{ MW}_L/(Va_v\,\rho_L)] \times (D_w/D_{O_2,w})^{0.5}$, where POWR(hp) = (total power to aerators)V $Va_v(\text{ft}^2)$ = (fraction of area agitated)A
4	$k_g(\text{m/s}) = 1.35 \times 10^{-7}(\text{Re})^{1.42}\text{P}^{0.4}(Sc_G)^{0.5}(\text{Fr})^{-0.21}(D_a\,MW_a/d)$, where $\text{Re} = d^2 w \rho_a/\mu_a$ $P = [0.85\text{ POWR}(550\text{ ft-lb}_f/\text{s-hp})/N_I]g_c/[\rho_L(d^*)^5 w^3]$ $Sc_G = \mu_a/(\rho_a D_a)$ $\text{Fr} = d^* w^2/g_c$
5	$k_l(\text{m/s}) = f_{air,l}Q/(3600\text{ s/min } h_c \pi d_c)$, where $f_{air,l} = 1 - 1/r$ $r = \exp[0.77(h_c)^{0.623}(Q/(\pi d_c))^{0.66}(D_w/D_{O_2,w})^{0.66}]$
6	$k_g(\text{m/s}) = 0.001 + 0.0462U^*(Sc_G)^{-0.67}$ where $U^*(\text{m/s}) = (6.1 + 0.63U_{10})^{0.5}U_{10}/100$ $Sc_G = \mu_a/(\rho_a D_a)$
	Overall Mass-Transfer Coefficients for water (K) and oil (K_{oil}) Phases and for Weirs (K_D)
7	$K(\text{m/s}) = k_l K_{eq} k_g/(K_{eq} k_g + k_l)$ where $K_{eq} = H/(RT)$
8	$K(\text{m/s}) = [\text{MW}_L/(k_l \rho_L(100\text{ cm/m})) + \text{MW}_a/(k_g \rho_a H\ 55{,}555\ (100\text{ cm/m}))]^{-1}\text{MW}_L/[(100\text{ cm/m})\rho_L]$

Table 5-13. (*Continued*)

Equation Number	Equation
9	$K_{oil} = k_g K_{eq,oil}$ where $K_{eq,oil} = P^* \rho_a MW_{oil}/(\rho_{oil} MW_a P_o)$
10	$K_D = 0.16\,h(D_w/D_{O_2,w})^{0.75}$
	Air Emissions (E)
11	$E\,(g/s) = (1 - Ct/Co) V\,Co/t$ where $Ct/Co = \exp(-KAt/V)$
12	$E(g/s) = KC_L A,$ where $C_L(g/m^3) = Q\,Co/(KA + Q)$
13	$E(g/s) = (1 - Ct/Co) V\,Co/t,$ where $Ct/Co = \exp[-(KA + K_{eq}Q_a)t/V]$
14	$E(g/s) = (KA + Q_a K_{eq}) C_L,$ where $C_L(g/m^3) = Q\,Co/(KA + Q + Q_a K_{eq})$
15	$E(g/s) = (1 - Ct/Co) KA/(KA + K_{max} b_i V/K_s) V\,Co/t,$ where $Ct/Co = \exp(-K_{max} b_i t/K_s - KAt/V)$
16	$E\,(g/s) = KC_L A,$ where $C_L\,(g/m^3) = [-b + (b^2 - 4ac)^{0.5}]/(2a)$; $a = KA/Q + 1$; $b = K_s(KA/Q + 1) + K_{max} b_i V/Q - Co$; $c = -K_s\,Co$
17	$E\,(g/s) = (1 - Ct_{oil}/Co_{oil}) V_{oil}\,Co_{oil}/t,$ where $Ct_{oil}/Co_{oil} = \exp(-K_{oil} t/D_{oil})$; $Co_{oil} = K_{ow}\,Co/(1 - FO + FO\,K_{ow})$; $V_{oil} = FO\,V$; $D_{oil} = FO\,V/A$
18	$E\,(g/s) = K_{oil} C_{L,oil} A,$ where $C_{L,oil}\,(g/m^3) = Q_{oil}\,Co_{oil}/(K_{oil} A + Q_{oil})$; $Co_{oil} = K_{ow}\,Co/(1 - FO + FO\,K_{ow})$; $Q_{oil} = FO\,Q$
19	$E\,(g/s) = (1 - Ct/Co)(KA + Q_a K_{eq})/(KA + Q_a K_{eq} + K_{max}\,b_i V/K_s) V\,Co/t,$ where $Ct/Co = \exp[-(KA + K_{eq}Q_a)t/V - K_{max} b_i t/K_s]$
20	$E\,(g/s) = (KA + Q_a K_{eq}) C_L,$ where $C_L\,(g/m^3) = [-b + (b^2 - 4ac)^{0.5}]/(2a)$; $a = (KA + Q_a K_{eq})/Q + 1$; $b = K_s[(KA + Q_a K_{eq})/Q + 1] + K_{max} b_i V/Q - Co$; $c = -K_s\,Co$

Table 5-13. *(Continued)*

Equation Number	Equation
21	$E\,(\text{g/s}) = [1 - \exp(-K_D)]Q\,\text{Co}$
22	$E\,(\text{g/s}) = K_{oil}C_{L,oil}A$, where $C_{L,oil}\,(\text{g/m}^3) = Q_{oil}(\text{Co}^*_{oil})/(K_{oil}A + Q_{oil})$; $\text{Co}^*_{oil} = \text{Co}/FO$; $Q_{oil} = FO\,Q$
23	$E\,(\text{g/s}) = (1 - \text{Ct}_{oil}/\text{Co}^*_{oil})V_{oil}(\text{Co}^*_{oil})/t$, where $\text{Ct}_{oil}/\text{Co}^*_{oil} = \exp(-K_{oil}t/D_{oil})$; $\text{Co}^*_{oil} = \text{Co}/FO$; $V_{oil} = FO\,V$; $D_{oil} = FO\,V/A$
24	$E\,(\text{g/s}) = [1 - \exp(-K\pi d_e h_c/Q)]Q\,\text{Co}$

[a]Parameters are defined in Table 5-14.
Source: US EPA (1985a).

The effective diameter of the separator in this example is

$$d_e = 2\left(\frac{84\ \text{m}^2}{\pi}\right)^{0.5} = 10.3\ \text{m}$$

and the gas-side Schmidt number for benzene is

$$\text{Sc}_G = \frac{1.81 \times 10^{-4}\ \text{g/cm/s}}{1.2 \times 10^{-3}\ \text{g/cm}^3 \times 0.077\ \text{cm}^2/\text{s}} = 2.0.$$

For benzene, this makes the gas-phase mass-transfer coefficient

$$k_g\ \text{m/s} = 4.82 \times 10^{-3}(4.47\ \text{m/s})^{0.78}2.0^{-0.67}(10.3\ \text{m})^{-0.11} = 0.0075\ \text{m/s}.$$

From equation 9 in Table 5-13, the overall mass-transfer coefficient for transfer of a compound from the oil phase to the gas phase is

$$K_{oil} = k_g K_{eq,oil}.$$

In this equation the equilibrium constant for the partitioning of the compound between the gas phase and the oil phase is given by

$$K_{eq,oil} = \frac{P^*\rho_a \text{MW}_{oil}}{\rho_{oil}\text{MW}_a \text{P}_o}.$$

For benzene in the separator of this example, this partition coefficient is

Table 5-14. Definitions of and units for parameters in wastewater mass-transfer equations

Parameter	Definition	Units	Code[a]
A	Wastewater surface area	m^2 or ft^2	A
b_i	Biomass concentration (total biological solids)	g/m^3	B
C_L	Concentration of constituent in the liquid phase	g/m^3	D
$C_{L,oil}$	Concentration of constituent in the oil phase	g/m^3	D
Co	Initial concentration of constituent in the liquid phase	g/m^3	A
Co_{oil}	Initial concentration of constituent in the oil phase considering mass-transfer resistance between water and oil phases	g/m^3	D
Co^*_{oil}	Initial concentration of constituent in the oil phase considering no mass-transfer resistance between water and oil phases	g/m^3	D
Ct	Concentration of constituent in the liquid phase at time = t	g/m^3	D
Ct_{oil}	Concentration of constituent in the oil phase at time = t	g/m^3	D
d	Impeller diameter	cm	B
D	Wastewater depth	m or ft	A,B
d^*	Impeller diameter	ft	B
D_a	Diffusivity of constituent in air	cm^2/s	C
d_c	Clarifier diameter	m	B
d_e	Effective diameter	m	D
D_{ether}	Diffusivity of ether in water	cm^2/s	(8.5×10^{-6})[b]
$D_{O_2,w}$	Diffusivity of oxygen in water	cm^2/s	(2.4×10^{-5})[b]
D_{oil}	Oil film thickness	m	B
D_w	Diffusivity of constituent in water	cm^2/s	C
$f_{air,l}$	Fraction of constituent emitted to the air, considering zero gas resistance	Unitless	D
F/D	Fetch to depth ratio, d_e/D	Unitless	D
FO	Fraction of volume that is oil	Unitless	B
Fr	Froude number	Unitless	D
g_c	Gravitation constant (a conversion factor)	lb_m-ft/s^2-lb_f	32.17
h	Weir height (distance from the wastewater overflow to the receiving body of water)	ft	B

h_c	Clarifier weir height	m	B
H	Henry's law constant of constituent	atm-m^3/gmol	C
J	Oxygen-transfer rating of surface aerator	lb O_2/hr-hp	B
K	Overall mass-transfer coefficient for transfer of constituent from liquid phase to gas phase	m/s	D
K_D	Volatilization–reaeration theory mass-transfer coefficient	Unitless	D
K_{eq}	Equilibrium constant or partition coefficient (concentration in gas phase/concentration in liquid phase)	Unitless	D
$K_{eq,oil}$	Equilibrium constant or partition coefficient (concentration in gas phase/concentration in oil phase)	Unitless	
k_g	Gas-phase mass-transfer coefficient	m/s	D
k_l	Liquid-phase mass-transfer coefficient	m/s	D
K_{max}	Maximum biorate constant	g/s-g biomass	A,C
K_{oil}	Overall mass-transfer coefficient for transfer of constituent from oil phase to gas phase	m/s	D
K_{ow}	Octanol–water partition coefficient	Unitless	C
K_s	Half-saturation biorate constant	g/m^3	A,C
MW_a	Molecular weight of air	g/gmol	29
MW_{oil}	Molecular weight of oil	g/gmol	B
MW_L	Molecular weight of water	g/gmol	18
E	Emissions	g/s	D
N_I	Number of aerators	Unitless	A,B
O_t	Oxygen-transfer correction factor	Unitless	B
P	Power number	Unitless	D
P^*	Vapor pressure of the constituent	atm	C
P_o	Total pressure	atm	A
POWR	Total power to aerators	hp	B
Q	Volumetric flow rate	m^3/s	A
Q_a	Diffused air flow rate	m^3/s	B
Q_{oil}	Volumetric flow rate of oil	m^3/s	B

Table 5-14. *(Continued)*

Parameter	Definition	Units	Code[a]
r	Deficit ratio (ratio of the difference between the constituent concentration at solubility and actual constituent concentration in the upstream and the downstream)	Unitless	D
R	Universal gas constant	atm-m^3/gmol-K	8.21×10^{-5}
Re	Reynolds number	Unitless	D
Sc_G	Schmidt number on the gas side	Unitless	D
Sc_L	Schmidt number on the liquid side	Unitless	D
T	Temperature of water	°C or Kelvin	A
t	Residence time of disposal	s	A
U^*	Friction velocity	m/s	D
U_{10}	Wind speed 10 m above the liquid surface	m/s	B
V	Wastewater volume	m^3 or ft^3	A
Va_v	Turbulent surface area	ft^2	B
V_{oil}	Volume of oil	m^3	B
w	Rotational speed of impeller	rad/s	B
ρ_a	Density of air	g/cm^3	$(1.2 \times 10^{-3})^b$
ρ_L	Density of water	g/cm^3 or lb/ft^3	1^b or 62.4^b
ρ_{oil}	Density of oil	g/m^3	B
μ_a	Viscosity of air	g/cm-s	$(1.81 \times 10^{-4})^b$
μ_L	Viscosity of water	g/cm-s	$(8.93 \times 10^{-3})^b$

[a] *Codes:* A—site-specific parameter; B—site-specific parameter, default values available in US EPA (1985a); C—parameter can be obtained from literature, data on ~150 compounds available in US EPA (1985a); D—calculated value.

[b] Reported value at 25°C (298 K).

Source: US EPA (1985a).

Table 5-15. Diffusivities in air and vapor pressure

Compound	Diffusivity in Air, cm^2/s	Vapor Pressure at 25°C, mm Hg
Benzene	0.077	95.19
Toluene	0.079	28.4
Ethylbenzene	0.0658	9.53
Xylene[a]	0.059	8.3

[a]Property data for *m*-xylene.

$$K_{\mathrm{eq,oil}} = \frac{(95.19\,\mathrm{mmHg})(1.2 \times 10^{-3}\,\mathrm{g/cm^3})(282\,\mathrm{g/mol})}{(0.92\,\mathrm{g/cm^3})(29\,\mathrm{g/mol})(760\,\mathrm{mmHg})} = 1.6 \times 10^{-3}\,,$$

so the overall oil to gas mass-transfer coefficient is

$$K_{\mathrm{oil}} = 0.0075\,\mathrm{m/s} \times 1.6 \times 10^{-3} = 1.2 \times 10^{-5}\,\mathrm{m/s}.$$

Finally, emissions are calculated from

$$E = K_{\mathrm{oil}} C_{\mathrm{L,oil}} A\,,$$

where the concentration of a particular compound in the oil phase is

$$C_{\mathrm{L,oil}} = \frac{Q_{\mathrm{oil}} \mathrm{Co}_{\mathrm{oil}}}{K_{\mathrm{oil}} A + Q_{\mathrm{oil}}}$$

and the volumetric flow rate of the oil is

$$Q_{\mathrm{oil}} = FO\,Q.$$

The inlet concentration of a particular compound in the oil phase, $\mathrm{Co}_{\mathrm{oil}}$, is found by converting the information on mass fraction in the volatile organic compound portion of the oil layer to concentration in the oil layer in g/m^3. The equation for making this conversion is

$$\mathrm{Co}_{\mathrm{oil}} = R_i \left(0.87 \frac{\mathrm{g\ volatiles}}{\mathrm{cm^3\ volatiles}}\right)\left(0.1 \frac{\mathrm{cm^3\ volatiles}}{\mathrm{cm^3\ oil}}\right)\left(100 \frac{\mathrm{cm}}{\mathrm{m}}\right)^3.$$

For benzene this becomes

$$\begin{aligned}\mathrm{Co}_{\mathrm{oil}} &= (0.00926\,\mathrm{g\ benzene/g\ volatiles})(0.87\,\mathrm{g/cm^3})(0.1)(100\,\mathrm{cm/m})^3 \\ &= 810\,\mathrm{g/m^3}.\end{aligned}$$

Table 5-16. Intermediate and final results for diffusion modeling

	Compound			
Parameter	Benzene	Toluene	Ethylbenzene	Xylene
Sc_G	2.0	1.9	2.3	2.6
k_G, m/s	0.0075	0.0078	0.0069	0.0063
$K_{eq,oil}$	0.0016	0.00047	0.00016	0.00014
K_{oil}, m/s	1.2×10^{-5}	3.7×10^{-6}	1.1×10^{-6}	8.8×10^{-7}
Co_{oil}	810	14,000	11,000	60,000
$C_{L,oil}$	720	14,000	11,000	60,000
E, g/s	0.7	4	1	4
E, kg/hr	3	20	4	20

For the separator of this example, the volumetric flow rate of the oil is

$$Q_{oil} = 0.1(7600\ \text{m}^3/\text{day})(\text{day}/24\ \text{hr})(\text{hr}/3600\ \text{s}) = 0.0088\ \text{m}^3/\text{s}.$$

Therefore, the concentration of benzene in the oil phase is

$$\begin{aligned} C_{L,oil} &= (0.0088\ \text{m}^3/\text{s})(800\ \text{g/m}^3)/(1.2 \times 10^{-5}\ \text{m/s} \times 84\ \text{m}^2 + 0.0088\ \text{m}^3/\text{s}) \\ &= 720\ \text{g/m}^3 \end{aligned}$$

and benzene emissions are

$$E = (1.2 \times 10^{-5}\ \text{m/s})(720\ \text{g/m}^3)(84\ \text{m}^2) = 0.7\ \text{g/s} = 3\ \text{kg/hr}.$$

Values of Sc_G, k_G, $K_{eq,oil}$, K_{oil}, Co_{oil}, $C_{L,oil}$, and E for all four compounds are given in Table 5-16.

The equations of Table 5-13 are based on completely mixed systems, as opposed to those that exhibit plug flow. An oil–water separator would actually not be perfectly mixed, but because of low flow rates, resembles a perfectly mixed system more than a plug-flow system. Plug-flow models, for high flow rates in narrow channels, can be found in another reference (US EPA, 1989). Mass transfer models for emissions from drains, manholes, and trenches are currently under development. However, there is a published emission factor for hydrocarbon emissions from drains in refineries. It is 0.76 kg/day/source (US EPA, 1985a).

Measurement techniques for quantifying secondary emissions can be either

Table 5-17. Techniques for measuring secondary emissions

Measurement Method	Sources for Additional Information	Suitable Applications
Isolation flux chamber	Balfour and Schmidt, 1984; Gholson et al., 1987	Surface impoundments, land treatment, landfills, large open-top tanks
Broadband-infrared		Open-top tanks, clarifiers, aeration units
Dual-beam laser	McClenny et al., 1974; Measures, 1984	Surface impoundments, land treatment, holding tanks, clarifiers, equalization basins, aeration units
Fourier transform infrared (FTIR)	Herget, 1978; Herget and Brasher, 1979, 1980	Surface impoundments, land treatment, holding tanks, clarifiers, equalization basins, aeration units
Reviews of remote-sensing techniques	Baker and McCready, 1988; Saeger et al., 1988; Minnich et al., 1989	

Source: CMA (1990).

direct or indirect. Table 5-17 lists several methods of directly measuring secondary air emissions, along with additional sources of information and suitable applications. When choosing a method, factors such as the strength of the emission source, the type and size of the source, the surroundings, and cost must be considered. The following example describes one direct measurement technique for estimating the emissions from an oil–water separator.

Example 5-5 Direct Measurement of Secondary Emissions Using an Isolation Flux Chamber An isolation flux chamber, pictured in Figure 5-7, can be used to directly measure hydrocarbon emissions as it rests on the surface of a wastewater treatment body's surface. When the contents of the chamber reach steady state they are sampled and analyzed to determine the concentration of the compounds of interest. Determine the emissions of benzene, toluene, ethylbenzene, and xylene in kg/hr from the oil–water separator of Examples 5-3 and 5-4. Recall that the area of the separator is 84 m^2. Data from flux chamber measurements are given in Table 5-18. First you will need to develop an equation relating the measured concentration [C, in ppmW (parts per million by weight) of carbon] to the mass flux, J, from the separ-

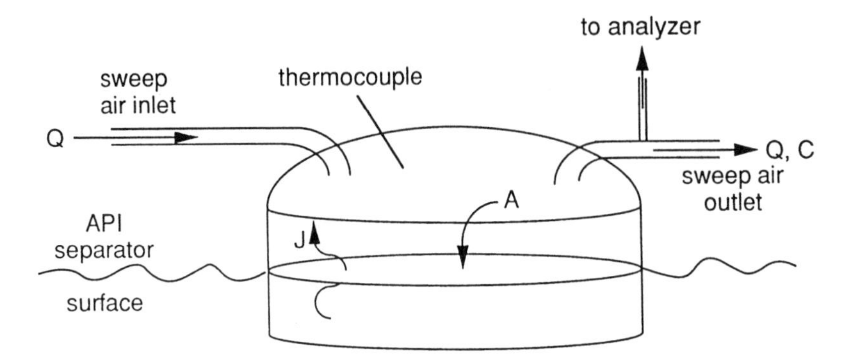

Figure 5-7. An isolation flux chamber.

Table 5-18. Analysis results of flux chamber effluent

Compound	Concentration, ppmW[a] of Carbon
Benzene	21
Toluene	120
Ethylbenzene	31
Xylene	142

[a]Parts per million by weight (ppmW) of carbon is the mass of carbon in the species of interest per million mass units of material sampled.

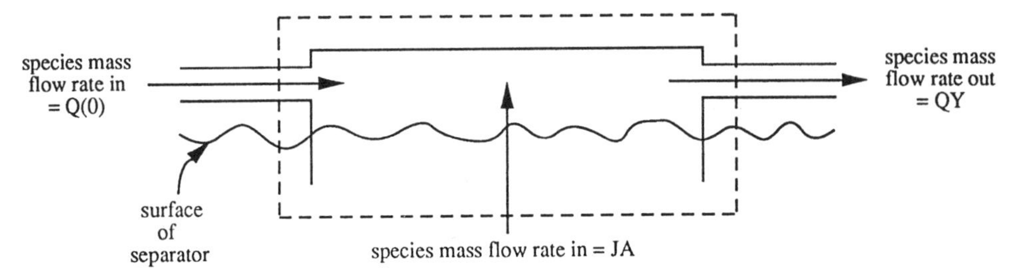

Figure 5-8. Mass-balance diagram for a flux chamber.

ator. Assume the gases behave ideally. Q, the sweep airflow rate of the flux chamber, is 5 L/min and A, the area the flux chamber shares with the surface of the separator, is 0.13 m^2. The ambient temperature is 25°C.

Solution Let Y be the species concentration in g/L in the chamber. Perform a mass balance around the dashed line in Figure 5-8 to obtain the equation

$$JA_{\text{chamber}} = QY.$$

Table 5-19. Emissions measured using an isolation flux chamber

Compound	n	MW	Emissions, kg/hr
Benzene	6	78.11	0.005
Toluene	7	92.13	0.03
Ethylbenzene	8	106.16	0.008
Xylene	8	106.16	0.04

To convert C (ppmW of carbon) to Y (g/L), let n be the number of carbon atoms per mole of species to obtain

$$Y = \left(\frac{C\text{ g carbon}}{10^6\text{ g air}}\right)(\text{MW}_{\text{air}})\left(\frac{1}{\text{molar volume of air}}\right)\left(\frac{\text{mol species}}{12n\text{ g carbon}}\right) \times \text{MW}_{\text{species}}.$$

Substitution gives

$$J = \left(\frac{C\text{ g carbon}}{10^6\text{ g air}}\right)(\text{MW}_{\text{air}})\left(\frac{P}{RT}\right)\left(\frac{\text{mol species}}{12n\text{ g carbon}}\right)(\text{MW}_{\text{species}})\left(\frac{Q}{A_{\text{chamber}}}\right).$$

Emissions are equal to J, the flux per area, multiplied by the area of the separator. Emissions, n, and the molecular weights of the compounds are given in Table 5-19.

Measurements taken using flux chambers are affected by the positioning of the chamber (whether it is before or after the skimmer), by surfacing gas bubbles, and by the position of the skimmer on the surface of the separator. Also, flux chambers perturb conditions such as convective airflow at the surface of the separator by their presence. However, they represent the best method of estimating emissions from an oil–water separator. A flux chamber would not be suitable for measuring emissions from a large wastewater body of varied composition, or from one with a highly agitated surface (CMA, 1990).

Indirect measurement methods include the transect technique, where emission estimates are made based on information obtained about the downwind plume of the secondary emission source (Balfour and Schmidt, 1984; Esplin, 1988), and the concentration-profile method, illustrated in the following example.

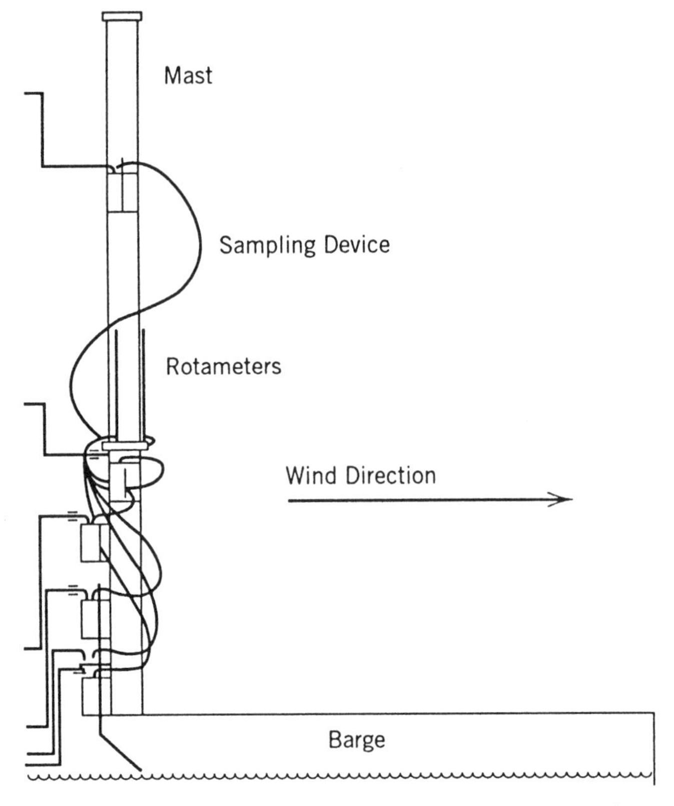

Figure 5-9. Concentration mast for diffusion model measurements (CMA, 1990).

Example 5-6 Indirect Measurement of Secondary Emissions Using Concentration-Profile Measurements In the concentration-profile method of indirectly measuring emissions from a wastewater treatment body, measurements of compound concentration, wind speed, and temperature are taken using a device like the one pictured in Figure 5-9. The measurements must be taken in the reestablished boundary layer pictured in Figure 5-10 so results are not affected by upwind disturbances. Also, the measurements must be taken at intervals within 2 m of the surface. Emissions of VOC can be estimated using a diffusion model (US EPA, 1982). Emissions per unit area are given by

$$J_i = -0.16\left(\frac{D_i}{D_{\mathrm{w}}}\right)^{2/3}\frac{S_{\mathrm{v}}S_{ci}}{\phi_{\mathrm{m}}^2 S_{\mathrm{cw}}^{(\mathrm{t})}}.$$

The denominator in this equation is a stability correction factor that will be taken as unity for this example. The term S_{v} characterizes the velocity

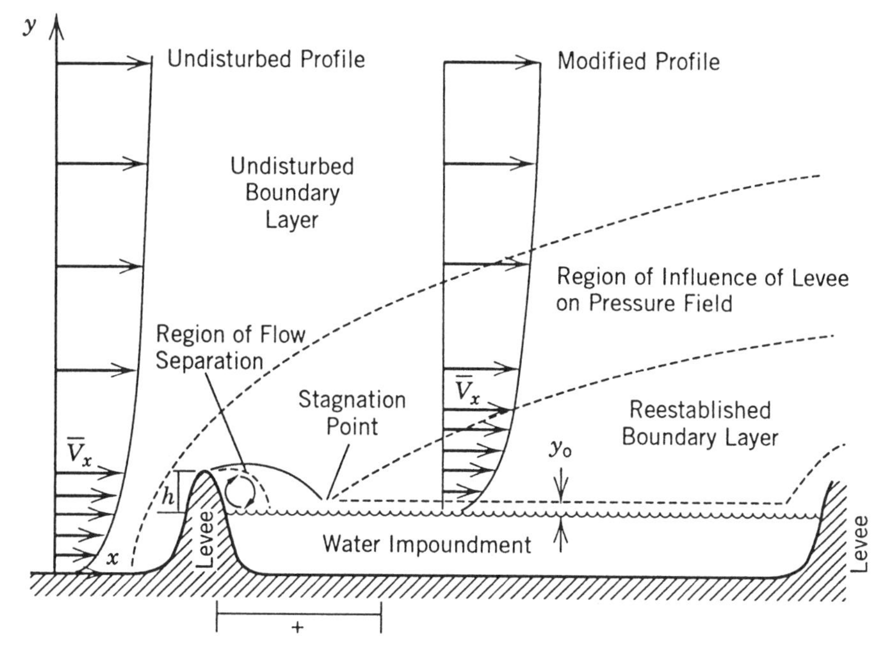

Figure 5-10. Boundary layers above a wastewater treatment pond (US EPA, 1982).

Table 5-20. Data for the concentration-profile method

Compound	Diffusivity in Air, cm^2/s	Concentration at 10 cm, ng/L	Concentration at 100 cm, ng/L
Benzene	0.077	470	390
Toluene	0.079	890	590
Ethylbenzene	0.0658	240	190
Xylene	0.059	900	650
Water	0.22		

distribution and is defined as the slope of the line from a plot of air velocity versus the natural log of the measurement height in cm; S_{ci} is the slope of the line from a plot of the concentration of compound i versus the natural log of the measurement height in cm. Subscript w denotes water, and D represents diffusivity in air. In practice, at least six observations of each variable would be made, but in the interest of simplifying the calculations, data from two heights will be assumed to provide an approximate picture. The wind velocity at 10 cm is 230 cm/s and at 100 cm is 700 cm/s. Concentration data and diffusion coefficients are given in Table 5-20. Remember that the area of the oil–water separator is 84 m^2. Find the estimated releases of benzene, toluene, ethylbenzene, and xylene in kg/hr for the oil–water separator of Examples 5-3 to 5-5 using this indirect measurement technique.

Solution First, calculate S_v as follows:

$$S_v = \frac{230\ \text{cm/s} - 700\ \text{cm/s}}{\ln(10) - \ln(100)} = 204\ \text{cm/s}.$$

For benzene,

$$S_{cB} = \frac{470\ \text{ng/L} - 390\ \text{ng/L}}{\ln(10) - \ln(100)} = -34.74\ \text{ng/L}.$$

Values of S_{ci} for toluene, ethylbenzene, and xylene are found similarly and are −130.3, −21.71, and −108.6 ng/L, respectively. Emissions are equal to the flux (J) times the surface area of the separator, and for benzene the equation to find them is

$$E = -0.16\left(\frac{0.077}{0.22}\right)^{2/3}(204\ \text{cm/s})(-34.4\ \text{ng/L})(84\ \text{m}^2)(\text{m}/100\ \text{cm})$$
$$\times\ (1000\ \text{L/m}^3)(\text{kg}/10^{12}\ \text{ng})(3600\ \text{s/hr}) = 0.002\ \text{kg/hr}.$$

Emission estimates for toluene, ethylbenzene, and xylene are 0.006, 0.001, and 0.004 kg/hr, respectively.

As with fugitive emissions, different methods for estimating secondary emissions can give results that are different by several orders of magnitude. The concentration-profile method generally yields estimates that are 2–10 times lower than actual emissions (CMA, 1990). Tests conducted to validate emission measurements from isolation flux chambers show that these estimates are also consistently lower than actual emissions, but that they are usually within a factor of 5 of actual emissions (CMA, 1990). Emission factors for sources of secondary emissions are generally believed to substantially overestimate emissions. In the case of the API oil–water separator of Example 5-5, the emission factor estimates are 400–3000 times higher than the flux chamber estimates. Mass-transfer theory estimates are generally more accurate than emission factor estimates but can still be several orders of magnitude higher than actual emissions. Government and trade organizations are currently researching methods for estimating secondary emissions. Table 5-21 contains a sampling of sources for more information on estimating secondary emissions.

Table 5-21. Sources of information on secondary emissions

Source of Information	Content	Availability
"Development of Refinery Plot Plans," prepared for the United States Environmental Protection Agency by D. Powell, P. Peterson, K. Luedtke, and L. Levanas, Publication EPA-450/3-78-025, June 1978	Description of wastewater treatment facilities for refineries	Through NTIS as PB80-159288
"Compilation of Air Pollutant Emission Factors, Volume I with Supplement A," from the United States Environmental Protection Agency, Publication AP-42, 5th Ed.	Emission factors, reduction strategies, modeling of emissions	Through US GPO as Stock Numbers 055-000-00500-1 and 055-000-00551-6
"Industrial Wastewater Volatile Organic Compound Emissions—Background Information for BACT/LAER Determinations," from the United States Environmental Protection Agency, Publication EPA-450/3-90-004, Jan. 1990	Estimating and controlling volatile organic compound emissions from wastewater collection, treatment, and storage	From the Office of Air Quality Planning and Standards, Chemicals and Petroleum Branch, Research Triangle Park, NC
SIMS (Surface Impoundment Modeling System), from the United States Environmental Protection Agency	Software for estimating emissions from wastewater collection, treatment, and storage systems	Can be downloaded from the Clearinghouse for Inventories and Emission Factors electronic bulletin board (CHIEF BB). For information, call (919) 541-5384
CHEMDAT7, from the United States Environmental Protection Agency	Software for estimating VOC emissions from wastewater land treatment systems, open and closed landfills, waste storage piles, and more	Contact Emissions Standards Division, Chemicals and Petroleum Branch, Research Triangle Park, NC

5.3 RANKING WASTE STREAMS

A large manufacturing facility may have hundreds to thousands of waste streams, and time and budget constraints make it impossible for such a facility to pursue reduction alternatives for each stream. Instead, the streams are evaluated and the most important ones are selected for evaluation of pollution prevention options. Once the options are developed, they often must themselves be ranked, as discussed in Chapter 9. The subject of this section is the evaluation of streams for the purpose of prioritization.

It may be tempting to rank the waste streams from a facility by the amount of waste generated. In chemical manufacturing facilities and petroleum refineries, process wastewaters would almost always be selected as the subject of focus if waste streams were prioritized this way. Wastewaters constitute over 80% of total hazardous waste mass at refineries and 97% by mass of the total hazardous waste streams in the chemical manufacturing industries (CMA, 1992). Indeed, as mentioned in Chapter 2, wastewaters make up over 90% of RCRA hazardous waste in the nation (Baker and Warren, 1992). However, it must be remembered that wastewaters are made up mostly of water, and even though there may be large quantities of these streams, the mass of hazardous constituents they contain may be similar in magnitude to other sources of waste. Further, these wastes are generally extensively treated before release to the environment. Atmospheric emissions of hazardous materials are at least 100 times smaller nationally than the mass of hazardous waste generation. However, these much smaller emissions generally pose the major risk to public and environmental health from a facility because atmospheric emissions are direct releases to the environment, in contrast to solid and liquid wastes, which are generally treated and sent to secure disposal. Thus, it is inappropriate to rank waste streams for pollution prevention efforts solely on their quantity.

To produce a broader and more acceptable prioritization of waste streams at a large facility, the environmental impacts, treatment costs, and safety concerns associated with each stream must be evaluated. In addition, the technical challenges associated with reducing or eliminating different streams might be considered. Choosing which streams to focus on in waste reduction projects is not a straightforward task, not only because of the number of streams involved, but also because widely varying criteria are used in establishing priorities. Table 5-22 lists a set of criteria, proposed by the United States Environmental Protection Agency, that can be used to rank streams. Note that evaluations for some of these criteria, such as compliance with current and future regulations, would not be stated in quantitative terms. Even for the criteria that can be evaluated quantitatively, the units of measurement are not uniform, making comparisons between the criteria difficult. For example, for safety and health risks, the unit of measure might be avoided accidents or avoided cancer deaths. In contrast, waste management costs are measured in dollars.

Table 5-22. Typical criteria used in prioritizing waste streams

Compliance with current and future regulations
Costs of waste management (treatment and disposal)
Potential environmental and safety liability
Quantity of waste
Hazardous properties of the waste (including toxicity, flammability, corrosivity, and reactivity)
Other safety hazards to employees
Potential for (or ease of) minimization
Potential for removing bottlenecks in production or waste treatment
Potential recovery of valuable byproducts
Available budget for the waste minimization assessment program and projects

Source: US EPA (1988b).

When there are a large number of streams to be ranked using a diverse set of criteria, the only practical way to proceed is to devise a systematic procedure that produces quantitative results. The most common approach is to assign a dimensionless score to each criterion. For example, a waste stream that creates high risks might be given a high score for the health–safety criterion. The same stream might receive a low score for a waste quantity and frequency criterion if the waste is generated only occasionally and at very low levels. Using dimensionless scores allows an overall rating based on multiple criteria to be developed, but the simplification generates necessarily imperfect results.

An example of such a procedure can be taken from the experiences at a large petroleum refinery (Balik and Koraido, 1991). At this refinery, 660 waste streams were identified in a waste survey. As shown in Figure 5-11, these waste streams were released to multiple environmental media. Streams were chosen for further study based on evaluations for the following criteria:

- Waste quantity and frequency
- The cost of managing existing waste
- Possible regulatory impact in the future
- Safety and health risks to the employees and the community
- Ease and cost of implementing pollution prevention options
- Demonstrated effectiveness of pollution prevention options

A rating of 0–5 points (5 indicating the most important streams to target for prevention) was assigned to each stream for each criterion. The points for each stream were added and streams with the highest scores were assigned the highest priority for reduction. The range of possible scores is thus 0–30. Figure 5-12 shows the distribution of total scores for the refinery waste streams. The ranking resulted in a small population of high-priority waste

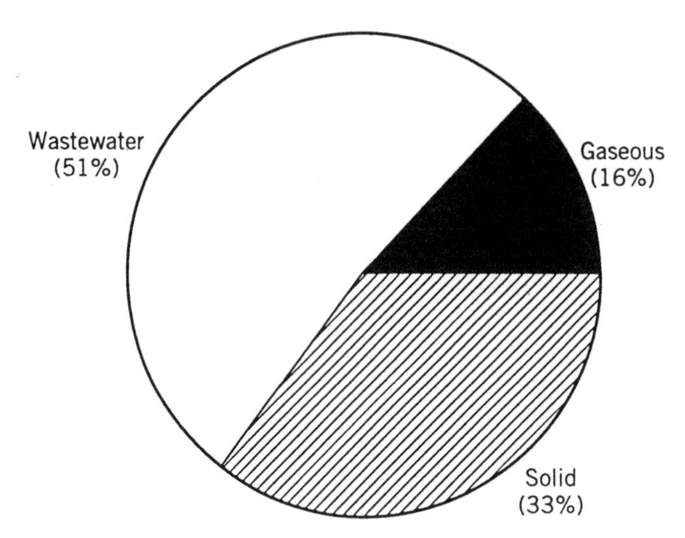

Figure 5-11. Distribution of waste streams across environmental media in a large mid-western petroleum refinery. (From "Identifying Pollution Prevention Options for a Petroleum Refinery," J. A. Balik and S. M. Koraido, *Pollution Prevention Review*, Copyright @ 1991 Executive Enterprises, Inc. Reprinted by permission of John Wiley & Sons, Inc.)

streams, a small population of low-priority waste streams, and a large number of streams that are ranked somewhere in the middle.

Although this method is extremely simplified, it still requires that nearly 4000 ratings from 0 to 5 be assigned. Even if each rating could be developed in only 10 min by a single individual, this represents 660 worker-hours. A variation on the theme might be to select a single criterion and perform an initial screening. Criteria evaluations for streams that received a low score on the screening criterion would not be performed. At the refinery, for example, the safety and health risks ratings for all the streams might be determined first, and streams with a score of 0 or 1 might be left off the list for future ratings. The implicit assumption in this procedure is that safety and health risks are more important than any other single criterion. The time required to screen the waste streams might also be reduced by eliminating some of the criteria. The last two criteria, ease and cost of option implementation and demonstrated effectiveness, are particularly questionable because they are more relevant to the pollution prevention strategies for the waste streams than they are to the significance of the waste streams themselves.

Another way to streamline the process would be to assign scores of 0–3 instead of 0–5 for the criteria. It is generally obvious when to assign very high scores or very low scores, but intermediate cases are more difficult to assign. Sensitivity analyses of prioritization schemes for pollution prevention options have shown that such coarse evaluations may produce the same results as finer evaluations, as discussed in Chapter 9.

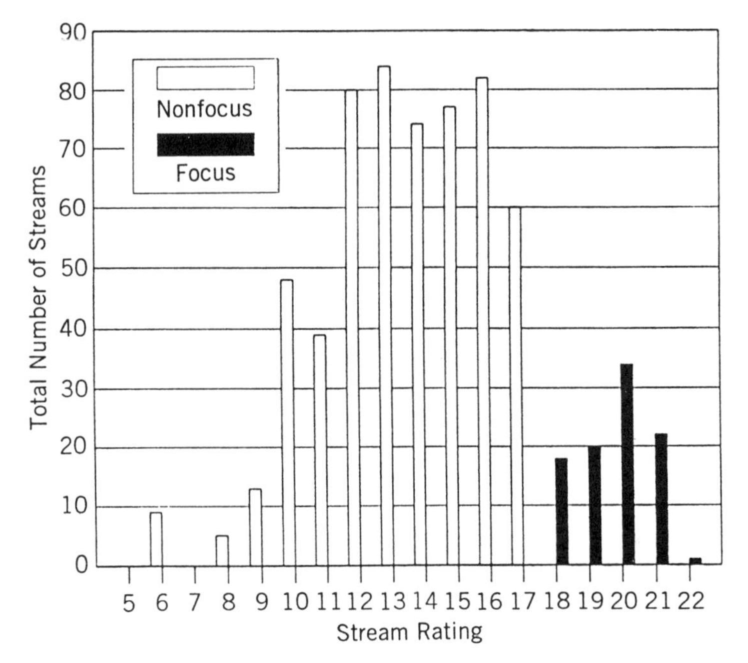

Figure 5-12. Distribution of ratings for waste streams at a large oil refinery in the midwestern United States. (From "Identifying Pollution Prevention Options for a Petroleum Refinery," J. A. Balik and S. M. Koraido, *Pollution Prevention Review*, Copyright @ 1991 Executive Enterprises, Inc. Reprinted by permission of John Wiley & Sons, Inc.)

It might be considered questionable to assign all the criteria equal weight when screening the streams at the refinery discussed above. A potential improvement of the ranking process would be to assign weights to the criteria based on their perceived importance. For example, if the health–safety criterion were scored from 0 to 10, 10 being high, and all other criteria were scored from 0 to 5, the health–safety factor would be given twice the weight of all the other criteria in the analysis.

Despite the potential flaws, this semiquantitative system of ranking can be used to reduce the number of waste streams seriously evaluated for pollution prevention options. For large facilities wishing to incorporate a variety of concerns in their pollution prevention projects, a procedure such as this for systematically screening waste streams improves the decision-making process.

5.4 SUMMARY

This chapter has described the first step in engineering design for pollution prevention at a facility: determination of the wastes and emissions that are generated. While it is not possible in a single chapter to describe all the relevant techniques for estimating rates of waste and emission generation, the case studies of fugitive and secondary emissions illustrate many of the engineering challenges.

Key issues concerning waste audits and fugitive and secondary emissions are summarized below.

Waste audits:

- *While determining what wastes are generated is relatively straightforward, determining why the wastes are generated is difficult.* Yet, determining the mechanism of waste generation (why rather than what) is often the key to pollution prevention.
- *Wastewaters are the dominant component of total waste mass.* This is true not only for the petroleum refining industry, for which detailed data were presented in this chapter, but also of industrial wastes generally.
- *Direct emissions to the environment, such as fugitive emissions, are released at rates several orders of magnitude lower than the rates at which waste and wastewaters are generated, but these wastes must be characterized particularly well.* Nationally, on the order of three million tons of atmospheric releases are reported annually through the Toxic Chemical Release Inventory while more than 700 millions tons of RCRA hazardous waste are generated. These direct emissions, while representing significantly less mass, are frequently the focus of reduction programs. The reason for this focus is that releases often result in greater environmental risks than RCRA solid wastes and wastewaters. As a consequence, determining the extent of direct releases can be very important.

Fugitive emissions:

- *A number of methods are available for estimating rates of fugitive emissions.*
- *Emission rates calculated using acceptable procedures can vary by several orders of magnitude.* More time-consuming estimation methods, involving data collection, frequently result in lower emission estimates.
- *Fugitive emission estimation methods are generally not reliable for industries outside those for which the methods were developed.* The example of fugitive emissions from acrolein manufacturing, presented in this chapter, which is not well served by average SOCMI factors, is a good case in point.

Secondary emissions:

- *A number of methods are available for estimating rates of secondary emissions.*
- *Emissions from secondary sources are difficult to measure accurately.* Emissions from each treatment component can vary according to lo-

cation within the component, seasonal variations, process conditions, and other factors.

- *Of the many methods for estimating secondary emissions, direct measurement techniques yield the most accurate results.* However, the direct measurement technique chosen must be suitable for the treatment unit whose emissions are being measured.
- *As with fugitive emissions, estimates of emission rates made using different methods can vary by several orders of magnitude.*
- *Emission factors are not available for many sources of secondary emissions.*

At a large facility with many waste streams, the results of waste audits and emission inventories are used in the process of ranking the streams so that the most important streams can be selected for potential pollution prevention options. The most systematic way to produce such a ranking is to identify the criteria that are important and assign dimensionless scores to the criteria for the different waste streams. The scores can be combined to develop a ranking of the streams. Waste stream quantities alone cannot be used to produce an acceptable ranking of many diverse waste streams.

QUESTIONS FOR DISCUSSION

1. Why is it difficult to estimate fugitive emissions accurately? Why is it important that they be measured accurately?
2. What are the incentives for developing facility-specific correlation curves for estimating fugitive emissions? What are the disincentives?
3. The uncertainties involved in estimating two significant sources of hydrocarbon air emissions were shown to be as much as several orders of magnitude in this chapter. What are the implications of this uncertainty when studying emissions at a national level, as was done in Chapter 2?
4. Which of the methods for quantifying secondary and fugitive emissions discussed in this chapter are direct measurement techniques? Which are indirect measurement techniques? Which do not rely on measurement?

PROBLEM STATEMENTS

1. Estimate the emissions of benzene in pounds per year from a popula tion of 10 valves at a refinery. Data for the valves are given in Table

Table 5-23. Process fluid data and screening concentrations for ten hypothetical valves

Valve	Service	Mass Fraction VOC in Process Fluid	Mass Fraction Benzene in VOC Portion of Process Fluid	Screening Concentration
1	Gas	1.0	0.1	11,000
2	LL[a]	0.5	0.1	9,000
3	Gas	1.0	0.2	7,000
4	LL	0.5	0.2	5,000
5	Gas	1.0	0.1	3,000
6	LL	0.5	0.1	2,000
7	Gas	1.0	0.2	1,000
8	LL	0.5	0.2	500
9	Gas	1.0	0.1	Not detectable
10	LL	0.5	0.1	Not detectable

[a]Light liquid.

5-23. Calculate the total emissions from this set of valves using (a) average emission factors, (b) leak/no-leak emission factors, and (c) EPA-provided correlation curves. How do the results of the three methods vary? Which is the most accurate?

2. The fugitive emissions of a highly reactive organic compound, compound A, from the valves at a very small synthetic organic chemical production unit must be estimated. Valve counts, mass fraction VOC, mass fraction of A in the VOC portion of the components, and screening concentrations for this unit are given in Table 5-24. Estimate the fugitive emissions of A from the valves in pounds per year (lb/yr) using (a) the average SOCMI emission factors from Table 5-6, (b) the leak/no-leak emission factors from Table 5-8, and (c) the facility-specific correlations for valves developed by bagging the components. The facility-specific correlations are

$$E(\text{g/hr}) = 0.00033C^{0.86} \qquad (C > 0.1)$$

and

$$E(\text{lb/hr}) = 3 \times 10^{-7} \qquad (C \leq 0.1)\,,$$

where E is the hydrocarbon emission rate for a valve and C is the screening concentration value of that valve in ppm. Note that the equation for estimating emissions from a component using average SOCMI emission factors is

$$E_{\text{A}} = m_{\text{voc}} m_{\text{A}} f_{\text{av}}\,,$$

Table 5-24. Valve descriptions at a small SOCMI unit

Stream No.	Type of Service	Number of Valves	Mass Fraction of VOC in Fluid Serviced by Valve (m_{VOC})	Mass Fraction of A in VOC Portion of Fluid Serviced by Valve (m_A)	Screening Value, ppm (C)
1	Gas	1	0.95	0.6	130
2	Gas	3	0.95	0.6	0
3	LL[a]	1	0.95	0.6	11,000
4	LL	1	0.95	0.6	500
5	LL	1	0.95	0.6	75
6	LL	1	0.95	0.6	25
7	LL	1	0.99	0.9	10
8	LL	1	0.99	0.9	5
9	LL	1	0.99	0.9	1
10	LL	8	1.000	0.999	0
11	LL	7	0.999	0.99	0
12	LL	4	0.99	0.9	0

[a]Light liquid.

where E_A is the emission rate of compound A, m_{VOC} is the mass fraction of VOC in the stream serviced by the component, m_A is the mass fraction of A in the VOC portion of the stream serviced by the component, and f_{av} is the average emission factor. The equation for estimating emissions from a component using leak/no-leak emission factors is

$$E_A = m_A f_{L/NL},$$

where E_A and m_A are as before and $f_{L/NL}$ is the appropriate leak/no-leak emission factor. There are 352 operating days per year. How do the results vary when different estimation methods are used? How sensitive are the estimated totals to the value of m_A for any single valve?

3. Repeat the calculations of Example 5-4 for an oil–water separator with an oil film that is <1 cm thick. How does your answer vary from the solution of the example? Is a large error introduced by using the average oil layer thickness of more than one centimeter?
4. The components of wastewater collection and treatment systems are often at least partially uncovered, which allows VOCs to escape. Some common wastewater collection and treatment components are junction boxes, sumps, lift stations, equalization basins, clarifiers, treatment tanks, and aeration basins. The emission mechanism from each of these components is similar; mass transfer driven by a concentration gradient with resistance occurring in both the air and water sides of the air/water

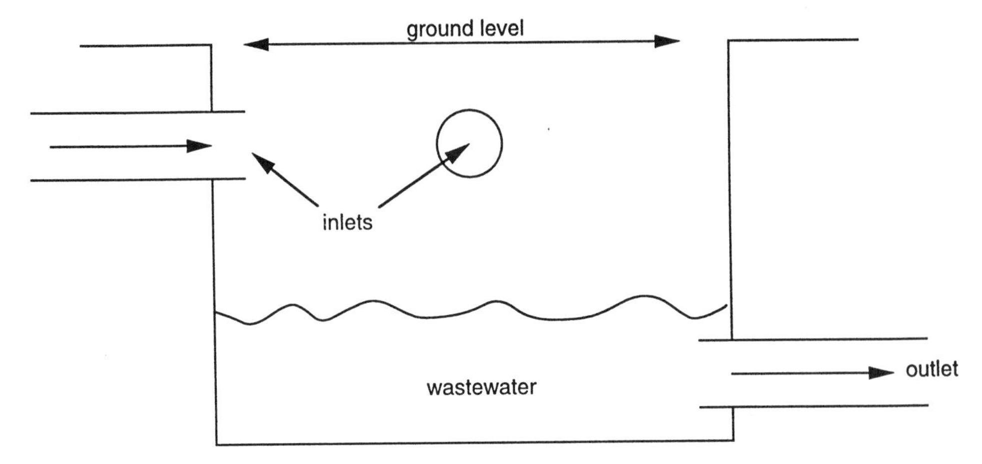

Figure 5-13. A junction box.

interface. Junction boxes such as the one pictured in Figure 5-13 are used to combine waste streams. Although they are not mechanically aerated, the wastewater flowing through them is turbulent, and a correlation for estimating the liquid-phase mass-transfer coefficient in mechanically aerated impoundments can be used. This correlation results in a value of 4.6×10^{-4} m/s for the liquid-phase mass-transfer coefficient of toluene in a typical junction box (US EPA, 1990). A correlation for toluene's vapor-phase mass-transfer coefficient in a typical junction box gives a value of 0.011 m/s (US EPA, 1990). Equilibrium is assumed to exist at the air/water interface, so the equilibrium partition coefficient of the compound between air and water must also be determined. It is found from

$$K_{eq} = H/RT,$$

where H is the Henry's constant for the compound, R is the ideal-gas-law constant, and T is the temperature in K (Kelvins). In this context, Henry's constant is defined by the equation

$$p_i = HC_i \ ,$$

where p_i is the partial pressure of compound i in equilibrium with the concentration of compound i in water, C_i. Once these coefficients have been determined, an overall mass-transfer coefficient can be calculated using the equation

$$1/K = 1/k_L + 1/(K_{eq}k_G).$$

Flux is then found from

$$J = KC_b,$$

where C_b is the bulk concentration of the compound in the waste stream. Find (a) the value of the overall mass-transfer coefficient, K, for toluene in this typical junction box, using values for the liquid and vapor phase mass-transfer coefficients of $k_L = 4.6 \times 10^{-4}$ m/s and $k_G = 0.011$ m/s (US EPA, 1990) and a Henry's constant for toluene of 0.011 atm-m^3/mol at 35°C; (b) an equation relating K, the overall mass-transfer coefficient; Q, the flow rate, A, the air/water interfacial area; C_0, the inlet concentration; and C_b, the bulk concentration in the junction box, using mass-balance techniques and the fact that because the contents of the junction box are well mixed, the bulk concentration in the box is equal to the effluent concentration; (c) the bulk concentration of toluene in the junction box if the inlet concentration of toluene is 5×10^{-4} g/cm^3, the air/water interfacial area is 0.656 m^2, and flow rate through the junction box is 2.52×10^{-3} m^3/s, and (d) the emissions of toluene from the junction box. What resistance dominates this mass transfer scenario? Why? What resistances are neglected?

OPEN-ENDED PROBLEMS

1. The waste streams in Table 5-3 are listed in order of quantity generated. List these wastes in order of the number of refineries that report their generation. Next, list the wastes in order of waste generation per refinery generating the waste. Assign dimensionless scores to the wastes using these three criteria (waste quantity, number of refineries, and waste per refinery) and produce a waste ranking. Weight the criteria as you find appropriate. Does the ranking you developed differ from the ranking for any single criteria?
2. Select an emission factor from one of the references given in this chapter and critically evaluate it.
3. A CD-ROM disk available from the National Pollution Prevention Center at the University of Michigan in Ann Arbor or from CAChE, Inc. at the University of Texas at Austin describes a vehicle chassis coating facility (the Ford-Wixom Phospate Coating System). Perform a waste audit of this facility using the data provided in the CD.

REFERENCES

American Petroleum Institute (API), "The 1982 Refinery Solid Wastes Survey," Washington, DC, 1983.

American Petroleum Institute (API), "Generation and Management of Residual Materials: Petroleum Refining Performance, 1991 Survey," API Publication 329, Washington, DC, 1994.

Baker, G. L. and D. I. McCready, "Plant-Wide Remote Detection of Chemical Emissions–One Company's Perspective," *Proceedings of 81st Annual Meeting of the Air Pollution Control Association*, Dallas, June 1988.

Baker, R. D. and J. L. Warren, "Generation of Hazardous Waste in the United States," *Hazardous Waste & Hazardous Materials*, **9**(1), 19–35, Winter 1992.

Balfour, W. D. and C. E. Schmidt, "Sampling Approaches for Measuring Emission Rates from Hazardous Waste Disposal Facilities," *Proceedings of 77th Annual Meeting of the Air Pollution Control Association*, San Francisco, June 1984.

Balik, J. A. and S. M. Koraido, "Identifying Pollution Prevention Options for a Petroleum Refinery," *Pollution Prevention Review*, **1**, 273–293, Summer 1991.

Berglund, R. L., D. A. Wood, and T. J. Covin, "Fugitive Emissions from the Acrolein Production Industry," paper presented at the Air and Waste Management Association International Specialty Conference: SARA Title III, Section 313, Industry Experience in Estimating Chemical Releases, April 6, 1989.

California Department of Toxic Substances Control (CA DTSC), "Hazardous Waste Minimization Bibliography," Document WR018, Sacramento, CA, 1991.

Chemical Manufacturers Association (CMA), "Improving Air Quality: A Guide to Estimating Secondary Emissions," Washington, DC, 1990.

Chemical Manufacturers Association (CMA), "Hazardous Waste Survey 1990," Washington, DC, 1992.

Dimmick, W. F. and K. C. Hustvedt, "Equipment Leaks of VOC: Emissions and Their Control," paper 84-62.1, presented at the 77th Annual Meeting of the Air Pollution Control Association, San Francisco, June 24–29, 1984.

Esplin, G. J., "Boundary Layer Emission Monitoring," *J. Air Pollution Control Assoc.* **38**, 1158–1161, Sep. 1988.

Gholson, A. R., J. R. Albritton, R. K. Jayanty, et al., "Evaluation of the Flux Chamber Method for Measuring Air Emissions of VOC from Surface Impoundments," *Proceedings of the 1987 Environmental Protection Agency/Air Pollution Control Association Symposium on Measurement of Toxic and Related Air Pollutants*, Research Triangle Park, NC, May 1987.

Heirigs, P. L., M.S. thesis, University of California, Los Angeles, 1991.

Herget, W. F., *American Lab*, **14**, 72–78, Dec. 1978.

Herget, W. F. and J. D. Brasher, *Appl. Optics*, **18**, 3404–3420, 1979.

Herget, W. F. and J. D. Brasher, *Optical Eng.* **19**, 508–514, 1980.

McClenny, W. A. et al., "Methodology for Comparison of Open-Path Monitors with Point Monitors," *J. Air Pollution Control Association*, **24**, 1044–1046, Dec. 1974.

Measures, R. M., *Laser Remote Sensing–Fundamentals and Applications*, Wiley, New York, 1984.

Minnich, T. R. et al., "Remote Sensing of Air Toxics for Pre-Remedial Hazardous Waste Site Investigation," Paper 108.1 presented at the Air and Waste Management Association Annual Meeting, Anaheim, CA, June 1989.

Pojasek, R. B. and L. J. Cali, "Contrasting Approaches to Pollution Prevention Auditing," *Pollution Prevention Review*, **1**, 225–235, Summer 1991.

Saeger, M. L., C. Sokol, et al., "A Review of Methods for Remote Sensing of Atmospheric Emissions from Stationary Sources," Environmental Protection Agency project 68-02-4442, March 1988.

United States Environmental Protection Agency (US EPA), "Assessment of Atmospheric Emissions from Petroleum Refining," Research Triangle Park, NC, Publication EPA-600/2-80-075a-d, 1980a.

United States Environmental Protection Agency (US EPA), "Problem-Oriented Report: Frequency of Leak Occurrence for Fittings in Synthetic Organic Chemical Plant Process Units," Research Triangle Park, NC, Publication EPA-600/2-81-003, Sep. 1980b.

United States Environmental Protection Agency (US EPA), "Measurement of Volatile Chemical Emissions from Wastewater Basins," available through NTIS as PB83-135632, Nov. 1982.

United States Environmental Protection Agency (US EPA), "Compilation of Air Pollutant Emission Factors, Volume I: Stationary Point and Area Sources," 4th ed., with Supplements A-D, Research Triangle Park, NC, Publication AP-42, 1985a.

United States Environmental Protection Agency (US EPA), *A Model for Evaluation of Refinery and Synfuels VOC Emission Data*, Volume I," Research Triangle Park, NC, Publication EPA-600/7-85-022a, 1985b.

United States Environmental Protection Agency (US EPA), "Emission Factors for Equipment Leaks of VOC and HAP," Research Triangle Park, NC, Publication EPA-450/3-86-002, Jan. 1986.

United States Environmental Protection Agency (US EPA), "Protocols for Generating Unit-Specific Emission Estimates for Equipment Leaks of VOC and VHAP," Research Triangle Park, NC, Publication EPA-450/3-89-010, 1988a.

United States Environmental Protection Agency (US EPA), "Waste Minimization Opportunity Assessment Manual," Publication EPA/625/7-88/003, available through NTIS as PB92-216985, 1988b.

United States Environmental Protection Agency (US EPA), "Hazardous Waste Treatment, Storage, and Disposal Facilities (TSDF)—Air Emission Models," Research Triangle Park, NC, Publication EPA-450/3-87-026, April 1989.

United States Environmental Protection Agency (US EPA), "Industrial Wastewater Volatile Organic Compound Emissions: Background Information for BACT/LAER Determinations," Control Technology Center, Research Triangle Park, NC, Jan. 1990.

United States Environmental Protection Agency (US EPA), "Facility Pollution Prevention Guide," EPA/600/R-92/088, 1992.

United States Environmental Protection Agency (US EPA), "Protocol for Equipment Leak Emission Estimates," available through NTIS as PB93-229219, June 1993.

6

POLLUTION PREVENTION FOR UNIT OPERATIONS

Once the wastes and emissions from a facility have been identified and a group of target streams has been selected for reduction, the development of design options for waste reduction can begin in earnest. Most presentations of waste reduction techniques have reported case histories for particular processes. This method of presentation has contributed to the notion that these techniques are process-specific. In practice, many waste reduction techniques can be broadly applied to the unit operations that make up most chemical processes. Devices such as reactors, storage vessels, distillation columns, and heat exchangers generate wastes through the same mechanisms regardless of the nature of the process in which they are embedded. In this chapter, successful waste reduction techniques are presented in a unit operations framework. Raw-material selection is discussed first. Next, methods for preventing solvent losses from cleaning operations and waste from storage and transport are presented, followed by techniques for reducing pollution from reactors, heat exchangers, and separation units. The chapter ends with a survey of industry-specific waste reduction manuals.

6.1 RAW-MATERIAL SELECTION

Careful selection of raw materials is an essential element of any process pollution prevention strategy. Waste reduction opportunities available in the selection of raw materials include (1) the elimination of feedstock impurities, (2) the use of less hazardous raw materials, (3) a reduction in the number of raw materials used, and (4) the utilization of waste materials from other processes (Nelson, 1990).

To illustrate the application of these pollution prevention approaches, consider the selection of raw materials at a petroleum refinery. Identifying material substitution opportunities when the primary raw material is crude oil may at first seem difficult. However, opportunities do exist. For example, the elimination of feedstock impurities might be pursued. Crude oil contains materials such as soils, metals, and sulfur- and nitrogen-containing compounds. Removing these materials during refining results in large volumes of waste. Some types of crude oil contain fewer impurities and therefore produce less waste than do others, so substitution of these costlier crude oils in place of a "dirtier" crude might be considered for pollution prevention. However, economic, technical, and market factors influence the choice of crude oil at each facility, and this strategy is not always possible.

Utilization of waste materials from other processes is another approach that has been implemented at some refineries where used motor oil supplements the refinery feedstock. Further, there are many inputs to refining other than crude oil, and these inputs can be considered as raw materials for which substitutes can be found or that could be purified in order to reduce waste. Solvents are used, for example, to clean fouled plant equipment, and deemulsifiers are used in waste treatment processes. Other chemicals are used to treat process water. Large volumes of catalyst are required for refinery processes such as cracking, where long-chain hydrocarbons are broken into smaller chains suitable for blending into gasoline. Process water is required for desalting, heating (in the form of steam from boilers), and cooling. Purifying process water can lead to substantial waste reduction, as illustrated in the following case study of a refinery that reduced waste generation rates at a profit by removing impurities in their process water (Griffin, 1994).

6.1.1 Upgrading Raw Materials to Prevent Pollution

Figure 6-1 shows the major uses of process water at a midsize refinery located in the southwestern United States (Griffin, 1994). Process water enters the desalter, where it is contacted with crude oil in order to remove impurities that can interfere with downstream processing. The water leaving the desalter is contaminated with oil and is routed to the wastewater treatment facility for oil recovery and removal of toxic constituents. Process water also enters ion exchange softeners. After being softened, the water is heated in boilers to make steam for heating refinery process streams. Much of this steam is lost, but about a third returns to the boiler as condensate. Finally, process water is used to cool some process streams in heat exchangers. After leaving the exchangers, the water is sent to cooling towers to reduce its temperature. Cooling towers use evaporation of water to cool the process water stream, so there are water losses to evaporation, and additional losses of water occur in the form of water droplets called *windage*.

The boiler and cooling towers both evaporate water. Impurities present in the process water evaporate much less readily and become concentrated

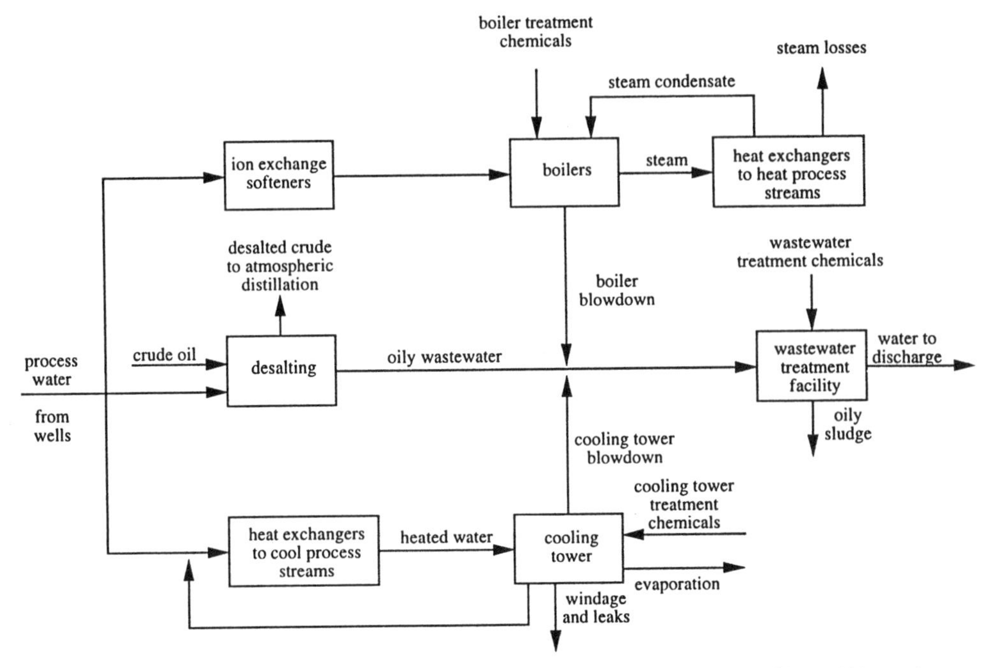

Figure 6-1. Conceptual block flow diagram of process water at a refinery. (From "Upgrading Process Water to Prevent Pollution in Petroleum Refining," K. S. Rosselot and D. T. Allen, *Pollution Prevention Review*, copyright © 1996 John Wiley & Sons, Inc. Reprinted by permission of John Wiley & Sons, Inc.)

in the boiler and in the cooling tower. These impurities can become so concentrated that scale formation begins or processes are upset. When this happens, a purge stream called *blowdown* is necessary to keep the concentration of impurities at an acceptable level. The blowdown from both the cooling tower and from the boiler at the refinery in this case study is routed to the oily wastewater sewer to be treated along with water from the desalter. In the past, when concentrated calcium salts from the cooling tower met with the alkaline boiler blowdown, salts precipitated. This precipitate plugged sewer lines, degraded wastewater treatment equipment, and resulted in greater sludge generation rates from wastewater treatment processes. Remember from the previous chapter that one pound of solids in oily wastewater treatment can create as much as 10 lb of oily sludge, because the surfaces of the solids provide sites to which oil and water can stabilize and adhere. The sludge is a RCRA hazardous (i.e., defined as hazardous by the Resource Conservation and Recovery Act) waste and is expensive to dispose of. It also contains oil that would have otherwise been recovered in the oily wastewater treatment unit and sent to refining processes.

The water supply for the refinery described in this case study is a well with extremely hard water containing high total dissolved solids, as shown

Table 6-1. Raw-feedwater quality at a refinery in the southwestern United States

Parameter	Value
pH, pH units	7.0
Specific conductance, μS/cm	1600
Total alkalinity, ppm as $CaCO_3$	180
Total hardness, ppm as $CaCO_3$	748
Sodium, ppm	86
Calcium, ppm	200
Magnesium, ppm	58
Potassium, ppm	1.0
Sulfate, ppm	560
Bicarbonate, ppm	227
Chloride, ppm	113
Silica, ppm	16
Nitrate, ppm	1.8

Source: Griffin (1994).

in Table 6-1. The predominant salt in the water is calcium sulfate, which forms a scale on process equipment that cannot be chemically removed. Instead, physical methods such as blasting and grinding are necessary. The blowdown rates from the cooling towers had to be very high to avoid formation of scale, and even then, considerable scale was formed in heat exchangers, necessitating frequent cleaning to regain heat-transfer efficiency.

Because of the nature of the water supply, conventional pretreatment processes were not applicable. Less conventional pretreatment options, such as demineralization by distillation and ion exchange, reverse osmosis, and electrodialysis were rejected in the past because of high costs. Improvements in membrane technology and favorable economic factors caused reverse-osmosis to be reconsidered and eventually chosen as a pretreatment option, despite the lack of documented cases of refineries using this technology. In the reverse-osmosis process, raw water under pressure is applied to a semipermeable membrane, forcing purified water through the membrane. The purified water at the refinery of this example contains 5% of the impurities of the raw water. Rejected water with concentrated impurities is suitable for use in irrigation.

It would have been desirable for the refinery to single out the most important streams for feedwater pretreatment, but piping and space constraints forced the pretreatment of all of the process feedwater. Even though reverse osmosis is expensive, it proved to be cost-effective at this refinery. The greatest savings are due to the reduction in hazardous wastewater sludge generation. In fact, savings in sludge disposal costs alone are greater than the cost of pretreatment. Consumption of chemicals for treating the boiler and cooling tower water have been reduced by more than 90%, and the

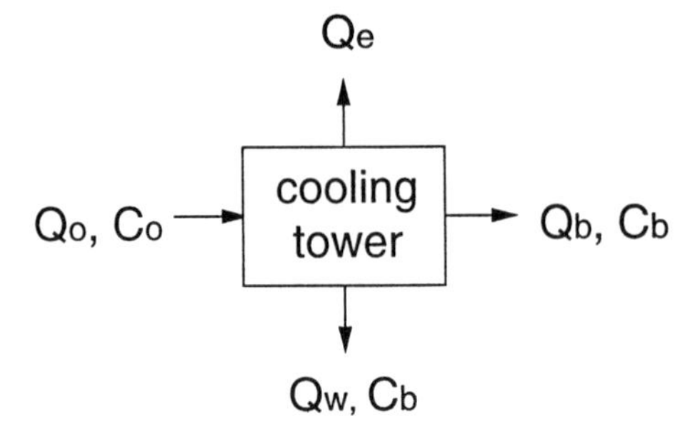

LEGEND:
C = hardness, ppm as $CaCO_3$
Q = flow rate, millions of gallons per day

subscripts:
e = evaporative
w = windage and leaks
o = flow to towers
b = blowdown

Figure 6-2. Mass-balance diagram for cooling tower at a refinery. (From "Upgrading Process Water to Prevent Pollution in Petroleum Refining," K. S. Rosselot and D. T. Allen, *Pollution Prevention Review*, copyright © 1996 John Wiley & Sons, Inc. Reprinted by permission of John Wiley & Sons, Inc.)

resultant savings amount to about one third of the cost of the reverse osmosis. Other savings have resulted from reduced maintenance needs due to less scale buildup in water-cooled heat exchangers.

Example 6-1 Advantages of Feedwater Pretreatment at a Southwestern Petroleum Refinery Figure 6-2 is a mass balance diagram of the cooling tower at the refinery, where Q denotes flow rates and C denotes hardness levels. Table 6-2 provides flow rate and hardness data for the tower prior to feedwater pretreatment. If pretreated feedwater hardness is 38 ppm as $CaCO_3$ and the water in the cooling towers is 34 times as hard, what is the new cooling-tower blowdown rate after feedwater pretreatment? If the total wastewater volume prior to feedwater pretreatment is 0.9 million gallons per day (MGD), what percentage reduction in wastewater volume is due to reduced cooling tower blowdown? If ion concentrations in pretreated feedwater are 5% of what they are in the raw well water, how much longer is it between regenerations for the sodium ion exchange softener for the boiler?

Table 6-2. Flow rates and hardness values for hypothetical refinery cooling towers prior to feedwater pretreatment

Symbol	Value
Flow rates, MGD[a]	
Q_e	1.4
Q_0	2.0
Q_w	0.041
Q_b	0.54
Hardness, ppm as $CaCO_3$	
C_o	750
C_b	2600

[a]MGD = million gallons per day.

Source: "Upgrading Process Water to Prevent Pollution in Petroleum Refining," K. S. Rosselot and D. T. Allen, *Pollution Prevention Review*, copyright © 1996 John Wiley & Sons, Inc. Reprinted by permission of John Wiley & Sons, Inc.

Solution

Mass balance for hardness:

$$Q_oC_o = Q_bC_b + Q_wC_b.$$

Mass balance for water:

$$Q_o = Q_e + Q_b + Q_w.$$

After pretreatment:

$$C_o = 38 \text{ ppm as } CaCO_3$$

$$C_b = 34 \times 38 \text{ ppm} = 1300 \text{ ppm as } CaCO_3.$$

Q_w and Q_e (windage and evaporation) are the same before and after pretreatment. Solve for Q_o:

$$Q_o \times 38 \text{ ppm} = Q_b \times 1300 \text{ ppm} + 0.041 \text{ MGD} \times 1300 \text{ ppm}$$

$$Q_o/34 = Q_b + 0.041 \text{ MGD}$$

$$Q_b = Q_o/34 - 0.041 \text{ MGD}$$

$$Q_o = 1.4 \text{ MGD} + Q_b + 0.041 \text{ MGD}.$$

Substitute to get:

$$Q_o = 1.4\,\text{MGD} + Q_o/34$$

$$33Q_o = 47.6\,\text{MGD}$$

$$Q_o = 1.4\,\text{MGD}.$$

Therefore, $Q_b = 0$ (there is no blowdown required from the cooling tower). This is 0.54 MGD less than before, and reduces total wastewater volume by

$$\frac{0.54\,\text{MGD}}{0.9\,\text{MGD}} \times 100\% = 60\%.$$

The ion-exchange softener will last 20 times longer between regenerations after feedwater pretreatment because there are only 5% ($\frac{1}{20}$) of the ions present.

6.1.2 Environmental Evaluation of Substitute Chemicals

When chemical substitution is the route to pollution prevention, tradeoffs between the environmental impacts of the current material and its potential substitute must be considered. For example, if a volatile organic compound (VOC) solvent is replaced by a less volatile solvent, less atmospheric emissions result but the heavier solvent will generate more contaminated solvent sludge wastes. If the VOC solvent is replaced by an aqueous cleaner, the air emissions are replaced by aqueous wastes. To evaluate the environmental tradeoffs offered by competing materials, the life cycle assessment tools described in Chapter 4 can be used. This section describes a complementary approach for evaluating environmental impacts, illustrated by a case study of chlorinated solvent alternatives. Although the focus of this case study is narrow, some aspects of the analysis are widely applicable in the environmental evaluation of chemicals. Two such aspects are (1) the usefulness of normalized indicators of environmental effects and (2) the potential for tradeoffs in environmental impacts that can occur when substitutes are chosen.

Roughly a million tons of chlorinated solvents are used annually in processes ranging from vapor degreasing to the fabrication of electronic components. These solvents have many useful properties that make them suitable for a wide range of industrial applications. For example, they are nonflammable and have normal boiling points slightly above room temperature. However, some solvents have been implicated in stratospheric ozone depletion and their use is being phased out. Finding substitutes is problematic. Many characteristics, such as flammability, volatility, toxicity, environmental fate and impact, and solubility, play a role in the selection of substitutes. Environmental fate and impact alone is extremely complex, and it is a small subset of this characteristic that is used in this section to illustrate some of the tradeoffs that can occur when substitute materials are chosen.

One important environmental impact factor for chlorinated solvents is their potential to cause stratospheric ozone depletion. In the stratosphere (15–50 km in altitude), there is an ozone "layer" that absorbs ultraviolet radiation that would otherwise cause harm to life on the surface of the planet. In 1974, Mario Molina and Sherwood Rowland of the University of California, Irvine proposed that chlorinated solvents photodissociate if they reach the stratospheric ozone layer, releasing chlorine atoms that catalyze ozone destruction (Molina and Rowland, 1974). Molina and Rowland were awarded the 1995 Nobel prize in chemistry for this discovery of the role that chlorinated hydrocarbons play in stratospheric ozone depletion. The chemistry is as follows (Stolarski and Cicerone, 1974; Wofsy and McElroy, 1974):

$$\begin{array}{l} Cl + O_3 \rightarrow ClO + O_2 \\ ClO + O \rightarrow Cl + O_2 \\ \hline \text{Net: } O_3 + O \rightarrow 2O_2 \end{array}$$

For a compound to cause stratospheric ozone depletion, it must have a lifetime in the atmosphere sufficient to reach the stratosphere and it must contain chlorine or bromine. Reactivity in the ozone cycle and atmospheric lifetime can be simplisticly represented by ${}^{O}k_i$ and t_i in the equation (Allen et al., 1992):

$$\mathrm{ODP}_i = \frac{{}^{O}k_i t_i}{{}^{O}k_{\text{CFC-11}} t_{\text{CFC-11}} (1 \text{ mol CFC-11})},$$

where ${}^{O}k_i$ is the rate constant for reaction of compound i with atomic oxygen, t_i is the atmospheric lifetime of compound i, and ODP_i is a simplified index of the approximate ozone-depletion potential for compound i, normalized by the ozone-depletion potential of CFC-11. This index provides a simplified relative measure of the extent to which a compound contributes to ozone depletion.

A second factor in the environmental fate of solvents is their potential to contribute to global warming. The overall energy balance of our planet is complex, but is known to depend strongly on the emission of infrared radiation. When atmospheric gases intercept this reradiated energy, the earth's energy balance can be upset, warming the planet. The global-warming potential of a chemical depends in part on both its ability to absorb infrared radiation and the length of time that it remains in the atmosphere. Thus, a simplistic index of the global-warming potential for compound i, normalized

Table 6-3. Environmental indices for the major chlorinated solvents

Compound	ODP Index, (gmol CFC-11)$^{-1}$	GWP Index, (gmol CFC-11)$^{-1}$	SFP Index, (gmol 1-octene)$^{-1}$
CH_2Cl_2 (methylene chloride)	0.009	0.007	0.0029
$Cl_3C—CH_3$ (1,1,1-trichloroethane)	0.15	0.057	0.00044
$HClC=CCl_2$ (trichloroethylene)	None	0.001	0.051
$Cl_2C=CCl_2$ (tetrachloroethylene)	None	0.009	0.0038
$Cl_2FC—CClF_2$ (CFC-113)	1.3	2.5	0.000011

Source: Allen et al. (1992).

by the global-warming potential of CFC-11, is given by (Allen et al., 1992):

$$\mathrm{GWP}_i = \frac{I_i t_i}{I_{\text{CFC-11}} t_{\text{CFC-11}} (1 \text{ mol CFC-11})},$$

where I_i is the infrared absorption band intensity of compound i and t_i is its atmospheric lifetime. This index provides a measure of the relative rise in global temperature to be expected due to atmospheric emissions.

Yet another environmental fate factor for solvents is their potential to cause smog. Smog is, in part, the result of the photochemical oxidation of hydrocarbons such as organic solvents. The smog-formation potential (SFP) of a chemical depends in part on the reaction rate for the oxidation of the compound by the hydroxyl radical, which is a measure of the tendency of the chemical to participate in photochemical reactions. Therefore, a simplistic smog formation potential index for compound i, normalized by the smog formation potential of 1-octene, is given by (Allen et al., 1992):

$$\mathrm{SFP}_i = \frac{{}^{\cdot\mathrm{OH}}k_i}{{}^{\cdot\mathrm{OH}}k_{\text{1-octene}} (1 \text{ mol 1-octene})},$$

where ${}^{\cdot\mathrm{OH}}k_i$ is the rate constant for the reaction of compound i with the hydroxyl radical.

The simplified indices of stratospheric ozone depletion, global warming, and smog-formation potential described above are given in Table 6-3 for the major chlorinated solvents. This table shows that CFC-113 has the highest ODP and GWP indices and that trichloroethylene has the highest SFP index.

Smog-formation, ozone-depletion, and global-warming potentials for five solvents based on emissions in 1979 and 1989 are given in Table 6-4. These impact potentials are arrived at by multiplying the indices in Table 6-3 by the atmospheric releases of these solvents. Atmospheric emissions were estimated to be 80% of production (Allen et al., 1992). The rankings of the solvents for the three potentials remain the same from 1979 to 1989 for the case of smog formation and global warming, and change slightly for ozone depletion. The sum of the potentials for this group of solvents for 1979 and 1989 shows that overall ozone-depletion and global-warming potentials increased during this period while smog-formation potential decreased. The following examples illustrate in further detail the tradeoffs that can be encountered when substituting one chlorinated solvent for another.

Example 6-2 Replacement of CFC-113 with 1,1,1-Trichloroethane Suppose that CFC-113 is to be replaced by an equal mass of 1,1,1-trichloroethane (TCA). Indices for the three potentials are given in Table 6-3, and data for the 1989 contributions to smog-formation, global-warming, and ozone-depletion potentials are given in Table 6-4. Production of CFC-113 and TCA in 1989 was 78,000 and 353,000 metric tons, respectively, and the molecular weight of TCA is 133 g/mol. Determine the total smog-formation, global-warming, and ozone-depletion potential for this set of compounds before and after the substitution. How will the substitution affect the environmental impacts? Assume that 80% of the solvents that are produced are eventually released to the atmosphere and that solvent use has been stable since 1989.

Solution CFC-113 production drops to zero while TCA production increases to 431,000 metric tons per year. Atmospheric emissions of TCA (in moles) are

$$(80\%)(431{,}000 \text{ metric tons})\left(\frac{10^6 \text{ g}}{\text{metric ton}}\right)\left(\frac{\text{mol}}{133 \text{ g}}\right) = 2.59 \times 10^9 \text{ mol}.$$

Multiplying this value by the indices in Table 6-3 gives the results in Table 6-5. Comparison of the totals from Tables 6-4 and 6-5 reveals that substitution of TCA for CFC-113 leaves overall smog-formation potential unchanged, reduces overall global-warming potential by 80%, and reduces overall ozone depletion potential by one-half.

Note that the smog-formation potential of TCA would rise if this substitution were made (because the releases of TCA would be larger), but even with the increase its smog-formation potential is not large enough to affect the overall SFP value of this group of chlorinated solvents.

Table 6-4. Atmospheric impact of solvent emissions in 1979 and 1989; ranking in parentheses

Compound	Smog Formation Potential		Global Warming Potential		Ozone Depletion Potential	
	1979	1989	1979	1989	1979	1989
CH_2Cl_2	(2)	(2)	(3)	(3)	(3)	(3)
(methylene chloride)	7.8×10^6	5.8×10^6	1.9×10^7	1.4×10^7	2.4×10^7	1.8×10^7
$Cl_3C—CH_3$	(4)	(4)	(2)	(2)	(1)	(2)
(1,1,1-trichloroethane)	8.6×10^5	9.3×10^5	1.1×10^8	1.2×10^8	2.9×10^8	3.2×10^8
$HClC{=}CCl_2$	(1)	(1)	(5)	(5)		
(trichloroethylene)	4.5×10^7	2.5×10^7	8.8×10^5	5.0×10^5	0	0
$Cl_2C{=}CCl_2$	(3)	(3)	(4)	(4)		
(perchloroethylene)	6.4×10^6	3.9×10^6	1.5×10^7	9.3×10^6	0	0
$Cl_2FC—CClF_2$	(5)	(5)	(1)	(1)	(2)	(1)
(CFC-113)	2.2×10^3	3.7×10^3	5.0×10^8	8.3×10^8	2.6×10^8	4.3×10^8
Total	6.0×10^7	3.6×10^7	6.5×10^8	9.8×10^8	5.8×10^8	7.7×10^8

Source: Allen et al. (1992).

Table 6-5. Atmospheric impacts of solvents on substitution of 1,1,1-trichloroethane for CFC-113

Compound	Smog-Formation Potential	Global-Warming Potential	Ozone-Depletion Potential
CH_2Cl_2 (methylene chloride)	5.8×10^6	1.4×10^7	1.8×10^7
$Cl_3C—CH_3$ (1,1,1-trichloroethane)	1.1×10^6	1.5×10^8	3.9×10^8
$HClC{=}CCl_2$ (trichloroethylene)	2.5×10^7	5.0×10^5	0
$Cl_2C{=}CCl_2$ (perchloroethylene)	3.9×10^6	9.3×10^6	0
$Cl_2FC—CClF_2$ (CFC-113)	0	0	0
Total	3.6×10^7	1.7×10^8	4.1×10^8

Example 6-3 Replacement of Trichloroethylene with 1,1,1-Trichloroethane Since the connection between trichloroethylene (TCE) and photochemical smog was established, TCA has been substituted for TCE. Evaluate this substitution not only on the basis of smog-formation potential but also on the basis of global-warming and ozone-depletion potential. As in Example 6-2, use 1989 production data. Also, assume that the two solvents are equivalent on a mass basis and that air emissions are 80% of production. In 1989, production of TCA and TCE was 353,000 and 82,000 metric tons, respectively. The molecular weight of TCA is 133 g/mol.

Solution TCE production drops to zero while TCA production increases to 435,000 metric tons per year. Atmospheric emissions of TCA (in mol) are

$$(80\%)(435{,}000 \text{ metric tons})\left(\frac{10^6 \text{ g}}{\text{metric ton}}\right)\left(\frac{\text{mol}}{133 \text{ g}}\right) = 2.62 \times 10^9 \text{ mol}.$$

Multiplying this value by the indices in Table 6-3 gives the results in Table 6-6. On substitution of TCA for TCE, overall smog-formation potential for the group of solvents drops to 31% of what it was in 1989, but global-warming potential is virtually unchanged and ozone-depletion potential increases by 10%.

Note that the solvents with high smog-formation potential (methylene chloride, trichloroethylene, and tetrachloroethylene) have low ozone-depletion and global-warming potentials. This is because they are quickly oxidized by

Table 6-6. Atmospheric impacts of solvents on substitution of 1,1,1-trichloroethane for trichloroethylene

Compound	Smog-Formation Potential	Global-Warming Potential	Ozone-Depletion Potential
CH_2Cl_2 (methylene chloride)	5.8×10^6	1.4×10^7	1.8×10^7
$Cl_3C—CH_3$ (1,1,1-trichloroethane)	1.2×10^6	1.5×10^8	3.9×10^8
$HClC{=}CCl_2$ (trichloroethylene)	0	0	0
$Cl_2C{=}CCl_2$ (perchloroethylene)	3.9×10^6	9.3×10^6	0
$Cl_2FC—CClF_2$ (CFC-113)	3.7×10^3	$8.3{\times}10^8$	4.3×10^8
Total	1.1×10^7	1.0×10^9	8.4×10^8

hydroxyl radicals in the lower atmosphere and do not remain in the atmosphere long enough to be a concern in global warming or stratospheric ozone depletion. Therefore, there is in general a tradeoff between the potential to cause global warming and depletion of ozone and the potential to cause smog formation in the use of organic solvents.

The example of chlorinated solvent substitution provides some preliminary insight into the tradeoffs and complexities of evaluating the environmental impacts of competing materials. As discussed in Chapter 4, an extensive set of literature is emerging around this issue. The general steps for evaluating the environmental impacts of a material include

- *Classification* of the environmental impacts into broad categories such as ozone-depletion potential, global-warming potential, smog-formation potential, generation of solid waste, and impacts on biodiversity.
- *Characterization* of the level of environmental impact using semiquantitative tools such as the indices described in Example 6-3 or more quantitative procedures such as risk assessments.
- *Evaluation* of the relative importance of the diverse environmental impacts of products.

More case studies of environmental assessments are provided in Chapter 4.

6.2 LOSSES FROM CLEANING OPERATIONS

When equipment becomes too fouled to operate properly and when parts are coated with materials that interfere with further processing, they must be cleaned. When solvents are used for cleaning, they may be emitted directly to the atmosphere through diffusion and convection. As discussed in the previous section on raw-material selection, some of the solvents used in cleaning parts and equipment are organic compounds linked to the formation of smog, and some chlorinated organic solvents have long atmospheric lifetimes and may pose a threat to the stratospheric ozone layer. Besides resulting in air emissions, cleaning with solvents produces dirty solvent that contains contaminants from the cleaning process. Sometimes it is cost-effective to reclaim waste solvent so that it can be used again, but in other cases waste solvent is destroyed. Mechanical cleaning methods and measures to reduce solvent evaporation can be used to prevent pollution from cleaning operations. If chemical cleaners are used, waste reduction through solvent recycling and proper inventory control can best be achieved if the variety of chemicals is minimized. Other measures for reducing solvent losses from parts and equipment cleaning are discussed below.

6.2.1 Parts Cleaning

Metals are often coated with thin films of light hydrocarbon oils in order to prevent oxidation of the raw metal feedstock during shipping and storage. Heavy oils, esters, and particulate matter tend to accumulate in these films during metal-forming operations, and these contaminants must be removed prior to surface treatment operations. Parts cleaning often consists of solvent degreasing, which can occur in cold cleaners, where the parts are soaked in liquid solvent, or in vapor degreasers, where cleaning is accomplished through condensation of hot solvent vapor on cold parts. Simple mass-transfer equations suitable for making rough estimates of emissions from parts cleaners can reveal which parameters have the most influence on emissions. The following example describes the vapor degreasing process in more detail and illustrates the development of pollution prevention measures for a vapor degreaser.

Example 6-4 Minimizing Solvent Emissions from Vapor Degreasers Vapor degreasers like the one pictured in Figure 6-3 are often left open to the atmosphere for ease of dipping and removing parts and because an explosive gaseous mixture might form if they are covered. In this example, estimates of solvent losses from a cylindrical degreaser with a diameter of 8 ft will be made in order to provide insight into methods for reducing solvent losses. The solvent used in this degreaser is 1,1,1-trichloroethane (TCA), which has a diffusivity in air of approximately $0.1\ cm^2/s$ and a molecular weight of

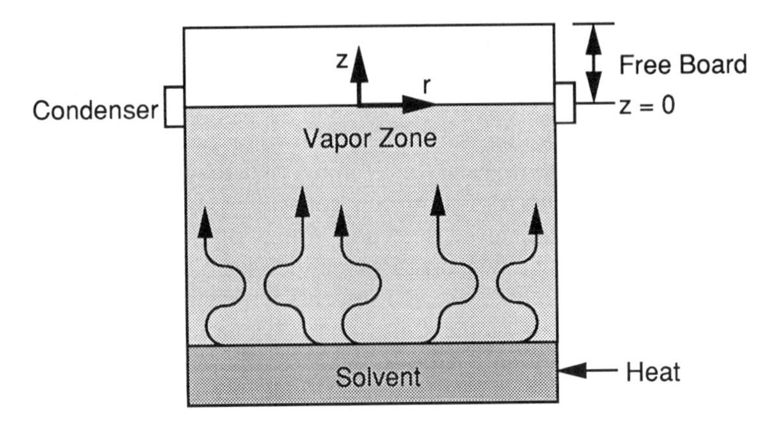

Figure 6-3. A vapor degreaser.

133 g/mol. The analysis is simplified by neglecting convective effects, assuming that the gases involved behave ideally, that the mole fraction of TCA in the air above the degreaser is effectively zero, that diffusivity is independent of concentration, and that temperature across the vapor surface is uniform. With these assumptions, emissions can be estimated using an equation describing the steady-state molar flux of TCA through a column of stagnant air, which (from Bird et al., 1960) is

$$N_{TCA_z} = -\frac{cD}{(1 - x_{TCA})}\frac{dx_{TCA}}{dz},$$

where N_{TCA_z} is the molar flux of TCA in the z direction, c is the total molar concentration of the air, D is the diffusivity of TCA in air, and x_{TCA} is the molar fraction of TCA. Combining this mass flux equation with the results of a mass balance over an incremental column height Δz gives

$$N_{TCA_z} = \frac{p_{TOT}D}{RTh}\ln\left[\frac{1}{1 - (p_{vap}/p_{TOT})}\right],$$

where h is the freeboard height, p_{vap} is the vapor pressure of TCA, p_{TOT} is the total pressure, R is the ideal-gas constant, and T is the temperature in K. Use this equation to estimate the emissions in lb/day from a degreaser with 5 ft of freeboard and a condenser that cools the vapor to 298 K. At this temperature, the vapor pressure of TCA is 127.3 mmHg. How much lower would the emissions be if the freeboard height were doubled? By how much are emissions reduced when the freeboard height is 5 ft and the condenser cools the vapor to 10°C, where the vapor pressure of TCA is approximately 62 mmHg?

Solution Emissions, E, are equal to flux multiplied by the vapor surface area. For 5 ft of freeboard and a condenser temperature of 298 K, emissions are

$$\frac{\pi(4\text{ ft})^2(1\text{ atm})(0.1\text{ cm}^2/\text{s})}{(298\text{ K})(0.08206\text{ L atm/mol K})(5\text{ ft})}\ln\left(\frac{1}{1-(127.3/760)}\right)\left(\frac{1000\text{ L}}{(100\text{ cm})^3}\right)$$

$$\times(30.48\text{ cm/ft})(3600\times 24\text{ s/day})(133\text{ g/mol})\left(\frac{2.2\text{ lb}}{1000\text{ g}}\right)=6\text{ lb/day}.$$

Emissions are inversely proportional to freeboard height, so if freeboard height is doubled, emissions are halved. When freeboard height is 5 ft, emissions (in terms of T and p_{vap}) are

$$E=9440\frac{\text{lb K}}{T\text{ day}}\ln\left[\frac{1}{1-(p_{\text{vap}}/760\text{ mmHg})}\right].$$

At a condenser temperature of 10°C or 283 K, emissions are

$$E=9440\frac{\text{lb K}}{283\text{ K day}}\ln\left(\frac{1}{1-\frac{62}{760}}\right)=3\text{ lb/day}.$$

Therefore, lowering the condenser temperature by 15°C to 283 K also halves the emissions.

Some additional generally recommended waste reduction practices for reducing solvent losses from parts cleaning are (Hunt, 1990): (1) enclose solvent cleaning units; (2) improve parts draining before and after cleaning; (3) use entirely closed-loop systems, such as supercritical CO_2 systems; and (4) use mechanical cleaning devices, such as plastic bead blasting, in place of solvents.

6.2.2 Equipment Cleaning

Equipment that can become fouled and require cleaning includes heat transfer surfaces, distillation column trays, and pipes. The use of a high-pressure rinse system, mechanical wipers, "pigs" (objects sent through pipes to scrape them clean), or compressed gas to clean lines can reduce solvent losses from the cleaning of plant equipment (Hunt, 1990). Blasting fouled surfaces with a material such as recyclable sand or dry ice is often a pollution prevention alternative to solvents. Some types of heat exchangers, such as gasketed plate exchangers, allow access to both sides of the exchanger so that mechanical methods are easily employed when fouling occurs (Carlson, 1992). When heat exchangers are cleaned, careful control of the cleaning process minimizes waste generation. This is important because heat exchanger bundle cleaning solids are a RCRA-listed hazardous waste. Some cleaning chemicals are reusable and may allow for product recovery from the accumulated solids (API, 1991a).

6.3 WASTE FROM STORAGE AND TRANSPORT

Waste from storage vessels takes many forms, from emissions due to vapor displacement during loading and unloading of storage tanks to wastes formed during storage to the storage containers themselves if they are discarded. Reducing waste from storage vessels therefore consists of a variety of activities, as shown in this section.

6.3.1 Storage Tanks

Tanks for storing organic liquids are found at petroleum refineries, organic chemical manufacturing facilities, bulk storage and transport facilities, and other facilities handling organic liquids. They are used to dampen fluctuations in input and output flow. Storage tanks can be a disastrous source of waste when weakly active undesired reactions run away, and it is important to monitor temperatures where this can occur and to design tanks so that heat dissipation effects dominate. Inadequate heat dissipation is of particular concern in the storage of bulk solids and viscous liquids (Gygax, 1988). Two other aspects of pollution prevention for storage tanks involving tank bottoms and standing and breathing losses are discussed below. Reduction of emissions due to the loading and unloading of storage tanks is discussed in the next section.

6.3.1.1 *Tank Bottoms* Tank bottoms are solids or sludges that accumulate at the bottom of large storage vessels. They are composed of rusts, soil contaminants, heavy feedstock constituents, and other heavy materials that are likely to settle out of the process fluid being stored. The bottoms generally accumulate slowly since the materials that create them are frequently present at low concentration in the process fluid. If the tank bottoms are composed primarily of heavy feedstock constituents (e.g., heavy oils in a refinery storage tank), then mixers can be used to prevent their formation (API, 1991a). Mixers can be used even if the tank bottoms are not heavy feedstock constituents, as long as the materials the bottoms contain are compatible with later processing steps and will not generate more waste downstream.

Agents that solubilize the materials that form tank bottoms can sometimes be added to storage tanks. However, these agents or the materials in the tank bottoms can at times generate more downstream waste than was prevented. The addition of emulsifying agents to the crude-oil storage tanks at a petroleum refinery illustrates this point. The emulsifiers reduce water and solids in tank bottom wastes, but they may or may not provide an overall reduction in waste generation, depending on the fate of the emulsifying agents in downstream processing and on the composition of the tank bottoms. According to the American Petroleum Institute (API, 1991b), the primary constituents of refinery tank bottoms are oil, water, and solids. This composition is compatible with desalting, which is the first processing unit after crude-oil storage. In desalting, the crude oil is contacted with water to remove dissolved salts. Since emulsions can become wastes in oil–water contacting operations, the emulsifying agents added to the crude-oil tank to reduce tank bottom sludges may generate additional waste in the desalting operation. Some refineries add deemulsifiers to circumvent this problem (API, 1991a). Clearly, evaluating this waste reduction option is not simple. A detailed analysis involving not only the tank operation but also desalter operation is required to determine whether net waste reduction can be economically accomplished by adding emulsifiers to crude-oil storage tanks.

6.3.1.2 *Standing and Breathing Losses* Standing and breathing losses are evaporative losses not associated with loading and unloading operations that occur from floating-roof and fixed-roof tanks, respectively. Techniques for reducing emissions from storage tanks depend on the type of tank. There are four main kinds of storage tanks: fixed-roof, floating-roof, variable-vapor-space, and pressurized. Pressurized tanks have very low emissions and are not discussed here.

Fixed-roof tanks have a permanent, rigid roof and are vented to the atmosphere. Breathing losses are vapor emissions from fixed-roof tanks that occur when the temperature or barometric pressure changes. When the pressure outside the tank decreases, vapors are expelled. When the pressure rises, fresh air is drawn into the tank. As this air becomes saturated with vapor, it occupies more volume and some of it must be expelled. The temperature dependence of vapor pressure causes similar mechanisms to occur when the temperature changes. In some cases, a pressure/vacuum vent that allows the tank to operate at slight internal pressure or vacuum can be used to reduce breathing losses from fixed-roof tanks (US EPA, 1985). Also, narrow tanks have lower breathing losses, as do insulated tanks whose contents undergo reduced temperature fluctuations. Vapor-recovery devices that trap and condense escaping vapors for return to the tank can be used to reduce emissions from fixed-roof storage tanks.

Example 6-5 Reducing Breathing Losses in Fixed-Roof Tanks The emissions from a fixed roof storage tank include both breathing losses (L_B) and working losses (L_W):

$$L_T = L_B + L_W.$$

This example examines breathing losses, estimated from the equation

$$L_B = 365 V_V W_V K_E K_S,$$

where L_B is in pounds per year, V_V is the vapor-space volume in cubic feet, W_V is the vapor density in lb/ft^3, K_E is the dimensionless vapor-space expansion factor, and K_S is the vented-vapor saturation factor. The vapor space, V_V, can be calculated from the equation

$$V_V = \frac{\pi}{4} D^2 H_{VO},$$

where H_{VO} is the vapor-space outage in feet and D is the diameter of the tank in feet. The vapor density can be calculated from

$$W_V = \frac{M_V P_{VA}}{R T_{LA}},$$

where M_V is the molecular weight of the vapor, P_{VA} is the vapor pressure of the liquid at the average liquid surface temperature in psia (lb/in.^{2} absolute), R is the gas constant, and T_{LA} is the average liquid surface temperature [Rankine units (°R)]. The average liquid surface temperature (°R) is given by

$$T_{LA} = 0.44 T_{AA} + 0.56 T_B + 0.0079 \alpha I,$$

where T_{AA} is the average daily ambient temperature (°R), T_B is the liquid bulk temperature (°R), α is the tank paint surface absorptance, and I is the daily total solar insolation factor in Btu/ft^2/day. T_B can be calculated from

$$T_B = T_{AA} + 6\alpha - 1.$$

The vented-vapor saturation factor is given by

$$K_S = \frac{1}{1 + 0.053 P_{VA} H_{VO}}$$

and accounts for the fact that the vapor leaving a tank is not completely saturated. The vapor-space expansion factor is given by

$$K_E = \frac{\Delta T_V}{T_{LA}} + \frac{\Delta P_V - \Delta P_B}{P_A - P_{VA}},$$

where ΔT_V is the daily vapor temperature range (°R), ΔP_V is the daily vapor pressure range in psia, ΔP_B is the breather vent pressure setting range (for this problem, assume this to be zero), P_A is the atmospheric pressure in psia, and P_{VA} is the vapor pressure at the average liquid surface temperature in psia. ΔP_V can be estimated using the equation

$$\Delta P_V = \frac{0.50 B P_{VA} \Delta T_V}{T_{LA}^2},$$

where B is a constant that depends on the fluid being stored, and ΔT_V is calculated from

$$\Delta T_V = 0.72 \Delta T_A + 0.028 \alpha I,$$

where T_A is the daily ambient temperature range (°R). Using these correlations, estimate the reduction in breathing loss emissions associated with painting a light gray tank that has paint in poor condition ($\alpha = 0.63$) with white paint ($\alpha = 0.17$) in Los Angeles, where the average annual value of I is 1500 Btu/ft^2/day. Assume that the tank is half full and contains crude oil whose vapor has an average molecular weight of 50. The average vapor pressure of the crude oil can be estimated from the correlation

$$\ln P_{VA} = A - \frac{B}{T_{LA}},$$

where P_{VA} is in psia, A is 11.3, B is 5300, and T_{LA} is in °R. (This value for B is also used in the correlation for estimating the average daily vapor-pressure range.) The tank holds one million gallons when full and is 60 ft in diameter. The average ambient temperature in Los Angeles is 65°F, with an average range of 15°F.

Solution First, calculate the vapor-space volume:

$$V_V = 0.5(10^6 \text{ gal}) \left(\frac{35.3 \text{ ft}^3}{264 \text{ gal}} \right) = 66{,}900 \text{ ft}^3.$$

From this, the vapor-space outage can be found:

$$H_{VO} = \frac{66{,}900\ \mathrm{ft}^3}{(60\ \mathrm{ft})^2}\frac{4}{\pi} = 23.7\ \mathrm{ft}.$$

To calculate W_V, T_B, T_{LA}, and P_{VA} must be found. For the gray tank:

$$T_B = 525 + 6(0.63) - 1 = 528°\mathrm{R},$$
$$T_{LA} = 0.44(525) + 0.56(528) + 0.0079(0.63)(1500) = 534°\mathrm{R},$$

and

$$P_{VA} = \exp\left(11.3 - \frac{5300}{534}\right) = 4.0\ \mathrm{psia},$$

and for the white tank:

$$T_B = 525 + 6(0.17) - 1 = 525°\mathrm{R},$$
$$T_{LA} = 0.44(525) + 0.56(525) + 0.0079(0.17)(1500) = 527°\mathrm{R},$$

and

$$P_{VA} = \exp\left(11.3 - \frac{5300}{527}\right) = 3.5\ \mathrm{psia}.$$

The vapor density for the gray tank is

$$W_V = \frac{(50\ \mathrm{lb/lb\cdot mol})4.0\ \mathrm{psia}}{(10.731\ \mathrm{psia\ ft^3/lb\cdot mol\ °R})534°\mathrm{R}} = 0.035\ \mathrm{lb/ft^3},$$

and for the white tank is

$$W_V = \frac{(50\ \mathrm{lb/lb\cdot mol})3.5\ \mathrm{psia}}{(10.731\ \mathrm{psia\ ft^3/lb\cdot mol\ °R})527°\mathrm{R}} = 0.031\ \mathrm{lb/ft^3},$$

To calculate K_E, ΔT_V, and ΔP_V must first be calculated. For the gray tank,

$$\Delta T_V = 0.72(15) + 0.028(0.63)(1500) = 37.3°\mathrm{R}$$

and

$$\Delta P_v = \frac{0.50(5300)(4.0)(37.3)}{534^2} = 1.39\ \mathrm{psia},$$

and for the white tank

$$\Delta T_{\mathrm{V}} = 0.72(15) + 0.028(0.17)(1500) = 17.9°\mathrm{R}$$

and

$$\Delta P_{\mathrm{V}} = \frac{0.50(5300)(3.5)(17.9)}{527^2} = 0.598 \text{ psia}.$$

The vapor-space expansion factor for the gray tank is

$$K_{\mathrm{E}} = \frac{37.3}{534} + \frac{1.39 - 0}{14.7 - 4.0} = 0.20$$

and for the white tank is

$$K_{\mathrm{E}} = \frac{17.9}{527} + \frac{0.598 - 0}{14.7 - 3.5} = 0.087$$

The vented-vapor saturation factor for the gray tank is

$$K_{\mathrm{S}} = \frac{1}{1 + 0.053(4.0)(23.7)} = 0.17$$

and for the white tank is

$$K_{\mathrm{S}} = \frac{1}{1 + 0.053(3.5)(23.7)} = 0.19$$

Finally, emissions can be estimated. For the gray tank, they are

$$L_{\mathrm{B}} = 365(66{,}900 \text{ ft}^3)(0.035 \text{ lb/ft}^3)(0.20)(0.17) = 29{,}000 \text{ lb/yr},$$

and for the white tank they are

$$L_{\mathrm{B}} = 365(66{,}900 \text{ ft}^3)(0.031 \text{ lb/ft}^3)(0.087)(0.19) = 13{,}000 \text{ lb/yr}.$$

Estimated breathing losses are therefore reduced by 55% when the tank is painted white.

Breathing losses can be prevented by using variable-vapor-space tanks, which might have a telescoping roof or a flexible diaphragm to allow for changes in vapor space. Breathing losses can also be prevented by floating a roof, called a *deck*, on the surface of the liquid in the tank. The floating roof can be external (exposed to the atmosphere), or it can be an internal floating roof mounted under a fixed roof. Internal floating-roof tanks have an advantage over external floating-roof tanks in that convective mechanisms for loss from the deck are virtually eliminated. This is especially advantageous in windy areas. Internal floating roofs have 60–99% fewer evaporative losses than fixed-roof tanks, depending on the type of roof, seals, and liquid being stored (US EPA, 1985). However, caution must be exercised in using internal floating roofs to store flammable liquids because of explosion hazards.

Floating-roof tanks have standing losses due to evaporation from rim seals, deck fittings, and deck seams. The following techniques for reducing standing losses from floating-roof tanks are derived from correlations for estimating these emissions (US EPA, 1985). Standing losses from floating-roof tanks are less when the seams are welded rather than bolted or riveted. Rim seal losses are minimized by choosing proper seals and using rim-mounted secondary seals. Deck fitting losses can be reduced by using fittings that are bolted and gasketed or fitted with a fabric seal. Also, minimizing the number of fittings reduces fitting losses.

Example 6-6 Reducing Standing Losses from Floating-Roof Tanks The standing loss in pounds per year from a single fitting in an external floating roof tank can be estimated using the equation

$$L_F = (K_1 + K_2 v^m) P^* M_v K_C,$$

where K_1 (lb mol/yr), K_2 (lb mol/yr/mph), and m are loss factors, v is the average wind speed in mi/hr, P^* is a dimensionless vapor-pressure function, M_v is the average vapor molecular weight, and K_C is a product factor. Estimate the losses from two types of slotted guide pole/sample wells, one with an ungasketed sliding cover and no float ($K_1 = 0$, $K_2 = 310$, and $m = 1.2$), and one with a gasketed sliding cover with a float ($K_1 = 0$, $K_2 = 8.5$, and $m = 2.4$). The tank is in Los Angeles, where the average wind speed is 6 mi/hr (mph). The liquid being stored is crude oil, and the tank is painted with light gray paint in poor condition, so that P^*, M_V, and K_C are 0.079, 50 lb/lb mol, and 0.4, respectively. How do the emissions for the two types of fitting differ?

Solution Losses from the ungasketed fitting are

$$L_F = [0 + 310(6)^{1.2}](0.79)(50)(0.4) = 4200\ \text{lb/yr}.$$

From the gasketed fitting, losses are

$$L_F = [0 + 8.5(6)^{2.4}](0.79)(50)(0.4) = 990 \text{ lb/yr}.$$

These emissions differ by a factor of 4.

6.3.2 Emissions from Loading and Unloading Operations

Vapors that are displaced in loading and off-loading operations can be a significant source of volatile hydrocarbon emissions from storage containers. It has been estimated that as much as 12 lb of organic compounds can be lost per 1000 gal of gasoline transferred during uncontrolled railcar and tank truck loading operations (US EPA, 1988). Less volatilization of storage fluids occurs during transfer operations when submerged loading as opposed to splash loading is done. Vapor recovery devices that trap and condense displaced gases reduce losses due to loading and unloading of fixed-roof storage tanks by 90–98% (US EPA, 1985). Vapor balance, where vapors from the container being filled are fed to the container being emptied, is another technique that can be applied in some cases to reduce emissions due to loading and unloading of fixed-roof tanks.

Unloading operations also create waste when the storage material clings to the surfaces of the container being emptied. Off-loading devices designed to completely empty shipping containers reduce waste, as does putting transport containers into dedicated service so that cleaning residual material out of the containers between loads is not necessary. Unloading of storage tanks with floating roofs exposes wet tank wall and column surfaces, which results in emissions called *withdrawal losses*. Tanks with a rusted interior have higher withdrawal losses because the rust provides surfaces to which the fluid being stored can easily cling. Also, reduction of wall and column area per fluid volume change decreases withdrawal losses.

Spills due to overfilling of storage containers are another source of emissions that occur during loading and unloading operations. These spills can be prevented through the use of appropriate overflow control equipment and/or overflow alarms.

6.3.3 Reusable Shipping Containers and Storage Bins

Shipping containers that are discarded can be a significant source of waste. Some suppliers will take back their shipping containers. Some containers can even be incorporated into the production process, such as soluble bags containing pigments or biocides (Hunt, 1990). Switching to internally reusable containers can be a cost effective way to reduce container waste, as shown in the following example.

Example 6-7 Reusable Bins for High-Volume Lubricants at a Refinery As part of a pollution prevention program at a large midwestern refinery, it was suggested that high-volume lubricating oils currently purchased in 55-gal drums be purchased in bulk and transferred to reusable tote bins instead (Balik and Koraido, 1991). The 55-gal drums were being sent to a vendor for recycling, but it was impractical to completely empty them. In addition, the bulk price for the lubricants is less than the price for lubricants in drums. If the capital cost of converting to reusable tote bins is $4000 and the annual savings in operating expenses is $25,000, calculate the capital payback period for this pollution prevention technique. There are no appreciable additional operating expenses as a result of using the tote bins. Neglect interest.

Solution Since interest can be neglected, the capital payback period is equal to capital cost divided by the annual net savings in operating expenses, or

$$\frac{\$4000}{(\$25{,}000/\text{yr}) - 0} = 0.16 \text{ year} = 9 \text{ weeks.}$$

6.3.4 Appropriate-Sized Containers

In some cases containers are not or cannot be reused and the containers become waste once they are emptied. Minimization of container waste can be achieved by ordering the largest-size container possible, but if the material being purchased has a finite shelf life, ordering too large a container may result in additional waste when the material degrades. Wasted raw material can be avoided by centralizing stocks and instituting careful inventory control. In addition to this, some materials are hazardous to store, which must be considered when choosing lot sizes. The following example illustrates how appropriate container sizes might be determined.

Example 6-8 Tradeoffs to Consider in Choosing Lot Sizes Consider the case of a chemical that is used at a rate of 500 lb/yr, has a shelf life of one year, and can be ordered in container sizes of five, 50, 100, 250, and 500 lb. What is the most appropriate lot size to order if the purchasing decision is based exclusively on waste generation? What if it is dangerous to store the material? What other issues might require consideration in choosing lot sizes?

Solution If the purchasing decision is based exclusively on waste generation, then the optimal container size is 500 lb. A smaller lot might be chosen to maximize worker safety and minimize the potential for leaks and spills if the material is a storage hazard. Storage space and facilities may also present problems for large lot sizes.

6.4 POLLUTION FROM PROCESS UNITS

In this section, methods for preventing pollution from specific process units are discussed. Waste reduction methods for reactors, separation processes, and heat exchangers are presented in a unit operations framework.

6.4.1 Reactors

Reactors are a key element in most chemical processes, and are particularly important in waste generation. The reactor is, after all, the unit where most of the undesired byproducts that eventually make up the waste streams are created. Five categories of analysis in examining reactor designs for waste reduction potential are discussed in this section. The first category is the consideration of selectivity to determine whether the reactor is producing the maximum amount of product and the minimum amount of by-product per unit mass of feed material. Next, examples of reconsidering reaction chemistry are provided, followed by a discussion of avoiding the storage of highly hazardous materials. Then, for catalytic reactors, a discussion of the ways in which changes in catalyst formulation may reduce attrition and deactivation, allowing for improved catalyst recycling and/or a reduction in catalyst waste is presented. Finally, new reactor configurations that reduce waste by combining reaction and separation operations in one unit are discussed. Methods for designing reactors in order to avoid the generation of trace contaminants, which is an issue of paramount importance in some processes and the optimization of chemical synthesis pathways, are discussed briefly in Chapter 11.

6.4.1.1 ***Methods for Maximizing Selectivity*** Although reactor designs have been continuously optimized for decades, the yield for some commodity chemicals is low, as shown in Table 6-7. Thus, in some instances, opportunities still exist for improving selectivity. The tools for optimizing selectivity are well known to chemical reaction engineers and have been described at length in reaction engineering texts (see, e.g., Fogler, 1992; Levenspiel, 1972). This section only briefly summarizes these tools. The analysis presented here focusses on the impact of reactor temperature, reactant concentration, and level of mixing on selectivity. Two simple reaction networks illustrate most of the basic concepts. These are a parallel reaction system,

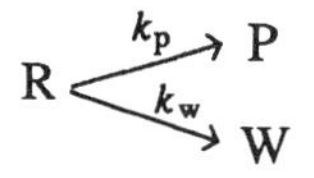

where R is the reactant, P is the product, W is a waste stream, and k_p and k_w are the rate constants for the reactions leading to products and wastes,

Table 6-7. Representative production and yield data for commodity chemical manufacturing

Chemical and Production Process	Annual Production, 1000 tons	Yield (%)[a]
Acrylonitrile	2,182	
Ammoxidation of propylene		65
Cyanation/oxidation of ethylene		76
1,3-Butadiene	2,546	
Dehydrogenation of *n*-butylenes		64
Oxidative dehydrogenation of *n*-butylenes		87
Dehydrogenation of *n*-butane		42
Ethylene	32,859	
Cracking of *n*-butane		34–41
Cracking of ethane		81
Cracking of propane		65
Maleic anhydride	359	
Oxidation of benzene		39
Oxidation of *n*-butane		48
Phthalic anhydride	863	
Oxidation of *o*-xylene		77
Oxidation of naphthalene		69
Vinyl chloride	8,439	
Chlorination and oxychlorination of ethylene		96
Dehydrochlorination of ethylene dichloride		98

[a]Calculated using data reported in Rudd et al. (1981).

Source: Adapted from Allen (1992).

respectively. A series reaction system

$$R \xrightarrow{k_p} P \xrightarrow{k_w} W$$

where the desired product is the intermediate is the other example reaction network. If the rate constants k_p and k_w are written in the Arrhenius form, then

$$k_P = A_P e^{-(E_P/RT)}$$

and

$$k_W = A_W e^{-(E_W/RT)},$$

where E_P and E_W are the activation energies and A_P and A_W are the frequency factors for the reactions.

Table 6-8. Values of rate ratios showing the effect of changes in temperature on parallel reactions

Direction of Temperature Change	Activation Energy Relationship	
	$E_P > E_W$	$E_P < E_W$
$T_0 > T_1$	S decreases. Lowering temperature lowers k_P more than k_W; overall conversion and selectivity are reduced	S increases. Lowering temperature reduces k_P less than k_W; overall conversion is reduced but selectivity increases
$T_1 > T_0$	S increases. Increasing temperature increases k_P more than k_W; overall conversion and selectivity both increase	S decreases. Increasing temperature increases k_P more than k_W; overall conversion increases but selectivity is decreased

Impact of Reactor Temperature on Selectivity The selectivity of the parallel network to the desired product is governed by the ratio of the reaction rates. If the reaction temperature is changed, the change in the ratio of the rates will be

$$S = \frac{e^{-(E_P/RT_1)}/e^{-(E_W/RT_1)}}{e^{-(E_P/RT_0)}/e^{-(E_W/RT_0)}},$$

where T_0 is the original temperature and T_1 is the new temperature. High values of S indicate a greater production of product and less waste. Table 6-8 shows the values of S for a variety of activation energies and temperature scenarios in a parallel reaction network.

For the series reaction network, the situation is more complex. The qualitative behavior of a series reaction network is shown in Figure 6-4. Initially, all the reactant forms product, but since product is not initially present, waste does not form. As product concentrations build up, substantial amounts of product react to form waste. The goal is to maximize the amount of product formed and minimize the amount of waste formed, and the optimum set of conditions to achieve this goal depends on the relative values of k_P and k_W. Consider the two limiting cases shown in Figure 6-5. When $k_P/k_W \rightarrow \infty$, then essentially no waste is formed. Conversely, if $k_P/k_W \rightarrow 0$, no net product is formed because all of the product is immediately converted to waste. The results for intermediate values of k_P/k_W are plotted in Figure 6-6.

The simple series and parallel reactions described above are interesting starting points, but by no means do they characterize all behaviors. Figure 6-7 shows the selectivity behaviors of a number of reversible series and

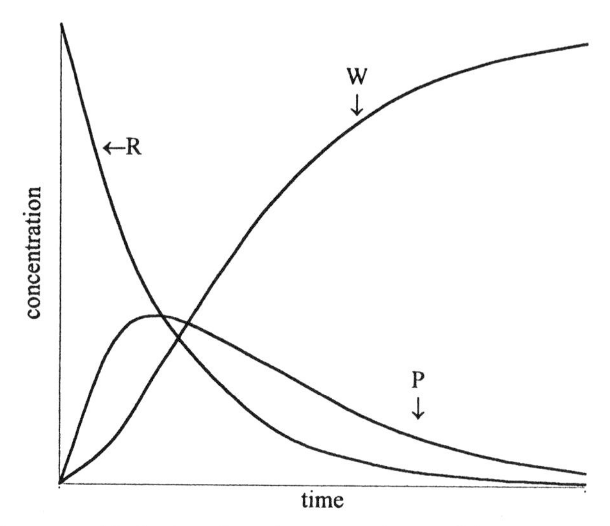

Figure 6-4. The general behavior of an elementary irreversible series reaction network.

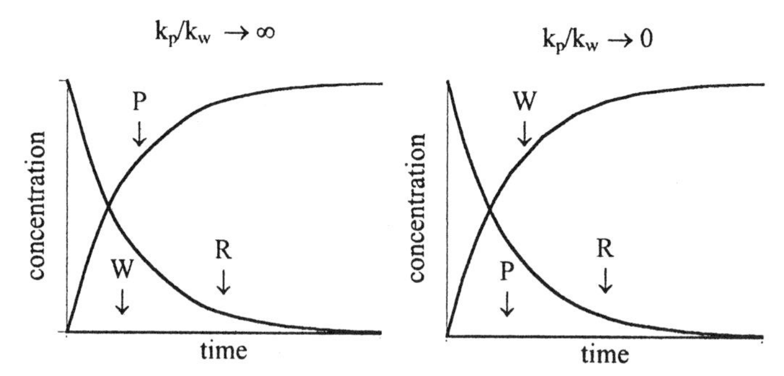

Figure 6-5. An elementary irreversible series reaction at the limiting ratios of the activation energies.

parallel reactions. Analysis of these and more complex reaction networks is presented in a number of standard reaction engineering texts.

Impact of Reactor Concentration on Selectivity The selectivity of the series and parallel reaction networks is sensitive to the starting concentration of the reactants as well as the temperature, since the overall rate of disappearance of the reactant is generally dependent on the concentration of the reactant. In the parallel network, the rate of formation of product and waste can be expressed as

$$\text{Rate of appearance of } P = A_{\mathrm{P}} e^{-(E_{\mathrm{P}}/RT)}[\mathrm{R}]^{n_{\mathrm{P}}}$$

and

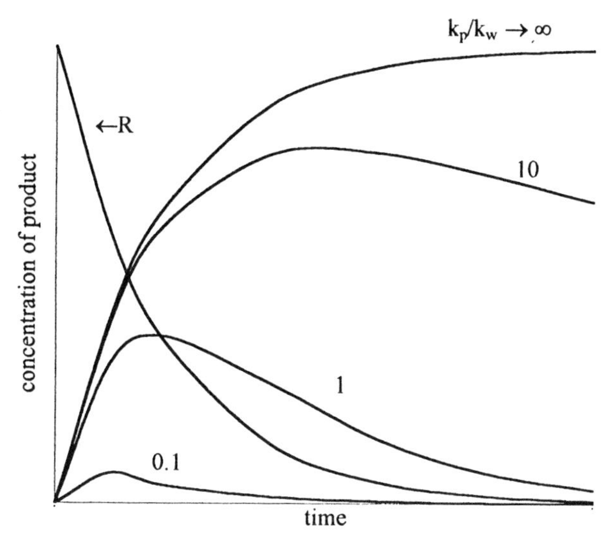

Figure 6-6. An irreversible elementary series reaction at various values of the activation-energy ratio.

$$\text{Rate of appearance of } W = A_W e^{-(E_W/RT)}[\text{R}]^{n_W}$$

where [R] is the concentration of reactant R. The selectivity ratio is

$$S = \frac{[\text{R}]^{n_P}}{[\text{R}]^{n_W}} = [\text{R}]^{n_P - n_W}.$$

If $n_P > n_W$, then selectivity will be improved by increasing the concentration of R. Conversely, if $n_P < n_W$, then increasing the concentration of R will reduce selectivity of the desired product.

As was the case with temperature, the analysis of the series reaction network $R \rightarrow P \rightarrow W$ is more complex than for the parallel network, and few generalizations can be made.

Impact of Mixing on Selectivity In the previous sections it was demonstrated that even for simple, first order series and parallel reactions, changes in temperature and reactant concentration can influence selectivity. Therefore, local inhomogeneities in the temperatures and concentrations found in reactors can lead to reduced selectivity. The impact of mixing is even more important for reactions that are not first order (where n, the concentration dependence of the reaction rate, is not equal to 1). Bimolecular and other nonfirst order reactions can be very sensitive to whether mixing occurs at a molecular level (micromixing) or at a more macroscopic level (macromixing).

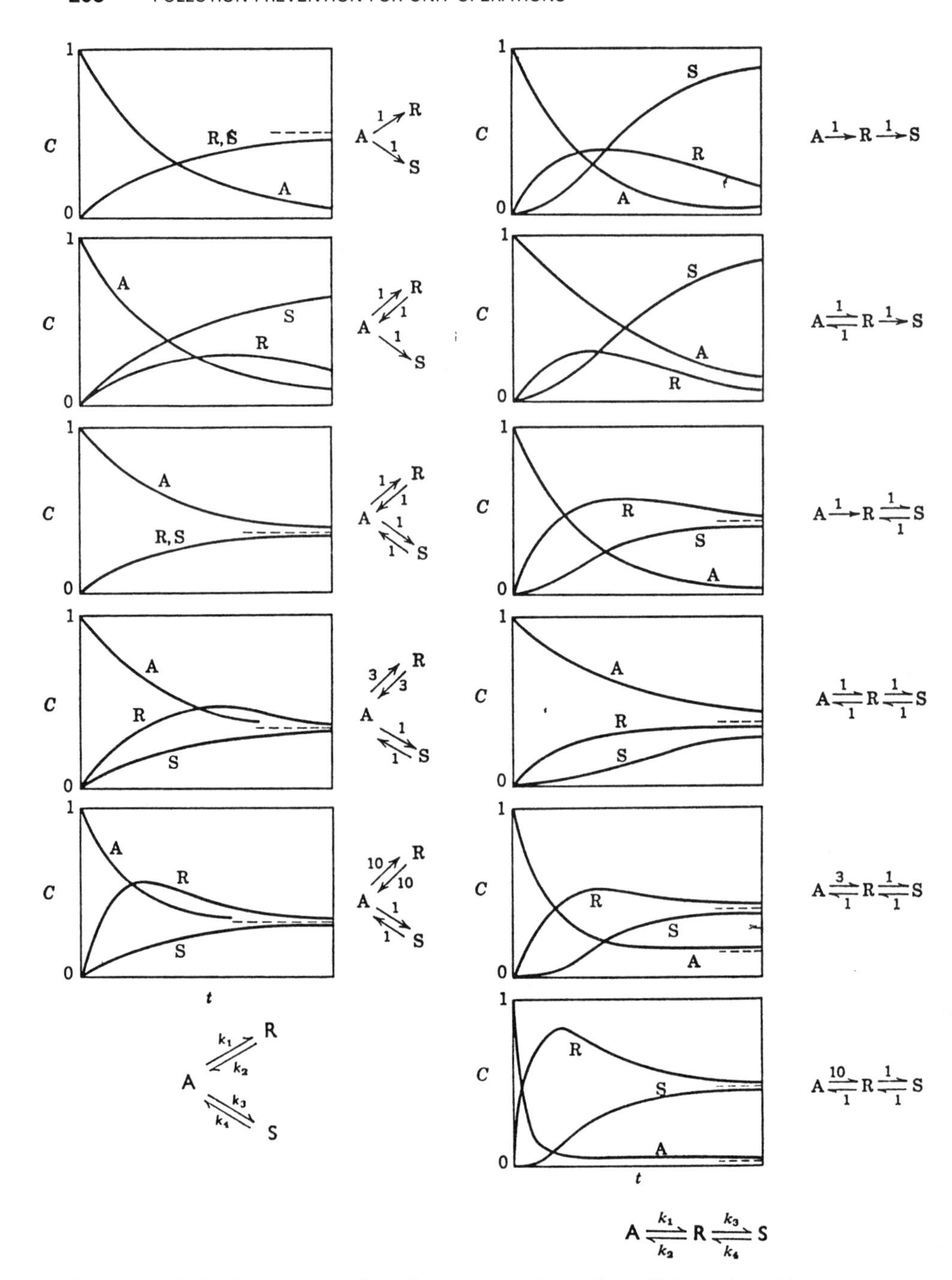

Figure 6-7. Selectivity in reversible elementary series and parallel reactions (Jungers et al., 1958).

A detailed description of micromixing and macromixing is beyond the scope of this text, but can be found in any standard reaction engineering textbook.

The discussion presented above provides an introduction to some of the qualitative and quantitative reactor design tools that can be used to improve selectivity. A more pragmatic review of some of the methods that can increase reactor selectivity and reduce waste has been summarized by Nelson (1990):

1. Improve physical mixing in the reactor, which will improve selectivity if the reaction order is greater than one (Levenspiel, 1972).
2. Distribute feeds better to avoid short-circuiting. This will ensure that as much of the feed as possible is in the reactor for the optimum residence time.
3. Premixing of reactants may result in better selectivity.
4. Provide a separate reactor for recycle streams. Three configurations of recycle reactors are shown in Figure 6-8. Configuration *a* shows a typical recycle reactor where unconverted reactants are recycled to a single-process reactor. This type of recycle reactor can be very effective in reducing wastes if undesired byproducts are produced in a reversible reaction that reaches equilibrium, so that recycling the byproduct inhibits the net formation of waste in the reactor (see Problem Statement 4 at the end of this chapter). If this is not the case, a separate recycle reactor can be used, as shown in parts *b* (Hagh and Allen, 1990; Kalnes and James, 1988) and *c* (Nelson, 1990) of Figure 6-8. The separate recycle reactor can be optimized for recycle conditions, which are often significantly different from conditions in the primary reactor.
5. Examine heating and cooling techniques, particularly focusing on avoiding cool spots and hot spots. Elimination of the use of direct steam contacting as a heating method is also likely to reduce wastes.

In the analysis presented above it was assumed that increasing product yield decreases waste. However, sometimes product yield must be sacrificed in order to minimize the generation of waste. The following example describes an instance in which the goals of maximizing product yield and minimizing waste generation are incompatible.

Example 6-9 Reaction Pathway Optimization for Waste Reduction Product P is formed using raw materials A, B, and C via the following pathways:

$$A + B \rightarrow D + H \qquad \text{(reaction 1)}$$

$$C + H \rightarrow E \qquad \text{(reaction 2)}$$

$$2D + E \rightarrow P \qquad \text{(reaction 3)}$$

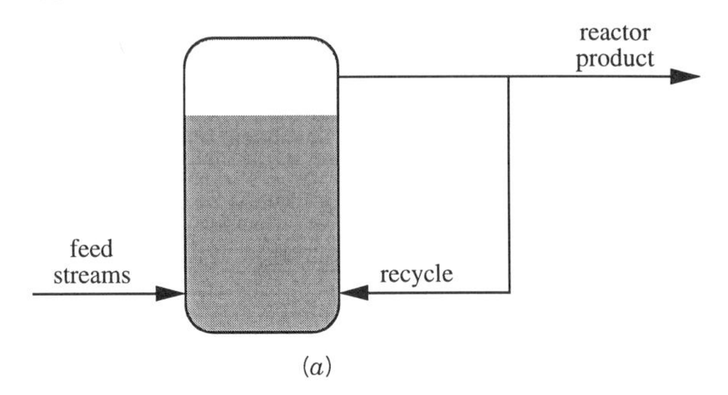

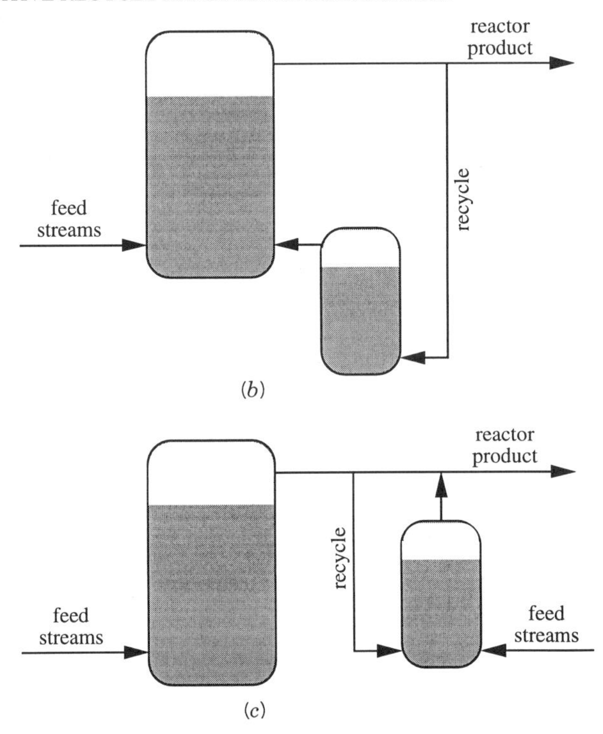

Figure 6-8. Recycle reactor configurations.

Intermediate H, formed in reaction 1 and consumed in reaction 2, is regulated under the Resource Conservation and Recovery Act (RCRA) as hazardous. Maximum yield of P is achieved when reaction 1 is run at twice the rate of reactions 2 and 3. Write the net reaction for these conditions. How much H per mole of P is created? If the rate of reaction 1 is 2 mol A reacted per second and no formation of H can be tolerated, what should the rates of reactions 2 and 3 be? Write the net reaction for these conditions.

Solution The net reaction for maximum yield of P per unit of A and B consumed is

$$2A + 2B + C \rightarrow P + H.$$

One mole of H is created for every mole of P under these conditions. If reactions 1 and 2 are run at the same rate, no net H is created. The rate of reaction 2 in this case is 2 mol C per second, and the rate of reaction 3 is 2 mol D per second. The net reaction for no net creation of H is

$$2A + 2B + 2C \rightarrow P + E.$$

6.4.1.2 ***Changing Process Chemistry*** Either the use of different precursors or the use of a different catalyst may be considered when analyzing a reactor for waste reduction. In some cases, waste created from undesirable side reactions makes a change in raw materials attractive. For example, when acrylonitrile is produced via the ammoxidation of propylene, side reactions create waste acrolein, acetonitrile, and hydrocyanic acid. Producing acrylonitrile through the cyanation/oxidation of ethylene might significantly reduce the wastes associated with acrylonitrile production.

Consideration of the reaction chemistry that results in the formation of waste can also lead to changes in raw materials. It has been shown that if oxygen is present during the chlorine catalyzed pyrolysis of methane, formation of carbonaceous deposits is avoided (Senkan, 1987). It has been suggested that this is because the oxygen reacts with the precursor to the waste, competing with the polymerization processes that create the carbonaceous deposits.

New catalyst technologies have considerable potential for reducing wastes (Allen, 1992). For example, a new catalyst for propylene polymerization has led to the reduction of nonlinear polypropylene (a waste that was landfilled or used as an auxiliary fuel) by 90%, or almost 200 million pounds per year (US DOE, 1991). Replacement catalysts for anhydrous hydrofluoric and fuming sulfuric acids could reduce the use of these toxic reagents and the generation of toxic intermediates in the chemical processing and petroleum refining industries (Cusumano, 1992). Toxic chloroorganic byproducts in the production of acetaldehyde can be reduced by more than a factor of 100 through the use of a retrofit catalyst system (Cusumano, 1992). Catalytic systems for reducing NO_x emissions from hydrocarbon combustion would reduce NO_x levels from 200 ppm to <1 ppm by reducing flame temperatures from 1800°C to 1300°C (Cusumano, 1992). For a systematic approach to considering alternative reaction pathways, see Chapter 11.

6.4.1.3 *Avoiding the Storage of Highly Hazardous Materials* If the reactants in a process are highly hazardous, their storage can be reduced by forming them where they are required and as they are needed. This is called *in situ, on-demand generation.* Problems associated with the synthesis of gallium arsenide (GaAs) and zinc selenide (ZnSe) in the electronics industry illustrate the potential utility of *in situ*, on-demand generation. A major hindrance to the development and implementation of GaAs and ZnSe materials is the fact that the precursor compounds, such as arsine (AsH_3) and hydrogen selenide (H_2Se) are highly volatile and extremely toxic. The large hydride gas flows used in GaAs and ZnSe synthesis systems necessitate the storage of large quantities of the toxic gases in high-pressure cylinders. An equipment failure could lead to the rapid discharge of the entire contents of a gas cylinder, and this is the primary concern with the use of the hydrides. Safety concerns have motivated the development of a number of less hazardous precursors such as tert-butyl arsine, dimethyl selenide, and methyl allyl selenide. Some of the organic precursors are less volatile, so they can be stored as liquids, reducing the rate of release to the gas phase in the event of an accident. However, because of source-induced carbon incorporation and high reaction temperatures, it has proved to be considerably more difficult to achieve high film quality with these precursors than with the hydrides. *In situ* generation of the hazardous hydrides offers an alternative to working with the carbon-containing precursors.

In situ production of H_2Se has been implemented in the synthesis of high-quality ZnSe films (Giapis, 1989). The process consists of (1) formation of selenium vapor, (2) reaction with hydrogen to produce H_2Se, and (3) condensation of excess H_2Se so that the hydride concentration can be precisely controlled. This *in-situ* production of the precursor represents a new level of process integration for microelectronics fabrication, with benefits in both product quality and waste minimization.

6.4.1.4 *Prolonging Catalyst Life* Catalysts for some processes, most notably those used in large quantities in the hydroprocessing of heavy distillate oils at petroleum refineries, lose their effectiveness during use but can be reused after undergoing a regeneration process. Careful control of the regeneration process can maximize the quality of the regenerated catalyst while minimizing the damage caused to the catalyst during regeneration (Dufresne, 1993). Catalysts can sometimes be modified so that they can last longer between changeouts, resulting in fewer waste-generating shutdowns for catalyst changeouts.

6.4.1.5 *Combined Reactor/Separators* Combining reactor and separator operations in a single unit can have many advantages. Catalytic distillation, used in the production of methyl *tert*-butyl ether (MTBE), is the most common example of combined reaction/separation in use today. Two industrial processes for synthesizing MTBE from methanol and isobutylene are in common use. In one process, a series of fixed-bed catalytic reactors send a mix of product, unreacted methanol, and unreacted isobutylene to a series of separation devices. A schematic of this process is shown in Figure 6-9*a*. The alternative process is to send the feed materials to a distillation column in which some of the packing material has been replaced by catalyst. This is called *catalytic distillation*, and is shown in Figure 6-9*b*. This single unit replaces a reactor and several distillation towers required in conventional MTBE manufacture, reducing both fugitive and process emissions. It also requires fewer heat exchangers, which can be a source of wastes. Conventional MTBE manufacture also requires water for separating the components, which is not necessary in the combined operation. Another advantage of catalytic distillation for this synthesis is that because MTBE is less volatile than the reagents, it moves down the distillation column and away from the reaction zone as it is produced, reducing thermodynamic limitations to conversion. It is important to note that the catalytic distillation of MTBE relies on large differences in volatility between the reactants and products.

Despite the advantages, which can include reduced capital costs, only a few commodity chemicals are manufactured using catalytic distillation. This is probably due in large part to the fact that the process configuration for catalytic distillation is very different from that of fixed-bed reactors, and retrofitting existing plants is not generally economical.

Membrane technology offers other new techniques for combining reaction and separation activities when the product molecules are smaller than the reactant molecules. Two such membranes are pictured in Figure 6-10. In the first configuration, reactants form products that are capable of permeating the membrane. In the second configuration, the membrane consists of two layers: a catalytic layer that reactants can permeate, and a second layer that is permeable to products only. The membranes can be formed into tubes, as shown in Figure 6-11, and bundled into hollow fiber reactors that provide a large permeable surface. Both types of membrane for Figure 6-10 hold particular promise for reversible reactions because the product is removed as it is formed, and can be quickly placed under conditions at which the reverse reaction is unlikely to occur. Removal of product also makes membrane reactors advantageous if the product can react with a reactant to form a waste.

6.4.2 Heat Exchangers

Heat exchangers can be a direct source of waste when high temperatures cause the fluids they contain to form sludges. These sludges increase heat-

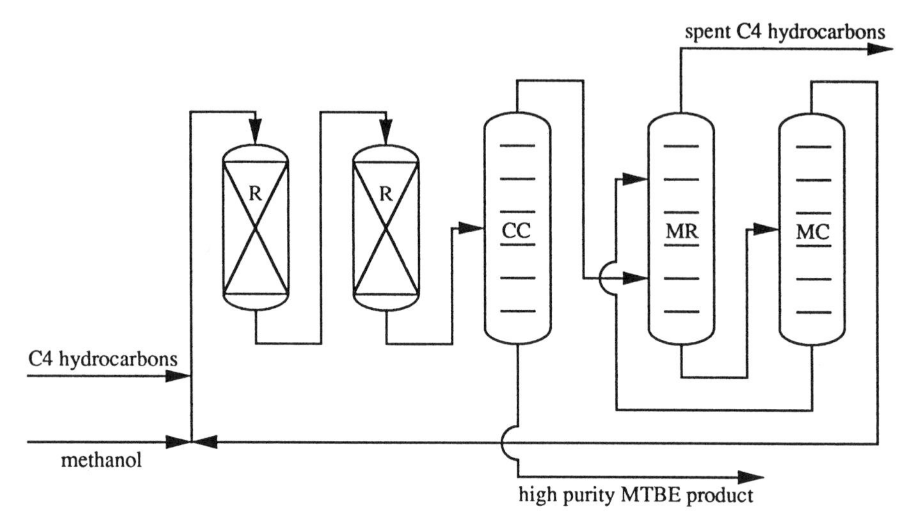

(*a*)

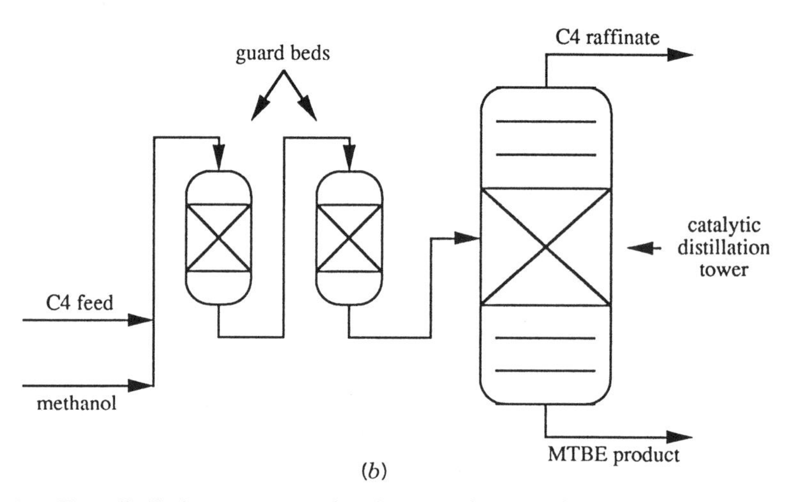

(*b*)

Figure 6-9. Two distillation processes for the manufacture of methyl *tert*-butyl ether (MTBE): (*a*) the conventional, fixed-bed process and (*b*) the catalytic distillation process.

transfer resistance and therefore must be periodically removed. Because it reduces efficiency and increases process energy requirements, sludge buildup in heat exchangers is an indirect source of combustion-related emissions. The methods for reducing heat-exchanger waste fall into two categories: reduction of sludge generation and improved cleaning techniques. Methods

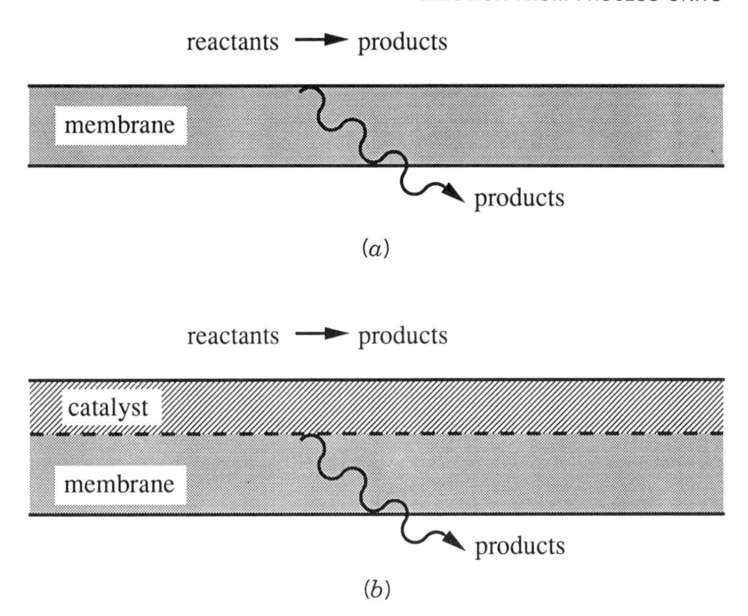

Figure 6-10. Reaction with membrane separation: (*a*) noncatalytic membrane; (*b*) catalytic membrane.

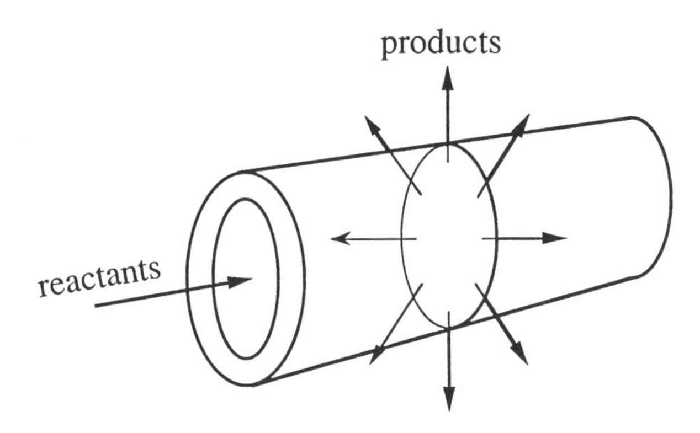

Figure 6-11. Single element of a hollow-fiber reactor.

that reduce sludge generation result in reduction of both direct and indirect wastes from heat exchangers. Improved cleaning techniques are discussed in a previous section.

One way to reduce sludge formation is to reduce the temperatures used in the heat exchanger. Excessively high temperatures are frequently used because of the availability of process steam at fixed pressures (Nelson, 1990). However, it is possible to tailor stream pressures to fit individual applications

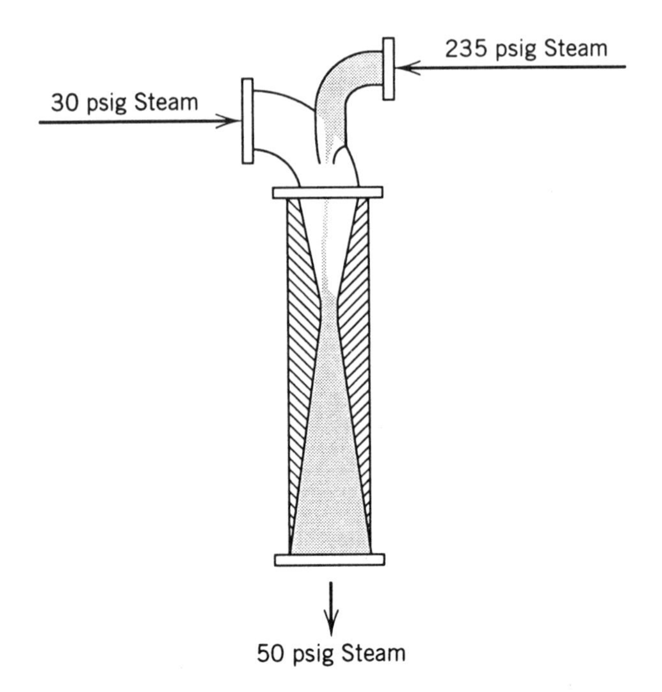

Figure 6-12. A device for upgrading low-pressure steam (Nelson, 1990). (Reprinted with permission from *Hydrocarbon Processing*, March 1990, copyright © 1990 by Gulf Publishing Co., all rights reserved.)

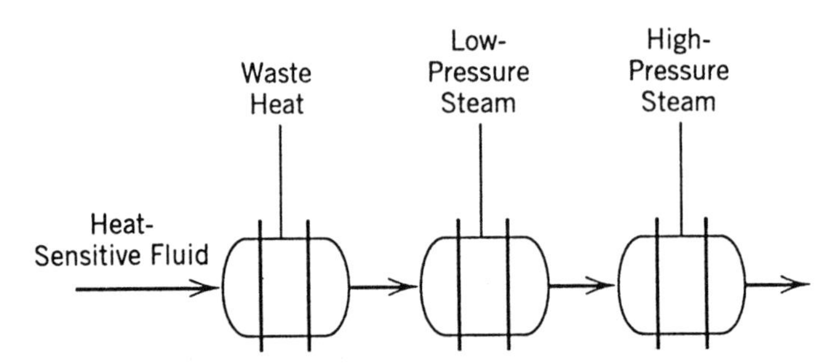

Figure 6-13. Staged heating may reduce waste generation (Nelson, 1990). (Reprinted with permission from *Hydrocarbon Processing*, March 1990, copyright © 1990 by Gulf Publishing Co., all rights reserved.)

with the use of a thermocompressor like the one pictured in Figure 6-12. Staged heating, as shown in Figure 6-13, can reduce sludge formation if high temperatures are unavoidable (Nelson, 1990).

Sludge buildup can also be minimized through the use of on-line cleaning techniques (Someah, 1992), scraped-wall exchangers (Carlson, 1992), and noncorroding tubes (Nelson, 1990). Antifoulants (scale inhibitors) prevent sludge buildup as well (API, 1991a). An alternative to shell-and-tube heat exchangers is the plate-and-frame heat exchanger that can reduce sludge waste in many ways. Because of the narrow space between plates in these exchangers, the heat-transfer fluids have high, uniform turbulence, enhancing heat transfer and necessitating a lower temperature hot side, which reduces sludge buildup. The narrow space also eliminates areas of little or no flow that allow buildup of dirt and debris in conventional heat exchangers. Non-corroding tubes result in less sludge generation in part because corroded areas slow down the heat transfer fluid, allowing sludge generation to begin.

6.4.3 Separation Equipment

In examining separation equipment for waste reduction, three levels of analysis can be considered. One level of analysis involves minimizing the wastes and emissions that are routinely generated in the operation of the equipment. A second level of analysis seeks to control excursions in operating conditions. The third level of analysis seeks to improve the design efficiency of the separation units. Waste reduction opportunities derived from each of these levels of analysis are presented below. However, the waste reduction options are not organized according to whether they minimize routine wastes, control excursions, or improve efficiencies.

6.4.3.1 *Distillation* Distillation columns produce wastes by inefficiently separating materials, through off-normal operation, and by generating sludges in heating equipment. The following solutions to these waste problems have been proposed (Nelson, 1990):

1. Increase the reflux ratio, add a section to the column, retray/repack the column, or improve feed distribution to increase column efficiency.
2. Insulate or preheat the column feed to reduce the load on the reboiler. A higher reboiler load results in higher temperatures and more sludge generation.
3. Reduce the pressure drop in the column, which lowers the load on the reboiler.

In addition, vacuum distillation reduces reboiler requirements, which reduces sludge formation.

Changes in tray configurations or tower packing may prevent pollution from distillation processes. At one facility, the conventional packing in an ethyl 3-ethoxy propionate (EEP) distillation column was replaced with high efficiency structured packing. The conversion resulted in 2.5 millions pounds per year more EEP available for sale and a decrease in still bottoms sent to incineration of 51% (Hazeltine, 1992).

Another method for preventing pollution from distillation columns involves reboiler redesign and is presented in the following example.

Example 6-10 The Design of Distillation Column Reboilers for Pollution Prevention When reboiler fouling causes excessive energy requirements or causes heat transfer to fall below design specifications, the heat exchanger must be shut down and cleaned. If this necessitates shutdown of the column, material in the column (called *holdup*) becomes off-specification product and may require disposal. A column reboiler and two alternative reboilers intended to reduce the holdup losses from reboiler shutdown are depicted in Figure 6-14. With the original reboiler, a column shutdown is required every 90 days. The first alternative is to purchase a larger reboiler that needs less frequent cleaning (once per year). Another alternative is to buy a second reboiler that is the same size as the original reboiler, so that the spare can be employed when reboiler cleanout is necessary, eliminating column shutdown for reboiler cleanouts entirely. What are the capital costs and annual shutdown costs of the two alternative designs if shutting the column down for reboiler cleanouts costs $50,000, installing the larger reboiler of alternative 1 costs $213,000, and installing the second reboiler of alternative 2 costs $150,000? Which of the three reboiler designs (original and two alternatives) is the most economically attractive? Which of the three results in the lowest pollution generation?

Solution The capital cost of alternative 1 is $213,000, and the annual shutdown costs are $50,000. The capital cost of alternative 2 is $150,000 with no shutdown costs. Since the original design has shutdown costs of $200,000/yr, the second alternative would pay for itself in less than a year and is the most economically attractive option. It is also the option that results in the most pollution prevention.

General methods for reducing waste from reboilers are discussed in the previous section on heat exchangers.

6.5 A SURVEY OF INDUSTRY-SPECIFIC WASTE REDUCTION GUIDES

Over the past decade, a number of states, as well as the federal government and industry trade organizations, have developed guides and handbooks for reducing wastes. These materials are generally in the form of analyses of individual facilities, although some guides attempt to be more general. A list of some of the documents that were available as this book went to press is compiled in Table 6-9. This list will very rapidly become incomplete. For the most up-to-date information on these very practical guides, a good starting point are the pollution prevention technical assistance offices in the states and EPA regions. These contacts are listed in Table 6-10.

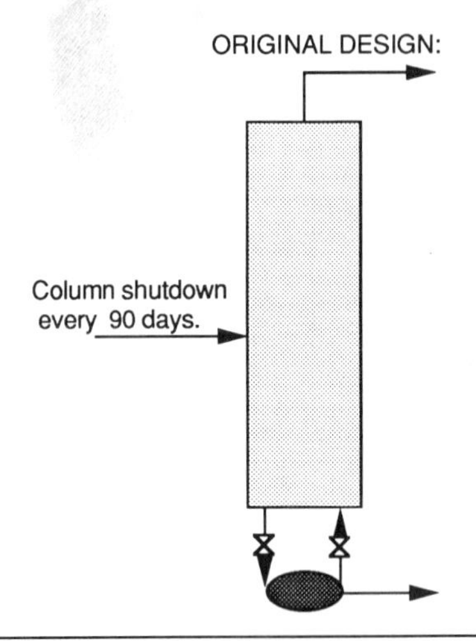

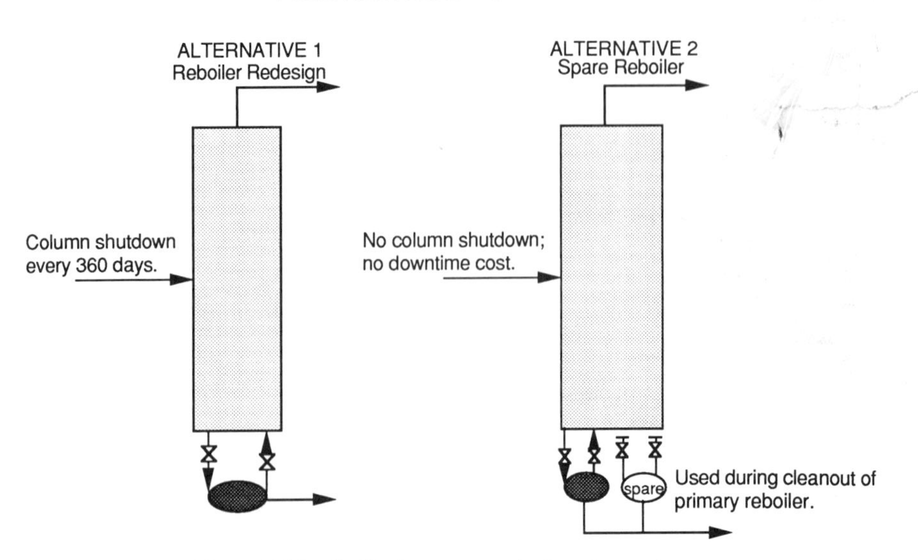

Figure 6-14. Distillation tower reboiler alternatives.

6.6 SUMMARY

Many techniques for preventing pollution are generally applicable if viewed from a unit operations standpoint. In this chapter, a matrix of approaches to pollution prevention in the chemical process industries was presented. This matrix is summarized in Table 6-11, where the rows are common unit operations and the columns are pollution prevention methods, organized into changes in operating practices, currently feasible changes in process

Table 6-9. Industry-specific waste reduction guides

Subject	Date of Issue	Report Number
Issued by the state of California		
Aerospace industry	1987	WR001
Arsenic wastes in the electronics industry	1988	WR002
Heavy metal waste sludges in ceramic products	1990	WR015
Solvent wastes in the electronics industry	1988	WR011
Printed-circuit-board industry	1987	WR014
Summaries issued by the EPA National Risk Management Research Laboratory in Cincinnati, OH		
Manufacture of automotive lighting equipment and accessories	1995	EPA/600/S-95/012
Manufacture of locking devices	1995	EPA/600/S-95/013
Metal fastener manufacture	1995	EPA/600/S-95/016
Outboard motor manufacture	1995	EPA/600/S-95/018
Manufacture of electroplated truck bumpers	1995	EPA/600/S-95/019
Manufacture of folding paperboard cartons	1995	EPA/600/S-95/021
Manufacture of rebuilt industrial crankshafts	1995	EPA/600/S-95/022
Manufacture of pressure sensitive adhesive tape	1995	EPA/600/S-95/023
Manufacture of wooden cabinets	1995	EPA/600/S-95/024
Manufacture of power supplies	1995	EPA/600/S-95/025
Metal parts coating	1995	EPA/600/S-95/027
Manufacture of gear cases for outboard motors	1995	EPA/600/S-95/028
Pharmaceuticals manufacture	1995	EPA/600/S-95/030
Manufacture of components for outboard motors	1995	EPA/600/S-95/031
Manufacture of aircraft landing gear	1995	EPA/600/S-95/032
Norfolk Naval Air Station	1995	EPA/600/SR-95/135
U.S. Army Corps of Engineers civil works facilities	1995	EPA/600/S-95/014

technologies, and process changes requiring technology breakthroughs. Selection of appropriate raw materials is a general approach to achieving pollution prevention in all process units. Solvent losses from both parts and equipment cleaning operations can sometimes be reduced through the use of mechanical cleaning methods. Methods for reducing waste from storage vessels consist of minimizing the container material that is wasted, reducing losses during loading and unloading, and minimizing waste such as tank bottoms that is formed during storage. Finally, pollution prevention from process units can be achieved by considering waste reduction techniques on a unit-by-unit basis.

Table 6-10. Government Technical and Financial Assistance Programs[a]

Alabama
Hazardous Material Management and Resource Recovery Program
University of Alabama
P.O. Box 870203
Tuscaloosa, AL 35487
(205) 271-7939

Alabama Department of Environmental Management
1751 Congressman W. L. Dickinson Drive
Montgomery, AL 36103
(205) 271-7939

Alaska
Alaska Health Project
Waste Reduction Assistant Program
1919 West Northern Lights, Suite 103
Anchorage, AK 99517
(907) 276-2864

Alaska Department of Environmental Conservation
P.O. Box O
Juneau, AK 99811-1800
(907) 465-2671

Arizona
Office of Water and Water Quality Management
Arizona Department of Environmental Quality
2005 North Central Avenue, Room 304
Phoenix, AZ 85004
(602) 257-6994

Arkansas
Arkansas Industrial Development Commission
One State Capitol Mall
Little Rock, AR 72201
(501) 371-1370

Hazardous Waste Division
Arkansas Department of Pollution Control and Ecology
P.O. Box 8913
Little Rock, AR 72210-8913
(501) 570-2861

California
Alternative Technology Section
Toxic Substances Control Division
California State Department of Health Services
P.O. Box 942732
Sacramento, CA 94234-7320
(916) 324-1807

Colorado
Colorado Department of Health
4120 East 11th Avenue
Denver, CO 80220
(303) 331-4830

CO Public Interest Research Group (CoPIRG)
1724 Gilpin
Denver, CO 80219
(303) 355-1861

Connecticut
Connecticut Hazardous Waste Management Service
Suite 360
900 Asylum Avenue
Hartford, CT 06105-1904
(203) 244-2007

Connecticut Department of Economic Development
210 Washington Street
Hartford, CT 06106
(203) 566-7196

Connecticut Department of Environmental Protection
165 Capital Avenue
Hartford, CT 06106
(203) 566-3437

Table 6-10. *(Continued)*

Delaware
Hazardous Waste Management Branch
Department of Natural Resources and Environmental Control
P.O. Box 1401
89 Kings Highway
Dover, DE 19903
(302) 739-3822
(302) 739-6400

District of Columbia
Office of Recycling
DC Department of Public Works, 8th Floor
2000 14th Street, N.W.
Washington, DC 20009
(202) 939-7116

Department of Environmental Programs
Metropolitan Council of Governments
777 N. Capital St., Suite 300
Washington, DC 20002-3200
(202) 962-3200

Florida
Hazardous Waste Reduction and Management Section
Florida Department of Environmental Regulation
Twin Towers Office Building
2600 Blair Stone Road
Tallahassee, FL 32399-2400
(904) 488-0300

Georgia
Hazardous Waste Technical Assistance Program
Georgia Institute of Technology
Georgia Technical Research Institute
Environmental Health and Safety Division
O'Keefe Building, Room 027
Atlanta, GA 30332
(404) 894-3806

Environmental Protection Division
Georgia Department of Natural Resources
Floyd Towers East, Suite 1154
2054 Butler Street, S.E.
Atlanta, GA 30334
(404) 656-2833

Hawaii
Solid and Hazardous Waste Branch
Hawaii Department of Health
645 Halekaulia Street, 2nd Floor
Honolulu, HI 96813
(808) 548-2270

Idaho
Idaho Hazardous Materials Bureau
450 West State Street
Boise, ID 83720
(208) 683-0710

Idaho Division of Environmental Quality
1410 North Hilton Street
Boise, ID 83706
(208) 683-0710

Illinois
Hazardous Waste Research and Information Center
Illinois Department of Energy and Natural Resources
One East Hazelwood Drive
Champaign, IL 61820
(217) 244-8905

Illinois Waste Elimination Research Center
Pritzker Department of Environmental Engineering
Alumni Building, Room 102
Illinois Institute of Technology
3200 South Federal Street
Chicago, IL 60616
(312) 567-3535

Table 6-10. *(Continued)*

Illinois Environmental Protection
Agency
P.O. Box 19276
Springfield, IL 62794-9276
(217) 785-8604

Indiana
Environmental Management and
Education Program
Young Graduate House, Room 120
Purdue University
West Lafayette, IN 47907
(317) 494-5036

Indiana Department of
Environmental Management
Office of Technical Assistance
P.O. Box 6015
105 South Meridian Street
Indianapolis, IN 46206-6015
(317) 232-8172

Iowa
Iowa Department of Natural
Resources
Air Quality and Solid Waste
Protection Bureau
Wallace State Office Building
900 East Grand Avenue
Des Moines, IA 50319-0034
(515) 281-8690

Center for Industrial Research and
Service
205 Engineering Annex
Iowa State University
Ames, IA 50011
(515) 294-3420

Kansas
Bureau of Waste Management
Department of Health and
Environment
Forbes Field, Building 730
Topeka, KN 66620
(913) 296-1607

Kentucky
Division of Waste Management
Natural Resources and
Environmental Protection
Cabinet
18 Reilly Road
Frankfort, KY 40601
(502) 564-6716

Louisiana
Department of Environmental
Quality
Office of Solid and Hazardous Waste
P.O. Box 44307
Baton Rouge, LA 70804
(504) 342-1354

Policy and Planning Division
Louisiana Department of
Environmental Quality
625 North 4th Street
Baton Rouge, LA 70804
(504) 765-0720

Maine
Maine Department of Environmental
Protection
State House Station #17
Augusta, ME 04333
(207) 289-2651

Office of Waste Reduction and
Recycling
Maine Waste Management Agency
State House Station #154
Augusta, ME 04333
(207) 289-5300

Maryland
Maryland Hazardous Waste Facilities
Siting Board
60 West Street, Suite 200A
Annapolis, MD 21401
(301) 974-3432

Table 6-10. *(Continued)*

Maryland Environmental Service
2020 Industrial Drive
Annapolis, MD 21401
(301) 269-3291
(800) 492-9188 (in Maryland)

Office of Waste Minimization and Recycling/Hazardous and Solid Waste Management
Maryland Department of the Environment
2500 Broening Highway, Building 40
Baltimore, MD 21224
(301) 631-3315

Massachusetts

Office of Safe Waste Management
Department of Environmental Protection
100 Cambridge Street, Room 1094
Boston, MA 02202
(617) 727-3260

Source Reduction Program
Massachusetts Department of Environmental Quality Engineering
1 Winter Street
Boston, MA 02108
(617) 292-5982

Massachusetts Department of Environmental Protection
75 Grove Street
Worcester, MA 01606
(508) 792-7692

Michigan

Resource Recovery Section
Department of Natural Resources
P.O. Box 30028
Lansing, MI 48909
(517) 373-0540

Minnesota

Minnesota Pollution Control Agency
Solid and Hazardous Waste Division
520 Lafayette Road
St. Paul, MN 55155
(612) 296-6300

Minnesota Technical Assistant Program
W-140 Boynton Health Service
University of Minnesota
Minneapolis, MN 55455
(612) 625-9677
(800) 247-0015 (in Minnesota)

Minnesota Waste Management Board
123 Thorson Center
7323 Fifty-Eighth Avenue North
Cystal, MN 55428
(612) 536-0816

Minnesota Office of Waste Management
1350 Energy Lane
St. Paul, MN 55108
(612) 649-5741

Mississippi

Hazardous Waste Division
Mississippi Department of Natural Resources
P.O. Box 10385
Jackson, MS 39209
(601) 961-5062

Environmental Protection Council
Office of Pollution Control
Mississippi Department of Environmental Quality
P.O. Box 10385
Jackson, MS 39209
(601) 961-5118

Table 6-10. *(Continued)*

Missouri
State Environmental Improvement and Energy
Resources Authority
P.O. Box 744
Jefferson City, MO 65102
(314) 751-4919

Missouri Department of Natural Resources
205 Jefferson Street
P.O. Box 176
Jefferson City, MO 65102
(314) 751-3176

Montana
Solid and Hazardous Waste Bureau
Montana Department of Health and Environmental Sciences
Cogswell Building, Room B0291
Helene, MT 59620
(406) 444-2821

Nebraska
Hazardous Waste Department
Nebraska Department of Environmental Control
301 Centennial Mall South
P.O. Box 98922
Lincoln, NE 68509-8922
(402) 471-4217

Nevada
Waste Management Program
Nevada Department of Environmental Protection
Capital Complex
123 West Nye Lane
Carson City, NV 89710
(702) 687-5872

New Hampshire
Waste Management Division
New Hampshire Department of Environmental Services
6 Hazen Drive
Concord, NH 03301-6509
(603) 271-2901

New Jersey
New Jersey Hazardous Waste Facilities Siting Commission
28 West State Street, Room 614
Trenton, NJ 08608
(609) 292-1459
(609) 292-1026

Hazardous Waste Advisement Program
Bureau of Regulation and Classification
New Jersey Department of Environmental Protection
401 East State Street, CN028
Trenton, NJ 08625
(609) 292-8341

New Jersey Department of Environmental Protection and Energy
Office of Pollution Prevention
401 East State Street, CN409
Trenton, NJ 08625-0402
(609) 984-5339

New Mexico
Hazardous and Radiation Waste Bureau
New Mexico Environmental Improvement Division
1190 St. Francis Drive
Santa Fe, NM 87503
(505) 827-2926

New York
New York State Environmental Facilities Corp.
50 Wolf Road
Albany, NY 12233
(518) 457-4138

Bureau of Hazardous Waste Program Development
NYSDEC
50 Wolf Road
Albany, NY 12233-7253
(518) 457-7267

Table 6-10. *(Continued)*

North Carolina

Pollution Prevention Pays Program
Department of Natural Resources and Community Development
P.O. Box 27687
512 North Salisbury Street
Raleigh, NC 27611
(919) 733-7015

Governor's Waste Management Board
325 North Salisbury Street
Raleigh, NC 27611
(919) 733-9020

Technical Assistant Unit
Solid and Hazardous Waste Management Branch
North Carolina Department of Human Resources
P.O. Box 2091
306 North Wilmington Street
Raleigh, NC 27602
(919) 733-2178

North Dakota

Division of Waste Management
Management and Special Studies
North Dakota Department of Health
1200 Missouri Avenue, Room 302
Bismarck, ND 58502-5520
(701) 224-2366

Ohio

Division of Solid and Hazardous Waste Management
Ohio Environmental Protection Agency
P.O. Box 1049
1800 Water Mark Drive
Columbus, OH 43266-0149
(614) 644-2917

Ohio Technical Transfer Organization
77 South High Street, 26th Floor
Columbus, OH 43255-0330
(614) 466-4286

Ohio's Thomas Edison Program
77 South High Street, 26th Floor
Columbus, OH 43215
(614) 466-3887

Ohio Department of Natural Resources
Fountain Square
Columbus, OH 43224-1387
(614) 265-6333

Oklahoma

Industrial Waste Elimination Program
Oklahoma State Department of Health
P.O. Box 53551
Oklahoma City, OH 73152
(405) 271-7353

Oregon

Oregon Hazardous Waste Reduction Program
Department of Environmental Quality
811 Southwest Sixth Avenue
Portland, OR 97204
(503) 229-5913

Pennsylvania

Pennsylvania Technical Assistance Program
501 F. Orvis Keller Building
University Park, PA 16802
(814) 865-0427

Bureau of Waste Management
Pennsylvania Department of Environmental Resources
P.O. Box 2063
Fulton Building
3rd and Locust Streets
Harrisburg, PA 17120
(717) 787-6239

Center for Hazardous Material Research
320 William Pitt Way
Pittsburgh, PA 15238
(412) 826-5320

Table 6-10. *(Continued)*

Rhode Island
Ocean State Cleanup and Recycling Program
Rhode Island Department of Environmental Management
9 Hayes Street
Providence, RI 02908-5003
(401) 277-3434
(800) 253-2674 (in Rhode Island)

Center of Environmental Studies
Brown University
P.O. Box 1943
135 Angell Street
Providence, RI 02912
(401) 863-3449

Tennessee
Center for Industrial Services
102 Alumni Hall
University of Tennessee
Knoxville, TN 37996
(615) 974-2456

Virginia
Office of Policy and Planning
Virginia Department of Waste Management
11th Floor, Monroe Building
101 North 14th Street
Richmond, VA 23219
(804) 225-2667

Washington
Hazardous Waste Section
Mail Stop PV-11
Washington Department of Ecology
Olympia, WA 98504-8711
(206) 459-6322

Wisconsin
Bureau of Solid Waste Management
Wisconsin Department of Natural Resources
P.O. Box 7921
101 South Webster Street
Madison, WI 53707
(608) 266-2699

Wyoming
Solid Waste Management Program
Wyoming Department of Environmental Quality
Herschler Building, 4th Floor, West Wing
122 West 25th Street
Cheyenne, WY 82002
(307) 777-7752

[a]The EPA's Office of Solid Waste and Emergency Response has set up a telephone call-in service to answer questions regarding RCRA and Superfund (CERCLA): (800) 424-9346 (outside the District of Columbia); (202) 382-3000 (in the District of Columbia). The EPA has also set up a Hotline for the Toxics Release Inventory: (800) 535-0202; (703) 920-9877 (in Virginia and Arkansas) and a Pollution Prevention Information Clearinghouse which uses the RCRA/Superfund Hotline listed above. The state programs listed above may also provide information.

QUESTIONS FOR DISCUSSION

1. Existing water management practices have to change to reflect the new value of water at the refinery described in the first section of this chapter. Standard practice prior to pretreatment was to crack open utility valves so they would dribble water to prevent the pipes from

Table 6-11. Process modifications for pollution prevention

Unit Operation	Changes in Operating Practices	Currently Feasible Modifications	Process Modifications that Require Technology Development
Storage vessels	Use of mixers to reduce sludge formation, add emulsifiers to dissolve bottoms when the bottoms and the emulsifiers do not interfere with downstream processing, maintain seals and paint in good condition	Floating-roof tanks, high-pressure tanks, insulated tanks, variable-vapor-space tanks	Process-specific changes to eliminate the need for storage of hazardous intermediates
Heat exchangers	Use of antifoulants, use of innovative cleaning devices for heat-exchanger tubes	Staged heat exchangers and use of adiabatic expanders to reduce heat-exchanger temperatures and reduce sludge formation, heat integration, on-line cleaning	
Reactors	Higher selectivity through better mixing of reactants, elimination of hot and cold spots, improved feed distribution, careful control of catalyst regeneration	Catalyst modifications to enhance selectivity or to prevent catalyst deactivation and attrition, separate recycle reactors when conditions warrant	New catalyst technologies, changes in process chemistry, integration of reaction and separation units
Separators	Reduced waste from reboilers	Improvements in separation efficiencies, insulate the column, change tray configuration or packing	New separation devices that efficiently separate dilute species

freezing during cold weather. What alternatives to this practice might be considered now that water is a more precious commodity?

2. Name some of the ways that raw materials in petroleum refining could be examined for waste reduction. Crude oil, water, solvents, and soaps are some of the raw materials you might want to consider.
3. Process water quality varies widely from one refinery to another and plays a larger role in waste generation at refineries whose cooling-tower and boiler blowdown is mixed with oily wastewaters prior to treatment. Also, pretreating the feedwater to a refinery can result in energy savings due to reduced boiler blowdown. However, some refineries generate an excess of process gas during operation that meets or exceeds their needs for boiler fuel, and any excess process gas is sent to a flare for thermal destruction, so energy conservation in boilers may lead to other emissions. Other refineries burn fuel in their boilers that would otherwise be sold. Is feedwater pretreatment a good pollution prevention strategy for refineries in general? Make a list of refinery characteristics that indicate a high potential for effective use of feedwater pretreatment.

PROBLEM STATEMENTS

1. For the vapor degreasing process described in Example 6-4, calculate the emissions reduction, assuming that the condenser is run at 15°C where the vapor pressure is 75 mmHg. Assuming that the heat of vaporization is 100 cal/g, calculate the minimum energy required by the condenser in kilowatts. Finally, using the cost of electricity in your region, calculate a lower bound on the recovery cost for the solvent in $/kg.
2. It has been estimated that as much as 12 lb of organic compounds can be lost per 1000 gal of gasoline transferred during loading and unloading operations. Assuming that gasoline vapor behaves as an ideal gas and the average molecular weight of the gasoline is 140, calculate the vapor pressure of the gasoline. How much would volatilization losses be decreased if vapor pressure were decreased by 2 psia?
3. Consider a parallel reaction network where a reactant R produces product P and waste W as follows:

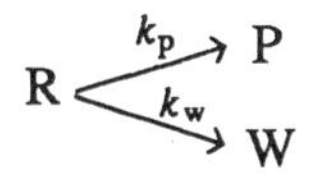

The rate constants k_P and k_W for the formation of product and waste have activation energies of 60 and 50 kcal/mol, respectively. Given that the ratio k_P/k_W at 100°C is 10, calculate the ratio of the rates at 90°C

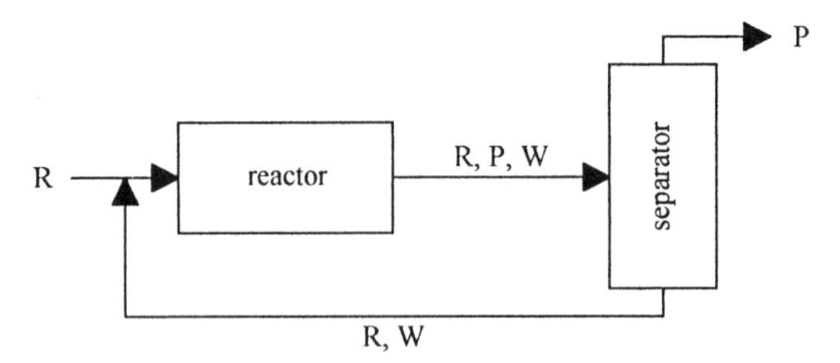

Figure 6-15. Process flow diagram for the parallel reaction network of Problem 4.

and 110°C. Do you recommend raising or lowering reactor temperature to improve selectivity? Are temperature fluctuations around the mean temperature desirable for this reaction?

4. The parallel reaction network

$$R \xrightarrow{k_P} P \qquad R \underset{k_{-W}}{\overset{k_W}{\rightleftharpoons}} W$$

is carried out in a well-mixed reactor. All the reactions are first-order, and the equilibrium constant for the reversible reaction forming waste W is

$$K_{eq} = \frac{k_W}{k_{-W}} = \frac{[W]}{[R]},$$

where [W] and [R] are the concentrations of waste and reactant. The process flow diagram for producing product P is shown in Figure 6-15. Reactant is fed to the reactor along with a recycle stream containing waste and unconverted reactant. The effluent from the reactor, containing unconverted reactant, product, and waste, is sent to an ideal separator that divides the reactor effluent into pure product and a stream containing waste and unconverted reactant.

a. Explain why there is no net production of waste in this process.
b. Given that the reactor converts 90% of the reactant fed to it, determine the flow rate of product and the flow rate of the recycle stream for every 100 mol of R fed to the process. Express your answer in terms of the rate constants and the equilibrium constant.

REFERENCES

Allen, D. T., N. Bakshani, and K. S. Rosselot, *Pollution Prevention: Homework and Design Problems for Engineering Curricula*, American Institute of Chemical Engineers, New York, 1992.

Allen, D. T., "The Role of Catalysis in Industrial Waste Reduction," in *Industrial Environmental Chemistry: Waste Minimization in Industrial Processes and Remediation of Hazardous Wastes*, A. E. Martell and D. Sawyer, eds., Plenum, New York, 1992.

American Petroleum Institute (API), "Waste Minimization in the Petroleum Industry: A Compendium of Practices," API Publication 849–00020, Washington, DC, 1991a.

American Petroleum Institute (API), "The Generation and Management of Wastes and Secondary Materials in the Petroleum Refining Industry: 1987–1988," API Publication 4530, Washington, DC, 1991b.

Balik, J. A. and S. M. Koraido, "Case Study: Identifying Pollution Options for a Petroleum Refinery," *Pollution Prevention Review*, **1**(3), 273–293, Summer 1991.

Bird, R. B., W. E Stewart, and E. N. Lightfoot, *Transport Phenomena*, Wiley, New York, 1960.

Carlson, J. A., "Understand the Capabilities of Plate-and-Frame Heat Exchangers," *Chem. Eng. Prog.*, **88**(7), 26–31, July 1992.

Cusumano, J. A., "Designer Catalysts: Hastemakers for a Clean Environment," *Proceedings of Pollution Prevention in the Chemical Process Industries*, April 6–7, 1992, compiled by McGraw-Hill, New York, 1992.

Dufresne, P., "Hydrotreating Catalysts Regeneration," preprints for the Symposium on Regeneration, Reactivation and Reworking of Spent Catalysts, 205th National Meeting of the American Chemical Society, Denver, CO, March 28–April 2, 1993.

Fogler, H. S., *Elements of Chemical Reaction Engineering*, 2nd ed., Prentice-Hall, Englewood Cliffs, NJ, 1992.

Giapis, K., Ph.D. thesis, University of Minnesota, 1989.

Griffin, D. G., "Feed Water Pretreatment to Reduce Process Wastewater," ENV-94–160, National Petroleum Refining Association, Washington, DC, 1994.

Gygax, R., "Chemical Reaction Engineering for Safety," *Chem. Eng. Sci.*, **43**(8), 1759–1771, 1988.

Hagh, B. and D. T. Allen, "Catalytic Hydrodechlorination," in *Innovative Hazardous Waste Treatment Technology*, Vol. 2, *Physical/Chemical Processes*, H. M. Freeman, ed., Technomic, Lancaster, PA, 1990, pp. 45–54.

Hazeltine, D. M., "Wastes: To Burn, or Not to Burn," *Chem. Eng. Prog.*, **88**(7), 53–58, July 1992.

Hunt, G. E., "Waste Reduction Techniques and Technologies," in *Hazardous Waste Minimization*, H. M. Freeman, ed., McGraw-Hill, New York, 1990.

Jungers, J. C. et al., *Cinétique chimique appliquée*, Technip, Paris, 1958.

Kalnes, T. N. and R. B. James, "Hydrogeneration and Recycle of Organic Waste Streams," *Environmental Progress*, **7**(3), 185–191, 1988.

Levenspiel, O., *Chemical Reaction Engineering*, 2nd ed., Wiley, New York, 1972.

Molina, M. J. and F. S. Rowland, "Stratospheric Sink for Chlorofluoromethanes: Chlorine Atom-Catalyzed Destruction of Ozone," *Nature*, **249**, 810, 1974.

Nelson, K. E., "Use These Ideas to Cut Waste," *Hydrocarbon Processing*, March 1990.

Rudd, D. F., S. Fathi-Afshar, A. A. Trevino, and M. A. Stadtherr, *Petrochemical Technology Assessment*, Wiley, New York, 1981.

Senkan, S. M., "Converting Methane by Chlorine-Catalyzed Oxidative Pyrolysis," *Chem. Eng. Prog.*, 58, Dec. 1987.

Someah, K., "On-Line Tube Cleaning: The Basics," *Chem. Eng. Prog.*, **88**(7), 39–45, July 1992.

Stolarski, R. S. and R. J. Cicerone, "Stratospheric Chlorine: A Possible Sink for Ozone," *Can. J. Chem.*, **52**, 1610, 1974.

U.S. Department of Energy (US DOE), Office of Industrial Technologies, "Industrial Waste Reduction Program: Program Plan," 1991.

U.S. Environmental Protection Agency (US EPA), "Compilation of Air Pollutant Emission Factors, Volume I: Stationary Point and Area Sources," 4th ed. with Supplements A to D, Research Triangle Park, NC, Publication No. AP-42, Sep. 1985.

U.S. Environmental Protection Agency (US EPA), "Waste Minimization Opportunity Assessment Manual," EPA/625/7–88/003, 1988.

Wofsy, S. C. and M. B. McElroy, "HO_x, NO_x and ClO_x: Their Role in Atmospheric Photochemistry," *Can. J. Chem.*, **52**, 1582, 1974.

7

PREVENTING FUGITIVE AND SECONDARY EMISSIONS

Fugitive and secondary emissions are significant sources of pollution at many chemical processing facilities. In Chapter 5, methods for measuring and estimating these emissions were discussed. In this chapter, strategies for preventing fugitive and secondary emissions are presented. Such strategies have been shown to be cost-effective pollution prevention alternatives at a number of facilities.

7.1 FUGITIVE EMISSIONS

Fugitive emissions are unintentional releases from process equipment. These releases are not trivial; it has been estimated that fugitive emissions account for approximately a third of the organic emissions from synthetic organic chemical manufacturing industry (SOCMI) facilities (US EPA, 1986). As was discussed in Chapter 5, the possible sources of fugitive emissions include pumps, compressors, valves, flanges, pressure-relief valves, sampling connections, and open-ended lines. These components can number in the thousands at a typical SOCMI facility and in the tens to hundreds of thousands at petroleum refineries.

Conventional technology components in good working order emit very low quantities of process fluid that, for most applications, are not a concern. However, when there is a mechanical failure of a seal, packing, or gasket, a leak results. The precise timing and location of leaks cannot be predicted, but they can be prevented or repaired, and leakless technology is available for applications where even minute releases cannot be tolerated.

The reduction of fugitive emissions is the topic of this section. First, typical profiles of component populations and their contribution to fugitive emissions

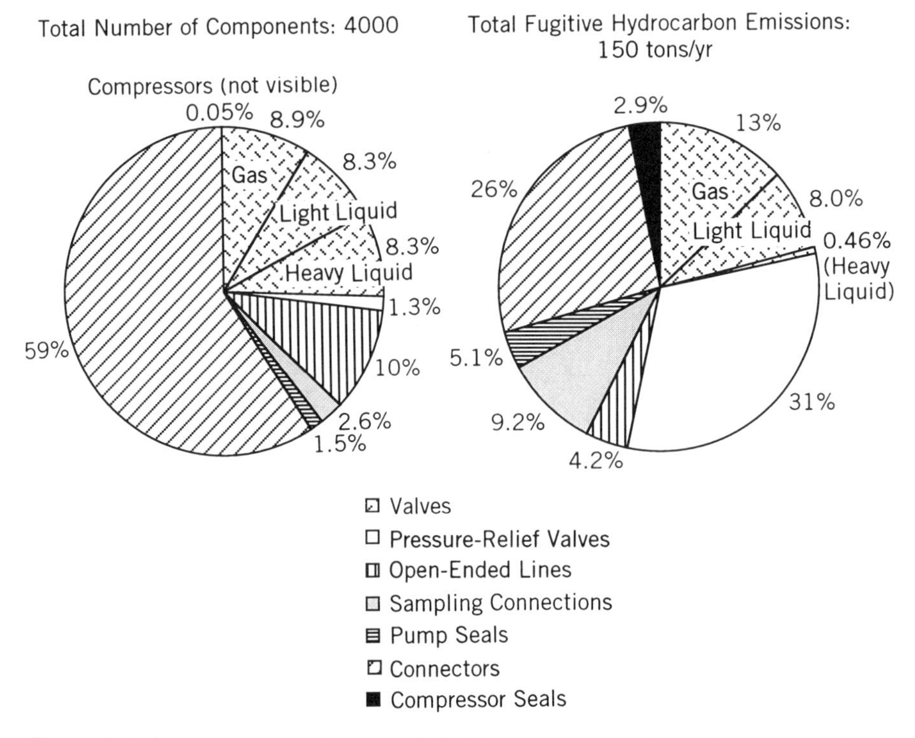

Figure 7-1. Component population and fugitive-emission profiles at a typical SOCMI facility.

are presented. Then, methods for reducing fugitive emissions are described, followed by case studies of fugitive-emission reduction programs. As discussed in this section, techniques for reducing fugitive emissions can be cost-effective and result in significant reduction of the total emissions from a facility.

7.1.1 Fugitive-Emission Profiles

A study of typical fugitive-emission profiles leads to a better understanding of the options for reducing such emissions, because the profiles reveal what the most common components are and what component types are generally responsible for the largest share of fugitive emissions. Profiles of the component population and fugitive emissions for a typical SOCMI facility and a refinery of moderate capacity are presented in this section.

7.1.1.1 ***Synthetic Organic Chemical Manufacturing Industry*** The component profile for a typical SOCMI facility is depicted in Figure 7-1. The values used to create this figure are weighted averages of three model SOCMI

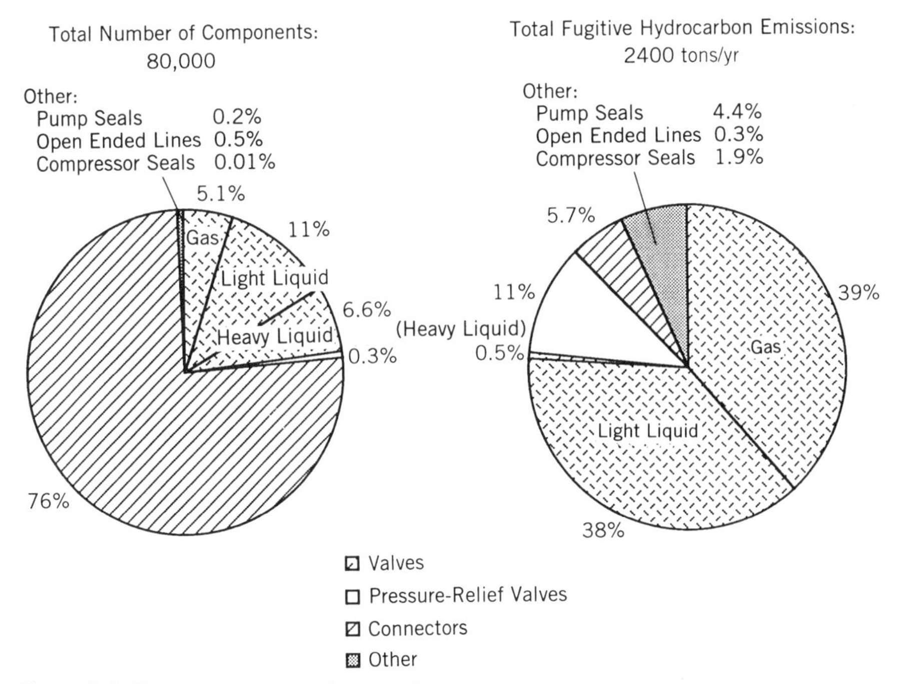

Figure 7-2. Component population and fugitive-emission profiles at a refinery of moderate capacity.

units used to make industrywide estimates of fugitive emissions (US EPA, 1982). Figure 7-1 includes a breakdown of fugitive hydrocarbon emissions by component type, as determined from the average SOCMI fugitive emission factors of Table 5-6, assuming that the mass fraction of VOC in the process fluids is 1.0. This figure shows that the contribution to fugitive emissions made by valves is roughly equivalent to their component contribution, and that pressure-relief valves and connectors are major contributors to SOCMI fugitive emissions.

7.1.1.2 *Petroleum Refineries* Figure 7-2 shows the distribution of emissions and the distribution of equipment for a refinery of moderate capacity. Although connectors (mostly flanges) account for ~75% of the components, they emit less than 6% of the fugitive hydrocarbons. Valves, on the other hand, represent ~20% of the component population and release ~75% of the fugitive hydrocarbons. This is in contrast to the profile for a SOCMI facility, where pressure-relief valves are the single largest contributor to fugitive emissions. The component breakdown for refineries differs from that of a typical SOCMI facility mainly in that there are proportionally more valves and fewer flanges. Also, there are more process components overall in a midsize refinery than in a typical chemical processing unit.

Table 7-1. Distribution of components in the cracking unit and hydrogen plant of a refinery of moderate capacity

		Cracker		H_2 Plant[b]
Component	Service[a]	Equipment Count	m_{VOC}	Equipment Count
Pump seals	LL	6	0.75	2
	HL	9	0.55	2
Compressor seals	HC gas	4	1.0	0
	H_2 gas	0		6
Valves	HC gas	200	1.0	70
	H_2 gas	0		80
	LL	196	0.75	427
	HL	294	0.55	427
Connectors	All	2277	0.75	3313
Relief valves	Gas	11	1.0	15
	Liquid	15	0.63	2
Open-ended lines	All	32	0.75	42
Sampling taps	All	17	0.75	24

[a]HL: heavy liquid, LL: light liquid, HC: hydrocarbon.
[b]$m_{VOC} = 1.0$ for all.
Source: Klee (1991).

Component and emission profiles vary within the different units of a refinery. This can be illustrated by comparing the fugitive emissions from the cracking unit and the hydrogen plant at a midsize petroleum refinery, as shown in the following example.

Example 7-1 Fugitive Emissions from Different Refinery Processes The average fugitive-emission factors for refineries from Table 5-6 can be used to estimate the distribution of fugitive hydrocarbon emissions from a cracking unit and a hydrogen plant at a refinery. Table 7-1 shows the types of equipment found in the two units and the fraction of the stream in service that is made up of volatile organic compounds. Use the average emission factors of Table 5-6 to develop profiles of the fugitive emissions by component type for these two units, and compare the contributions of the various components at the two units with each other and with the profile of fugitive emissions for a typical refinery. The relationship between fugitive hydrocarbon emissions and the factors in this table for a single piece of equipment is given by

$$E = m_{\mathrm{voc}} f_{\mathrm{av}} \,,$$

where E is the hydrocarbon emission rate, m_{VOC} is the mass fraction of volatile organic compounds plus the mass fraction of methane in the stream

Table 7-2. Fugitive emissions from a cracking unit and a hydrogen plant at a refinery

		Emissions, kg/hr			
Component	Service[a]	Cracker	(%)	H_2 Plant	(%)
Pump seals	LL	0.51	(4.0)	0.22	(1.9)
	HL	0.10	(0.81)	0.042	(0.36)
Compressor seals	HC gas	2.6	(20)	0	(0)
	H_2 gas	0	(0)	0.30	(2.6)
Valves	HC gas	5.3	(42)	1.9	(16)
	H_2 gas	0	(0)	0.66	(5.7)
	LL	1.6	(13)	4.7	(41)
	HL	0.037	(0.29)	0.85	(0.73)
Connectors	All	0.42	(3.3)	0.83	(7.2)
Relief valves	Gas	1.8	(14)	2.4	(21)
	Liquid	0.066	(0.52)	0.014	(0.12)
Open-ended lines	All	0.054	(0.43)	0.084	(0.72)
Sampling taps	All	0.19	(1.5)	0.36	(3.1)
Total	—	13	(100)	12	(100)

[a]HL: heavy liquid; LL: light liquid; HC: hydrocarbon.

(up to a maximum of 10%) serviced by the component, and f_{av} is the average emission factor. Use the SOCMI emission factor for sampling connections.

Solution To get the estimated emissions for the component types, multiply the estimated emissions per component by the number of components. Using pump seals in light liquid service in the cracker as an example:

$$\text{Emissions} = (6 \text{ sources})\left(\frac{0.75 \text{ kg VOC}}{\text{kg process fluid}}\right)\left(\frac{0.11 \text{ kg VOC}}{\text{hr} \cdot \text{source}}\right) = 0.51 \text{ kg/hr}.$$

Table 7-2 gives the estimated fugitive emissions for the two units. Although there are a third again as many components in the H_2 plant as in the cracking unit, the H_2 plant has lower estimated fugitive emissions. Figure 7-3 shows the distribution of emissions for the two units. For both units, the majority of emissions are from valves. Also, pressure-relief valves are important contributors to fugitive emissions for both units. The major emission distribution difference between the two units is in the compressor seals: they are a much more significant fraction of total emissions in the cracking unit than in the hydrogen plant. The hydrogen plant profile most closely resembles the profile at a typical refinery.

Overall, there are two points to be derived from these profiles. First, for the chemical manufacturing industries and for quite different processing

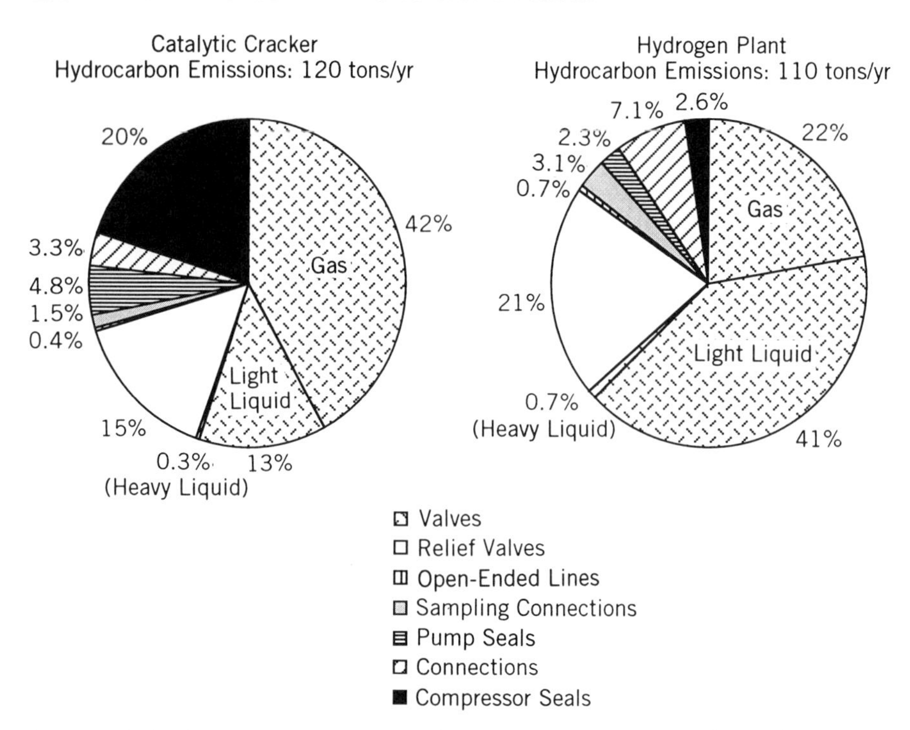

Figure 7-3. Fugitive-emission profiles for two units at a petroleum refinery of moderate capacity.

operations in petroleum refineries, valves are consistently a major source of fugitive emissions. The second point is that making generalizations about major sources of fugitive emissions other than valves is a difficult proposition.

7.1.2 Methods for Reducing Fugitive Emissions

There are two types of fugitive emissions: (1) low-level leaks from process equipment, which were profiled in the previous section, and (2) episodic fugitive emissions, where an event such as equipment failure results in a sudden large release. Often, methods for reducing low-level equipment leaks result in fewer episodes, and vice versa. Methods for reducing or eliminating both types of fugitive emissions can be divided into two groups: (1) leak detection and repair and (2) equipment modification.

7.1.2.1 *Leak Detection and Repair* In leak detection and repair (LDAR), process equipment such as pumps and valves is inspected periodically for leaks with an organic vapor analyzer (OVA). If the OVA detects hydrocarbon concentrations greater than some threshold level (e.g.,

10,000 ppm) around the equipment, the component is said to be leaking and repair is required.

Industrial LDAR programs vary widely in their inspection frequency and effectiveness. At one extreme is constant monitoring using area sensors. Area monitoring is especially effective for chemicals that can be detected at very low concentrations. For materials that either cannot be monitored or are too expensive to monitor constantly, manual detection is generally performed, typically on a monthly, quarterly, or annual basis. Although more costly, monitoring equipment at short intervals results in more effective emission reduction. For example, monitoring and repair of valves in light liquid service at monthly intervals is a third again as effective in reducing emissions as monitoring quarterly, and nearly three times as effective as monitoring every six months (US EPA, 1982).

When a component is found to be leaking, an appropriate method of repair must be chosen. Some repairs can be made to equipment in service, such as tightening bolts or lubricating packing. However, there are times when a repair cannot be made without removing the component from service. These repairs are usually made at normal shutdown times, because shutting down a process to repair a leaking component can result in larger emissions than long term emissions from the leak itself.

7.1.2.2 ***Equipment Modification*** Equipment modification to reduce fugitive emissions might include redesigning a unit so that fewer joints and other potential leaks exist, replacement of existing components with those that cause fewer or no emissions, or sealant injection. It is important to remember that seals on ordinary equipment allow some process fluid to escape, but when in good working order, these releases are very nearly zero. The following is a brief description of the types of equipment from which fugitive emissions are generated, where the emissions occur, and what equipment changes can be made to reduce or eliminate these emissions.

The most numerous component in a plant is the connector, which is used to connect piping to other piping or to equipment. Flanges are connectors consisting of gasket-sealed junctions that are used on pipe with a diameter of $\geq$2 in. Flanges may leak because of improperly selected gaskets or poor assembly. Other types of connectors are generally used on smaller-diameter pipe. Threaded connections, which leak when cross-threaded, are an example. Another type of small pipe connector is a nut-and-ferrule connection, which leaks when poorly assembled. All types of connectors are subject to thermal deformation and may leak as a result. Connectors and their associated emissions can be eliminated in some cases through the use of welded joints.

It is no surprise that in equipment with moving parts, releases generally occur around the moving part. Packing, which is subject to degradation, is often used around such parts to form a seal between the process fluid and the atmosphere. For valves, the moving part is the valve stem and emissions

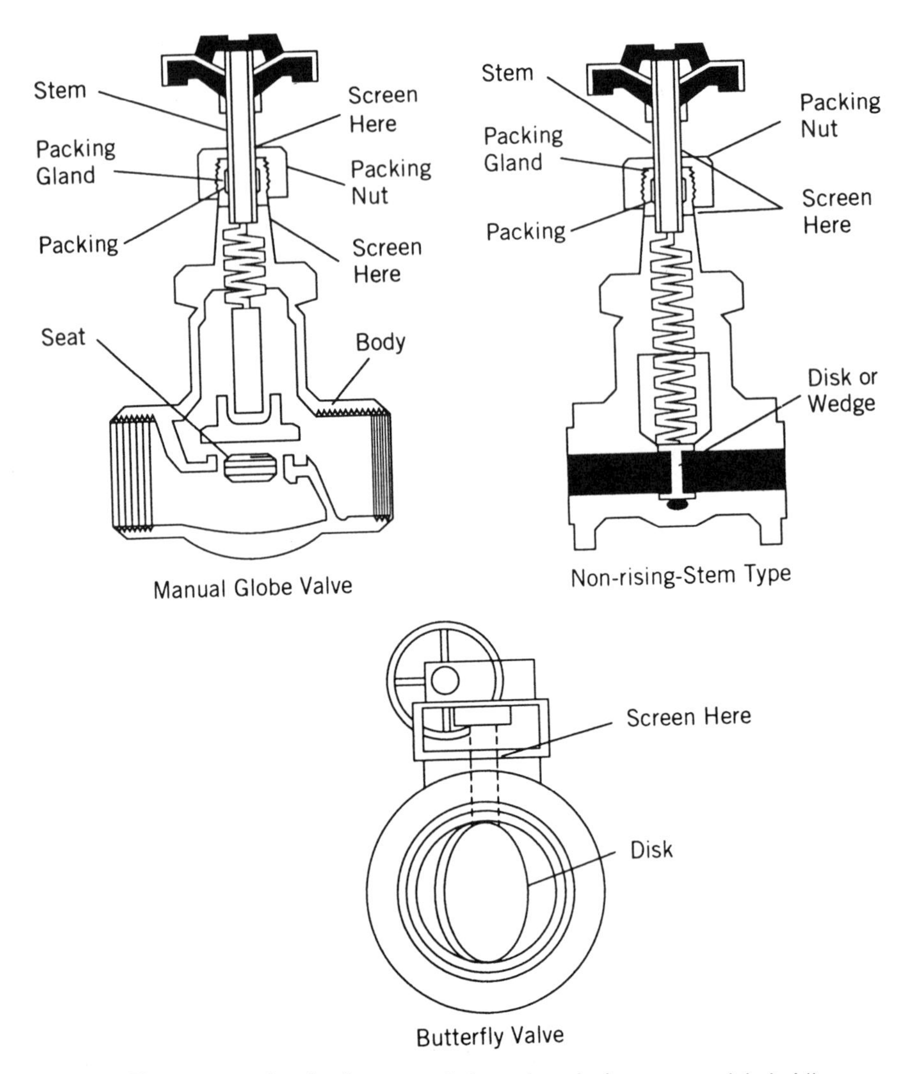

Figure 7-4. Three conventional valve types. Points where leaks occur are labeled "screen here" (US EPA, 1993).

usually occur through the packing gland around the stem. Three conventional valve types are shown in Figure 7-4. The points labeled "screen here" in this figure show where leaks occur and where screening with an OVA is required for LDAR programs. The expertise of the person performing the valve installation and the quality of the valve manufacturing process both affect valve leak rates. Valves with bent or nicked stems cannot be relied on to perform reliably in terms of leaks, and in these cases at least the stem must be replaced. Valve packing technologies that use rings to keep the packing

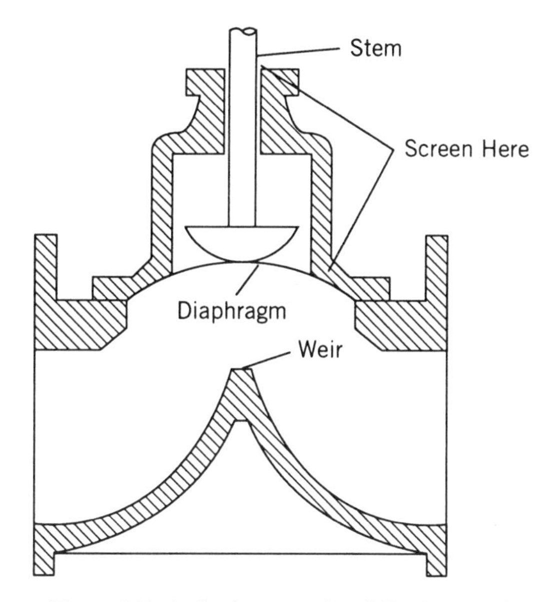

Figure 7-5. A diaphragm valve (US EPA, 1993).

from extruding and springs to maintain the packing under constant pressure (and thus in constant contact with the stem) have been developed. These systems can reduce leak rates and operate without maintenance for 10–50 times as long as valves with conventional packing (Brestel et al., 1991).

There are two main types of "sealless" or "leakless" valves that have no emissions through the stem. They are bellows valves, which are expensive and used mostly in the nuclear power industry, and diaphragm valves, which separate the valve stem from the process fluid through the use of a diaphragm. The diaphragm in some designs, such as the one of Figure 7-5, serves as the flow control device in addition to forming a barrier between the stem and the process fluid. If a diaphragm fails, emissions result, and packing is sometimes used as a backup for the diaphragm. Repair of a faulty diaphragm cannot be made without removing the valve from service. It can be difficult to replace conventional valves with leakless technology if there are significant space constraints.

As with valves, emissions from pumps occur largely around the moving parts; releases occur where the pump shaft meets the stationary casing. Two pump types are depicted in Figure 7-6. Packed seals are used, but well maintained mechanical seals generally leak less. However, mechanical seals are costly and time-consuming to repair, and sudden failure of a mechanical seal can result in large emissions. Because of this, mechanical seals are often backed up by either more mechanical seals or packed seals. When dual mechanical seals are used, a barrier fluid may be circulated between the seals to further reduce fugitive emissions. This barrier fluid must be treated to

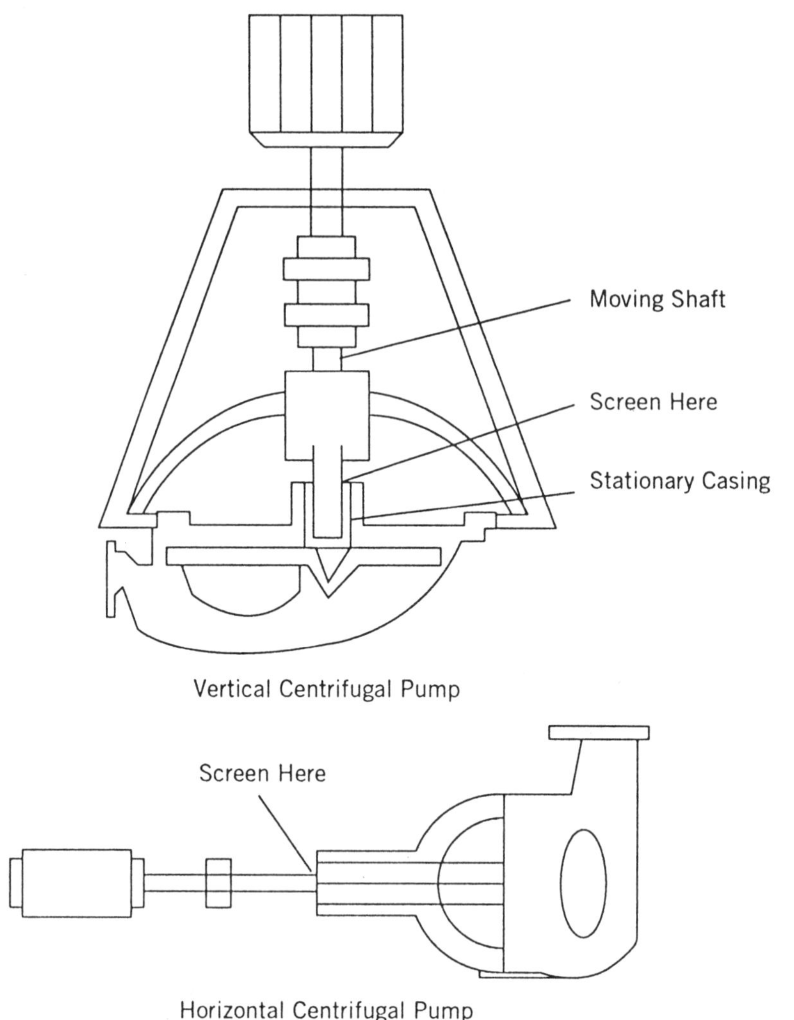

Figure 7-6. Two types of pump. Points labeled "screen here" are where leaks occur (US EPA, 1993).

remove process fluid. Table 7-3 shows that leakage indices for various types of pump seals vary by many orders of magnitude. Mechanical seals for pumps have improved greatly in the last few years and are a viable alternative to leakless technology in a wide variety of applications (Adams, 1994).

Sealless designs for pumps include the canned motor pump, where the pump bearings run in the process fluid; and the diaphragm pump, where a flexible diaphragm is used to drive the process fluid. There are also magnetic drive pumps in which the impeller is driven by magnets.

Compressors are similar to pumps in that they generally have rotating or

Table 7-3. Leakage indices for a centrifugal pump with different types of seals

Seal Type	Leakage Index
Packing with no sealant	100
Packing with sealant	10
Single mechanical seal, flushed	1.2
Tandem seal	0.15
Double seal	0.004

Source: W. V. Adams, "Control Fugitive Emissions from Mechanical Seals," *Chem. Eng. Prog.*, **87**(8), 36–41, Aug. 1991. Reproduced with permission of the American Institute of Chemical Engineers.

reciprocating shafts. Like pumps, they move process fluid, but it is in the form of a gas instead of a liquid. Again, packed and mechanical seals are used, but the use of packed seals is largely restricted to reciprocating compressors. Mechanical seals for compressors are not necessarily of the contact design used for pumps. Restrictive carbon rings and labyrinth-type seals that are composed of interlocking teeth are also used. Another type of seal used in compressors is a liquid film seal, in which an oil film is placed between the rotating shaft and a stationary gland.

Pressure relief devices are used to prevent operating pressure from exceeding the maximum allowable limit of the equipment. One type of pressure relief device is a valve that opens when the operating pressure exceeds a certain limit and closes when levels are safe again. These valves can leak because they did not reseat properly or because the operating pressure is near their limit and they are "simmering" (popping open and closed). Another type of pressure relief device is the rupture disk, which is leakless under normal operation. A rupture disk bursts when the operating pressure exceeds its limit, allowing process fluid to escape until a new disk is installed. Rupture disks can be mounted upstream of pressure-relief valves to eliminate emissions from poorly seated valves. Careful equipment design and proper process operation corrects the problem of simmering relief valves. Some pressure-relief valves have an improved "soft" seat that seals better on reseating.

Another class of components from which fugitive emissions originate is open-ended valves and lines. Drain valves, purge valves, and vent valves fall into this category. Process fluids leak when the valves are in poor repair or not fully closed. A pipe plug, cap, or blind flange can be installed over the open end to prevent emissions, or a second valve can be installed.

Sampling systems are used to verify that a process unit is operating properly. They must be purged before sampling in order to obtain a representative sample. If the purge stream is drained to the ground or sewer, it is an

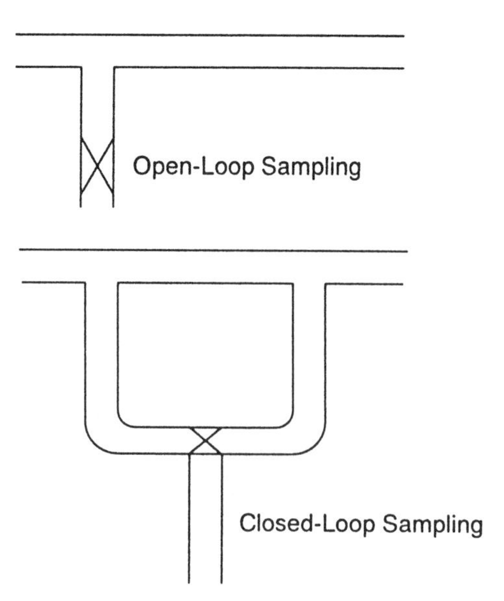

Figure 7-7. Closed-loop sampling.

episodic fugitive emission. The purge stream can be eliminated by modifying the sampling system so that the purge stream is routed back to the process, as shown in Figure 7-7. Such sampling systems are called *closed-loop sampling systems*.

7.1.2.3 *Effectiveness of Prevention Measures* The control effectiveness of various fugitive-emission reduction measures is given in Table 7-4 as a percentage of reduction. From the data of Figures 7-1 and 7-2 and the percentages in Table 7-4, it can be shown that by implementing the most effective of the control techniques, the fugitive hydrocarbon emissions from a moderate capacity refinery can be reduced by 71%, from 2400 to 690 tons/yr, and that the emissions from a typical SOCMI unit can be reduced by 66%, from 150 to 50 tons/yr. However, the control techniques most effective in reducing emissions are often much more expensive than techniques that reduce emissions by a smaller amount. For fugitive emissions, and many other types of emissions, eliminating the last few pounds costs considerably more than eliminating the majority of the emissions (Dimmick and Hustvedt, 1984). With pumps, for example, dual mechanical seals have been estimated to have an amortized annual expense roughly 10 times that of quarterly or monthly LDAR programs. Quarterly and monthly leak detection programs for valves can result in savings, with the monthly programs generating greater emission reductions than the quarterly programs. Leak detection and repair programs for pressure-relief devices do not have nearly the control effectiveness of venting the device to a flare and using a rupture disk,

Table 7-4. Effectiveness of various fugitive emission reduction techniques

Equipment	Control Technique	Control Effectiveness, %	
		SOCMI	Petroleum Refineries
Pumps, light liquid service	Dual mechanical seals	100	100
	Monthly leak detection and repair	60	80
	Quarterly leak detection and repair	30	70
Valves, gas/light liquid service	Monthly leak detection and repair	60	70
	Quarterly leak detection and repair	50	60
Pressure-relief devices	Tie to flare; rupture disk	100	100
	Monthly leak detection and repair	50	50
	Quarterly leak detection and repair	40	40
Open-ended lines	Caps, plugs, blinds	100	100
Compressors	Mechanical seals; vented degassing reservoirs	100	100
Sampling connections	Closed purge sampling systems	100	100

Source: Dimmick and Hustvedt (1984).

but they can result in annual savings. A common theme in fugitive-emission reduction programs throughout SOCMI facilities as well as in petroleum refineries is that good work practices and LDAR programs result in large emission reductions at a fraction of the cost of replacing equipment with that of leakless design.

7.1.3 Industry Experience in Reducing Fugitive Emissions

If average emission factors are used to estimate fugitive emissions from a typical SOCMI unit, they amount to 500–1500 g/Mg of product (Berglund and Hansen, 1990). However, the operators of plants handling chemicals that are toxic, explosive, or otherwise hazardous feel that these factors do not adequately describe fugitive emissions from their facilities. In addition to predicting incorrect emission quantities for these facilities, average SOCMI fugitive emission factors do not always accurately indicate the equipment type responsible for most fugitive emissions. The factors were developed to provide a realistic gauge of aggregate hydrocarbon emissions, but there are large differences in process operating conditions and procedures between different chemical manufacturing plants. In this section, a discussion of fugitive emissions for several industries, including some that manufacture hazardous chemicals, is presented. An emission reduction program at two SOCMI facilities is also presented.

7.1.3.1 *Low Fugitive Emissions in the Acrolein and Phosgene Industries* Hydrocarbon concentrations adjacent to equipment were measured in a study that included two acrolein plants, thirteen 1,3-butadiene plants, 12 ethylene oxide plants, and 13 phosgene plants (Berglund and Hansen, 1990). Figure 7-8 shows the distribution of screening concentrations (from OVA readings) by industry for the facilities in the study. Fugitive emissions as a fraction of total production and the percentage of emissions contributed by components with screening values >10,000 ppm are given in Table 7-5. Of the four industries, the 1,3-butadiene industry has the highest emissions and most resembles the facilities from which the average emission factors were developed. Table 7-5 and Figure 7-8 show that the phosgene and acrolein industries have achieved outstanding fugitive-emission control. In fact, no components were found to have a screening value of >1000 ppm in the phosgene plants, and only a small portion of components in the acrolein manufacturing facilities had screening values above this level.

When an examination was made to determine the reasons for the excellent performance in the acrolein and phosgene industries, it was discovered that the operators of these facilities are intolerant of leaks at a level far below the 10,000-ppm definition of leaking equipment. The staff are trained to be sensitive to leaks and to actively watch for them. Leaks are promptly repaired, even if unit shutdown is required. Procedures designed to minimize

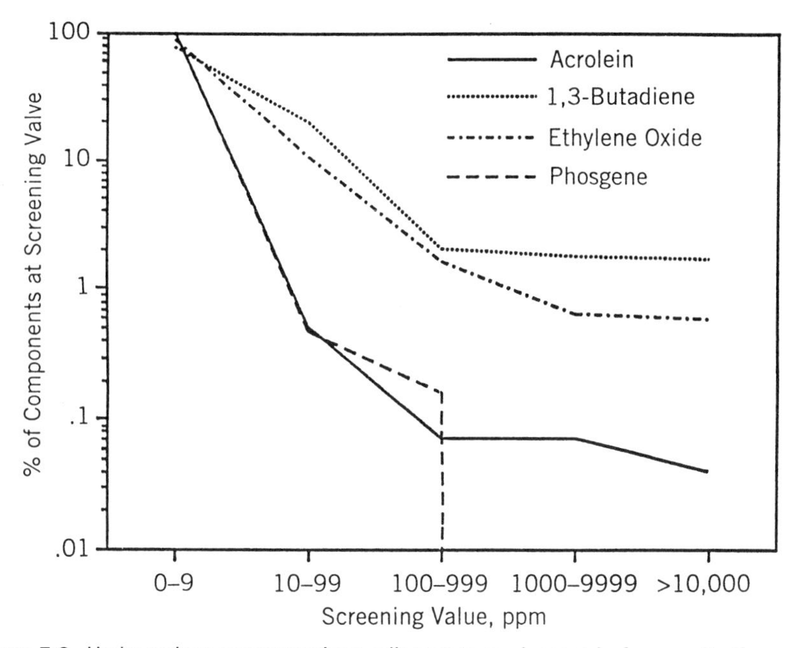

Figure 7-8. Hydrocarbon concentrations adjacent to equipment in four synthetic organic chemical manufacturing industries.

Table 7-5. Emission estimates for four SOCMI process units

Process Unit	Fugitive Emissions as a Fraction of Total Production, g/Mg[a]	Percentage of Emissions Attributable to Components with Screening Values >10,000 ppm
Typical process	500–1500	—
1,3-Butadiene	450	60
Ethylene oxide	35	70
Acrolein	0.5	90
Phosgene	0.1	0

[a]Estimated using EPA provided correlation curves, except for the typical process unit, which was estimated using average emission factors.

Source: Berglund and Hansen (1990).

fugitive emissions are followed: (1) units are designed for rapid startup and shutdown, (2) units are designed to contain a minimum number of components (the phosgene and acrolein facilities average 2000 components per unit, half as much as a typical SOCMI facility), (3) equipment vendors are evaluated and equipment is tested for leaks before and during installation, and (4) use of leakless technology is used sparingly but is focused on critical areas. The operators in these industries were able to apply years of experience to the reduction of fugitive emissions because they kept careful records about which manufacturers made reliable equipment, what were the best installation and test strategies, which areas of the unit required the most attention, and what might signal that a component is near failure. Low emissions are achieved not through the exclusive use of expensive leakless technology, but through careful application and maintenance of conventional technology. These industries serve as a role model for facility operators who wish to achieve low fugitive-emission rates.

7.1.3.2 *A Fugitive-Emission Reduction Program at Two Ethylene Oxide Plants* The findings of a fugitive-emission reduction study at two ethylene oxide facilities are applicable to many chemical process units. This project encompassed both episodic emissions, where an event such as equipment failure results in a sudden large release, and low-level leaks from process equipment (Berglund et al., 1990). Some of the reduction techniques discussed here apply to episodic emissions only, and some to low-level equipment leaks only. However, many of the techniques for reducing low-level equipment leaks also result in fewer episodic emissions.

Techniques followed in the program to reduce episodic emissions can be divided, as before, into two groups: work-practice changes and equipment modification. The history of episodic emissions at the ethylene oxide facilities showed that a large percentage of episodes occurred during startup and shutdown, so it was suggested that more personnel be on hand during these critical times to reduce the number of spills and leaks and to shorten response time. It was decided that some equipment modification was necessary as well, because records showed that some valve types had high failure rates or were difficult to close properly, indicating that they required replacement with a more suitable valve type. Also, single-seal pumps were often sources of episodic emissions.

Among the techniques instituted in the study for reducing both low-level leaks and episodic emissions were routine monitoring for leaks, which reduces episodic emissions because total seal failure usually does not occur suddenly. Improved gasketing techniques were followed, and the overall number of components was reduced to lessen the frequency of episodic emissions as well as the quantity of low-level equipment leaks. It was discovered that low-level equipment leaks were caused by the use of improper gasketing material, by casting flaws in the valves, by corroded parts, and by improper piping design that was putting the pipes and equipment under mechanical stress. To remedy this, faulty valves were replaced with high-quality conventional technology valves and improperly gasketed flanges were regasketed. As part of the program, new installations were tested for leaks before startup. Also, a monitoring program was instituted that included not only leak detection and repair, but record keeping of reduction progress, valve suppliers, installation and test strategies, areas of the unit requiring attention, and indicators of component failure as well.

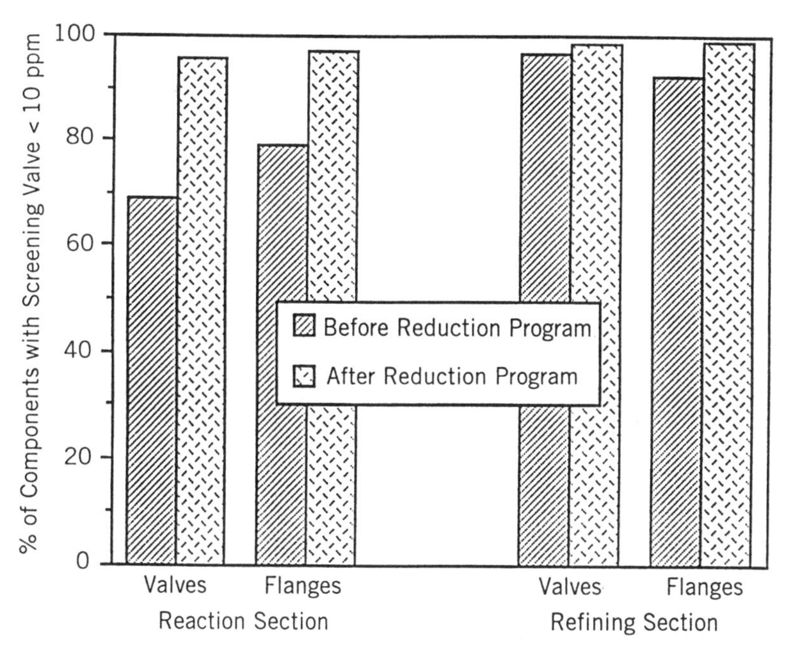

Figure 7-9. Fraction of valves and flanges with no detectable emissions in the two sections of an ethylene oxide plant before and after a fugitive-emission reduction program.

Progress in reducing low-level equipment leaks at one of the ethylene oxide plants in the study is given in Figure 7-9 for the two sections of the plant: the reaction section, where ethylene oxide concentrations are generally low, and the refining section, where concentrations are high. The figure shows the percentage of flanges and valves with screening values at or below background levels for the reaction and refining sections before and after the reduction program. The most dramatic reductions in low-level equipment leaks occurred in the reaction section, where ethylene oxide concentrations were relatively low and equipment changes were not made. Even in the refining section, equipment changes were intended more to eliminate episodic emissions than to reduce low-level equipment leaks.

This project cost $1.5 million over 2 years, with the spending broken down as follows: 15% for the screening and bagging of equipment, 55% for materials, and 30% for the labor costs of inspection, installation, and testing. No shutdown costs were incurred because replacement and repair of equipment was done during the annual turnaround periods. As is often the case, the most cost-effective control methods were those involving procedures instead of equipment.

7.2 SECONDARY EMISSIONS

As discussed in Chapter 5, secondary emissions are a result of the construction or operation of a major stationary source, but that do not come from the source itself. This includes emissions from all waste treatment operations. Reducing secondary emissions is complicated by the variety, number, and size of the sources.

Techniques for reducing secondary emissions can be divided into two groups: (1) the operating practices or equipment giving off secondary emissions are modified, and (2) the nature and/or quantity of the material responsible for secondary emissions is modified at its origin. The first of these strategies has traditionally been the focus of efforts to reduce secondary emissions. In this section, both of these methods for reducing secondary emissions are presented, but the discussion of secondary emission reduction is limited to wastewater treatment units, which are an important source of secondary emissions.

7.2.1 Wastewater Treatment

Wastewater is treated before release to make it less toxic and, in some cases, for recovery of materials. As discussed in Chapter 5, there are generally three stages of wastewater treatment: (1) primary treatment that uses physical operations to remove free oil and/or suspended solids, (2) secondary treatment for the removal of dissolved contaminants through chemical or biological action, and (3) tertiary treatment for the removal of residual contaminants.

7.2.1.1 ***Modifications to Wastewater Treatment Facilities*** As with secondary emissions in general, modification of existing treatment units or introduction of additional treatment units has traditionally been the focus of efforts to reduce emissions from wastewater treatment units. One of the most effective equipment modifications for reducing emissions from wastewater treatment plants is to add a pretreatment unit prior to the primary stages of the treatment facility to remove volatile organic compounds from the wastewater. This is commonly done through steam stripping, where fractional distillation of the wastewater stream is accomplished by directly contacting it with steam. If the volatile organics that are driven off are recycled, raw-material requirements are reduced. Another pretreatment process that can be used to remove volatile organics from wastewater is air stripping. The effectiveness of steam and air strippers is dependent on the volatility of the compounds to be removed, as shown in Table 7-6. Some pretreatment processes do not rely on a compound's volatility, including those that involve carbon and ion exchange, adsorption, chemical oxidation, membrane separation, and liquid extraction. Significant reduction in emissions can be accomplished by pretreatment of a small portion of the wastewater streams,

Table 7-6. Control effectiveness of steam and air stripping wastewater prior to treatment

Henry's Law Constant of Compound, atm-m^3/gmol	Effectiveness, %	
	Stream Stripper	Air Stripper
$>10^{-3}$	95–99	90–99
$<10^{-3}$ and $>10^{-5}$	90–95	$<$50–90
$<10^{-5}$	$<$50–90	$<$50–90

Source: US EPA (1985).

since 20% of the streams release 65% of secondary emissions (US EPA, 1990). Therefore, routing only those streams with a high potential for emission reduction through pretreatment is prudent.

Covering wastewater treatment units is another method of modifying equipment to reduce secondary emissions. The main objective of covering is to ensure that volatile organic compounds remain in the wastewater until they reach the stage of treatment at which they are destroyed, usually by biodegradation. Modifications include roofs, floating membranes, air-supported structures, and various collection system controls. If necessary, buildup of explosive mixtures is prevented by venting to a control device or to the atmosphere. Even if a unit that is covered with a roof is vented to the atmosphere, emissions are reduced because the surface is sheltered from wind, decreasing the mass transfer rate and maintaining a concentration of volatile species in the gas that is in contact with the wastewater, which inhibits additional volatilization.

The mass transfer equations of Table 5-13, some of which were used in Example 5-4 to estimate emissions from an API separator, can also be useful in developing preventive strategies for specific units. For example, if the emission rate from a unit is dominated by the liquid-phase mass-transfer coefficient, modifications intended to lower the gas-phase mass-transfer coefficient will not appreciably reduce the rate of emissions.

Example 7-2 Using Mass-Transfer Modeling of an API Oil–Water Separator to Identify Emission Prevention Opportunities Emissions for an API separator are modeled by

$$E = K_{\text{oil}} C_{\text{L,oil}} A,$$

where K_{oil} is the overall mass-transfer coefficient for transfer of a compound from the oil phase to the gas phase, $C_{\text{L,oil}}$ is the concentration of the compound in the oil phase, and A is the separator surface area. Explain what might be done to reduce emissions for the separator in terms of the above equation.

Solution Reducing area (A) would reduce emissions (E), but may not be possible. Reducing wind velocity by installing freeboard or a cover would reduce K_{oil} and therefore reduce E. (Many API separators are now enclosed, with precautions taken against explosion.) Reducing the concentration of the pollutant in the oil phase through pretreatment would reduce E.

Emission estimates for units in a wastewater treatment facility can also be used to identify which areas are responsible for the majority of emissions. This makes it possible to carefully consider the units whose modification would be the most effective in reducing emissions. It can be surprising where large emissions occur. At one facility for treating the wastewater from a large petrochemical plant, it was estimated that a single weir with a 4-ft drop was responsible for more than 15% of the emissions (Berglund and Whipple, 1987).

Although high levels of emission reduction can be achieved with the practices mentioned above, they can create hazardous "tertiary" waste. For example, if a wastewater treatment unit is covered and vented through activated charcoal, spent activated charcoal is a resultant waste. The option of pretreatment by steam stripping, very effective for reduction of highly volatile compound emissions, is costly and has energy requirements. One estimate suggests that a steam stripper designed to treat 300 L/min of wastewater flow has a total annual cost of $230,000 (1986 dollars) (US EPA, 1990). When burning of fossil fuels provides the heat required to create the steam, particulate matter, SO_2, NO_x, CO, and VOC emissions result.

7.2.1.2 *Modifying Wastewater Streams* The second group of reduction strategies is to alter the nature and/or quantity of the material responsible for secondary emissions by modifying the process responsible for its creation. This falls at the top of the waste management hierarchy, and requires a thorough waste audit as described in Chapter 5. By itself, physical characterization of the waste streams provides adequate information to implement the types of controls in the traditional reduction strategies, but the origin of the wastes must be understood in order to develop strategies that prevent generation of secondary emissions. The diversity of secondary-emission sources makes it impossible to provide general tactics for preventing these emissions. Each waste stream must be considered for options on a case-by-case basis.

Oily sludges are a high-volume secondary waste from wastewater treatment operations that are formed when solids in the wastewater electrostatically attract oil and water molecules to form heavy agglomerates that settle to the bottom of treatment vessels. If solids loading to wastewater treatment is reduced, sludge generation can be prevented, because solids are the limiting agent in sludge formation. If stormwater runoff and process wastewater can be collected and treated in separate systems, sludge formation is prevented because runoff generally has high solids content. Figure 7-10 shows 3 levels of sewer segregation, with part *a* depicting no segregation, and part *c* depicting the most effective level of segregation. Where segregated sewer systems are not possible, paving earthen areas in the plant and sweeping them to keep them free of dirt reduces the solids content of runoff entering the sewer during storms.

Small suspended solids and emulsifying chemicals can cause the formation of a lighter-than-water suspension of oil and water in wastewater treatment units like the DAF unit pictured in Figure 5-1. Reducing and eliminating the use of emulsifiers (such as detergents used in cleaning) that may find their way into the wastewater treatment processes therefore reduces the generation of waste emulsions. Emulsions require turbulence as well as emulsifying agents in order to form and limiting turbulence in locations where emulsifiers are present is another approach to source reduction of emulsions. For example, high-shear pumping of oily wastes can be avoided by using Archimedean screw pumps, and water sprays can be replaced with mud rakes or some other low-turbulence method of moving solids.

Mechanical, chemical, and thermal means of recovering the oil from sludges and emulsions have been developed. Figure 7-11 shows a recessed plate filter for recovery of oil from sludge. While techniques for such recovery can reduce sludge volume and return oil to the process at a profit, techniques for reducing the quantity of sludge generated in the first place are preferable. This is partly because recovery can require the use of filter aids or precoating materials that can be released with the filtrate and returned to oil–water separation steps, where they cause the formation of more sludge. Recovery methods can also require the use of chemicals such as ferric chloride and hydrated lime for destabilizing emulsions and increasing the filtration rate.

At one large refinery, a study aimed at developing pollution prevention options uncovered several specific tactics for reducing waste generated in the wastewater treatment unit (Balik and Koraido, 1991). The study team proposed that the coke fines and the water generated in the coke cutting operation be separated more efficiently, reducing the amount of solid waste generated in the refinery's oil–water separator and increasing recovery of coke. This could be accomplished by retrofitting the original coke solids separator with an inclined plate separator, and would generate annual cost savings of $250,000. Another pollution prevention proposal was to use reformer off-gas scrubber water as makeup water for the crude-desalting units. This would reduce consumption of city water and reduce the load to the wastewater treatment unit, generating an annual cost savings of $450,000. Capital costs for this option were projected to be paid back in approximately 6 months. A third pollution prevention proposal suggested by the study team involved reduction of spills where the coking unit loaded coke into railroad cars. Spills resulted both in degradation of the coke product and in increased solids loading to the wastewater treatment unit. Loading could be accomplished in a more controlled manner by crushing the coke and loading

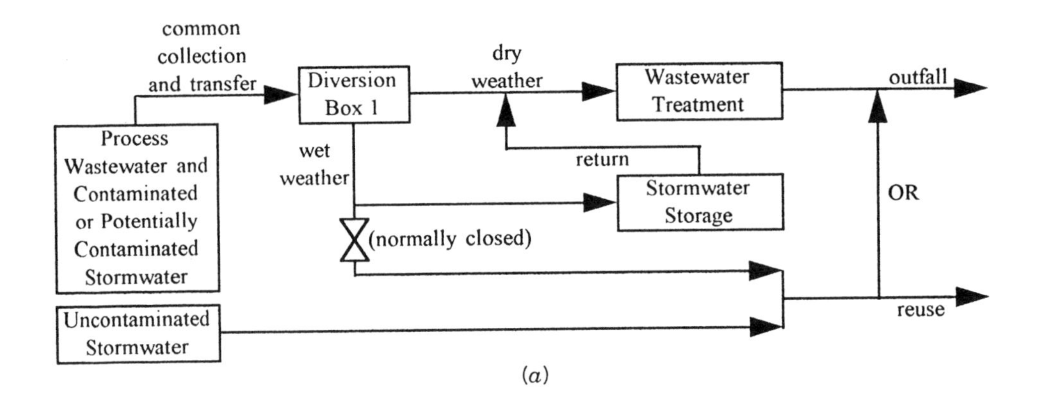

(*a*)

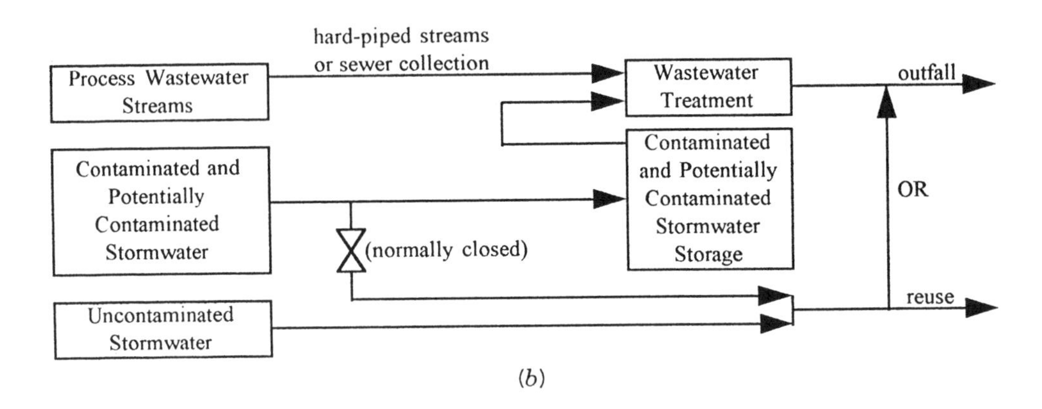

(*b*)

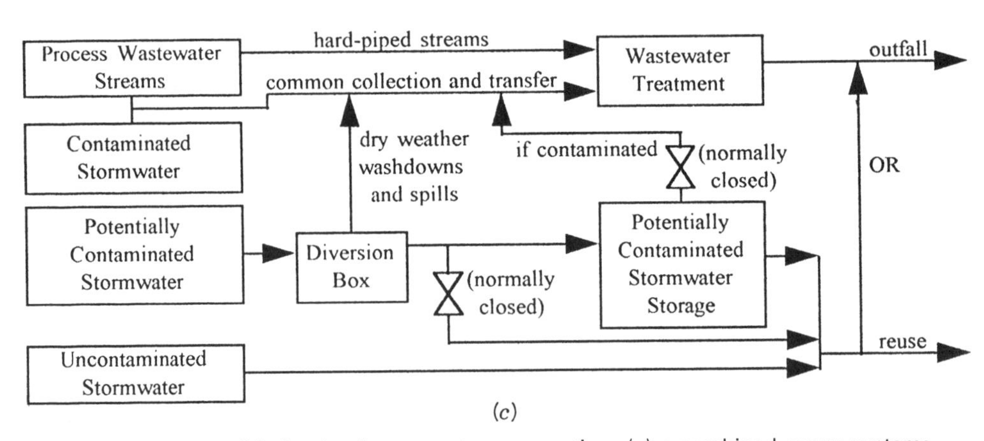

(*c*)

Figure 7-10. Possible levels of stormwater segregation: (*a*) a combined sewer system; (*b*) contaminated and potentially contaminated runoff are combined; (*c*) contaminated, potentially contaminated, and uncontaminated stormwater are segregated. (From D. Garg and R. B. Pair, Jr., "Effectively Manage Stormwater in a CPI Complex," *Chemical Engineering Progress*, **91**(5), 70–76, May 1995. Reproduced with permission of the American Institute of Chemical Engineers. Copyright © 1995 AIChE. All rights reserved.)

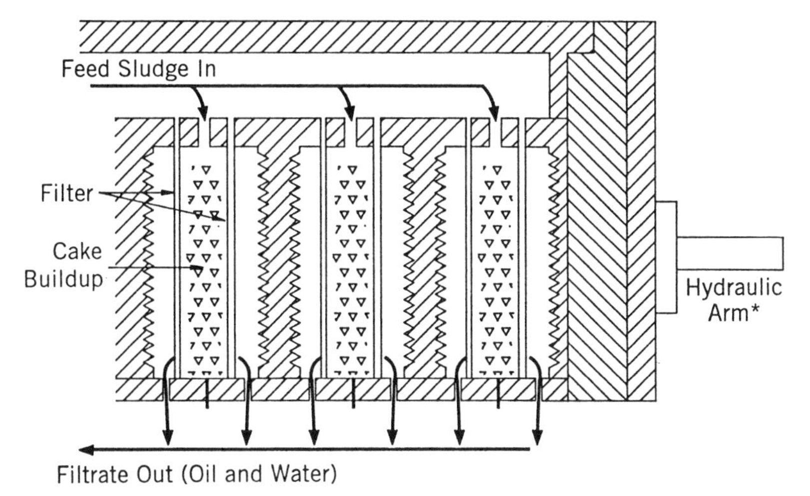

Figure 7-11. Operation of a recessed plate filter. (From *Waste Minimization in the Petroleum Industry: A Compendium of Practices*, Publication 849–00020, 1991. Reprinted courtesy of the American Petroleum Institute.)

it into an intermediate silo. Annual cost savings for this option as a result of a reduction in solid waste from the oil–water separator and an increase in coke recovery were projected to be $900,000 with a payback period of one year. These pollution prevention strategies could only be developed after a complete waste audit detailing the causes of pollution in each waste stream was performed.

7.3 SUMMARY

In reducing fugitive emissions, conscientious leak detection and repair programs have proven to be extremely effective at a fraction of the cost of replacing conventional equipment with "leakless" technology components. Besides being expensive, changing to leakless technology is not always feasible, and reduces emissions over well-maintained, high-quality conventional equipment only marginally. Reduction of secondary wastes requires an examination of upstream sources of the waste and an understanding of the mechanisms of formation. Reduction of secondary emissions can be accomplished through alteration of waste treatment equipment, but these reductions are often accompanied by cascading waste treatment technologies. Straightforward and substantial reductions in secondary emissions can result from modifications to the process responsible for the waste stream.

QUESTIONS FOR DISCUSSION

1. In Problem Statement 4 of Chapter 5 you were asked to estimate the emissions from a junction box. What resistance dominated this mass-transfer scenario? What resistances can be neglected? What would lower the mass-transfer coefficients, thus lowering emissions? Consider temperature, turbulence, the dimensions of the junction box, and air movement. How might emissions be reduced without affecting the mass-transfer coefficients?
2. How would the cost of reducing fugitive emissions behave as emissions approached zero? How would the quantity of fugitive emissions behave as the costs of reducing them approached zero?

PROBLEM STATEMENTS

1. Repeat the calculations of Example 5-4 for a covered oil–water separator (wind speed = 0 m/s). How does this compare to the estimate for an uncovered separator?
2. The estimated cost-effectiveness for various fugitive emission control techniques at refineries is given in Table 7-7. Determine the most cost-

Table 7-7. Cost-effectiveness of controls for equipment leaks at petroleum refineries

Equipment Type	Control Technique	Average $/Mg[a]
Pressure-relief valves	Quarterly leak detection and repair	(170)[b]
	Monthly leak detection and repair	(110)
	Rupture disks	410
Compressors	Controlled degassing vent	150
Open-ended lines	Caps on open-ended lines	460
Sampling systems	Close purge sampling	810
Valves	Quarterly leak detection and repair	(110)
	Monthly leak detection and repair	(60)
	Sealed bellows valves	4700
Pumps	Annual leak detection and repair	860
	Quarterly leak detection and repair	157
	Monthly leak detection and repair	158
	Dual mechanical seal systems vented to a flare	2000

[a]Average dollars per megagram (cost-effectiveness) = (annualized cost per component) ÷ (annual VOC emission reduction per component).
[b]Values in parens denote savings.
Source: Dimmick and Hustvedt (1984).

effective way to reduce fugitive emissions from the typical refinery of Figure 7-2 by 50% using the data in this table.

OPEN-ENDED PROBLEM STATEMENT

You work at a processing plant whose fugitive emissions are approximately 500 tons/yr. At this facility, components are inspected annually for leaks and are defined as leaking if the organic concentration around them is $\geq$10,000 ppm. Components found to be leaking are repaired following inspection. You have proposed that your facility institute a more aggressive leak detection and repair program to reduce fugitive emissions, saying that your company can reduce fugitive emissions dramatically while breaking even economically at a more frequent monitoring frequency and/or a lower leak rate definition. A colleague refutes this, saying that a leak-detection-and-repair (LDAR) study was performed at the facility in 1984, and the data indicated that the economically optimum LDAR program is one that is based on annual inspection with a leak definition of 10,000 ppm. Your management is convinced that more frequent inspections and/or a lower leak rate definition would be a poor use of your facility's pollution prevention dollars. You, however, are aware that the regulatory costs of emitting reactive organic gases in your area have increased by an order of magnitude in the last decade, as they have in the South Coast Air Quality Management District in California, whose approximate emission fees are shown in Figure 7-12.

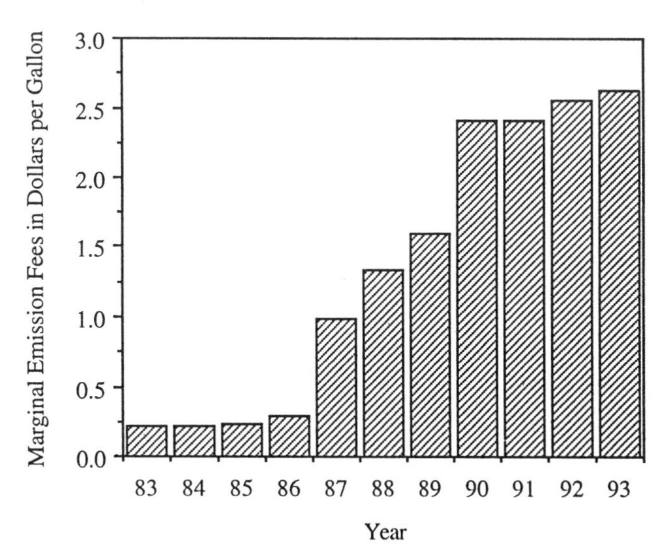

Figure 7-12. Historical emission fees for reactive organic gases in the South Coast Air Quality Management District in California (Fisher, 1994).

Table 7-8. Regulatory emission fees and product value for organic gases lost as fugitive emissions

	Year	
Cost, $/gal	1984	1994
Lost product	0.70	0.50
Emission fees	0.20	2.60
Total	0.90	3.10

Source: Fisher (1994).

Table 7-9. LDAR costs and fugitive emissions at a midsize refinery

LDAR Program			
Monitoring Frequency	Leak Definition, ppm	Annualized Cost, $1000/yr	Projected Fugitive Emissions, tons/yr
Annual	10,000	92	480
Quarterly	10,000	140	290
Quarterly	500	200	88

Source: Amoco/EPA (1992).

Your task is to propose an LDAR program, give the projected emissions, and show why the program you suggest is economically optimal. Use the emission fees and product values given for 1984 and 1994 in Table 7-8. The density of the product at your facility is close to that of water. Assume in your analysis that the cost of LDAR programs has not risen significantly since the 1980s. Because the component profile of your facility resembles the component profile at a midsize refinery, you can use the economic and emissions data in Table 7-9 in your analysis. With no LDAR program, this refinery had emissions of 800 tons/yr, but even with no LDAR program approximately $20,000/yr was spent on repairing and replacing leaking components. If emission reduction achieved through installation of leakless components costs roughly $20/lb (SOCMA, 1993), how do the costs of your LDAR program compare to replacement of components with leakless components? As part of your analysis, prepare a list of non-economic benefits that your program will provide. Do not use the values from Table 7-7 when solving this problem.

REFERENCES

Adams, W. V., "Control Fugitive Emissions from Mechanical Seals," *Chem. Eng. Prog.*, **87**(8), 36–41, Aug. 1991.

Adams, W. V., "Stop Leaks Economically with Advanced Mechanical Seals," *Chem. Eng. Prog.*, **90**(6), 24–31, June 1994.

American Petroleum Institute (API), "Waste Minimization in the Petroleum Industry: A Compendium of Practices," Publication 849-00020, Washington, DC, 1991.

Amoco/US EPA, "Pollution Prevention Project, Yorktown, VA: Project Summary," available through NTIS as PB92-228527, June 1992.

Balik, J. A. and S. M. Koraido, "Case Study: Identifying Pollution Options for a Petroleum Refinery," *Pollution Prevention Review*, **1**(3), 273–293, Summer 1991.

Berglund, R. L. and G. M. Whipple, "Predictive Modeling of Organic Emissions," *Chem. Eng. Prog.*, **83**(11), 46–54, Nov. 1987.

Berglund, R. L. and J. L. Hansen, "Fugitive Emissions: An Untapped Application for TQC," *ASQC Quality Congress Transactions*, San Francisco, CA, 1990.

Berglund, R. L., W. J. Mollere, C. R. Ritter, and L. W. Salathe, "A Quality Program to Reduce Episodic and Fugitive Emissions from an Ethylene Oxide Unit," paper for presentation at the Summer Meeting of the American Institute of Chemical Engineers, Aug. 19–22, 1990.

Brestel, R. et al., "Minimize Fugitive Emissions with a New Approach to Valve Packing," *Chem. Eng. Prog.*, **87**(8), 42–47, Aug. 1991.

Dimmick, W. F. and K. C. Hustvedt, "Equipment Leaks of VOC: Emissions and Their Control," Paper No. 84-62.1 presented at the 77th Annual Meeting of the Air Pollution Control Association, San Francisco, June 24–29, 1984.

Fisher, R., presentation at the Oil Industry Roundtable, EPA Region 9, San Francisco, April 13, 1994.

Garg, D. and R. B. Pair, Jr., "Effectively Manage Stormwater in a CPI Complex," *Chem. Eng. Prog.*, **91**(5), 70–76, May 1995.

Klee, H. (Amoco Yorktown Refinery), personal communication, 1991.

Synthetic Organic Chemical Manufacturers Association (SOCMA), "SOCMA Pollution Prevention Study," prepared by Woodward-Clyde Consultants, Project 90X4368, Wayne, NJ, Jan. 1993.

United States Environmental Protection Agency (US EPA), "Fugitive Emission Sources of Organic Compounds—Additional Information on Emissions, Emission Reductions, and Costs," Publication EPA-450/3-82-010, Research Triangle Park, NC, 1982.

U.S. Environmental Protection Agency (US EPA), "Compilation of Air Pollutant Emission Factors, Volume I: Stationary Point and Area Sources," 4th ed. with Supplements A to D, Publication AP-42, Research Triangle Park, NC, Sep. 1985.

United States Environmental Protection Agency (US EPA), "Emission Factors for Equipment Leaks of VOC and HAP," Publication EPA-450/3-86-002, Research Triangle Park, NC, Jan. 1986.

United States Environmental Protection Agency (US EPA), "Industrial Wastewater Volatile Organic Compound Emissions—Background Information for BACT/LAER Determinations," Research Triangle Park, NC, Publication EPA450/3-90-004, Jan. 1990.

United States Environmental Protection Agency (US EPA), "Protocol for Equipment Leak Emission Estimates," available through NTIS as PB93-229219, June 1993.

8

FLOWSHEET ANALYSIS FOR POLLUTION PREVENTION

Previous chapters in this text describe modifications to the design and operation of individual unit operations to reduce wastes and emissions. This approach implicitly assumes that a process with an established sequence of unit operations exists. The ways in which the sequencing and networking of unit operations influences waste generation are examined in this chapter, and a set of methods for synthesizing or restructuring process flowsheets are described. In the first section of this chapter, tools that are largely qualitative in concept and implementation are described. A more quantitative and rigorous flowsheet analysis tool called mass exchange network synthesis is discussed in the second section.

8.1 QUALITATIVE METHODS

In this section, two elements of qualitative flowsheet analysis for pollution prevention are described. The first is material flow analysis, where mass balances are used to identify pollution prevention opportunities. Next, frameworks for analyzing flowsheets are described. The analyses examine modifications to existing processes and the design of new processes.

8.1.1 Material Flow Analysis

The simplest analysis tool for waste minimization is the mass balance. While most process engineers are comfortable with the concept of a mass balance, a number of new issues arise when mass balances are applied to pollution prevention. Two key issues are defining system boundaries and determining the level of detail required in the mass flow analysis.

8.1.1.1 ***Defining System Boundaries*** The first issue that must be unambiguously addressed in performing a mass flow analysis of a flowsheet is the definition of system boundaries. In Section 6.3 (of Chapter 6), the discussion of storage tank wastes illustrated this important issue. A flowsheet for tank bottom waste is given in Figure 8-1. Recall that the American Petroleum Institute (1991) noted that some refineries reduce tank bottoms waste by adding emulsifying agents to the tanks. While this certainly reduces tank sludge, the emulsifying agents might increase downstream wastes. In this case, if the system boundary is just the storage tank, a single unit operation, then the net impact of adding an emulsifier is clear: less tank sludge is generated when the emulsifier is added. If the system that is chosen is larger, and includes both the tank and the downstream processing unit, then the impact of adding the emulsifier is less clear: the emulsifying agent may create more waste downstream than it prevents in the tank. In this case, the definition of the system boundary plays a key role in determining whether a strategy does, in fact, prevent pollution.

As another example of the impact of the choice of system boundaries,

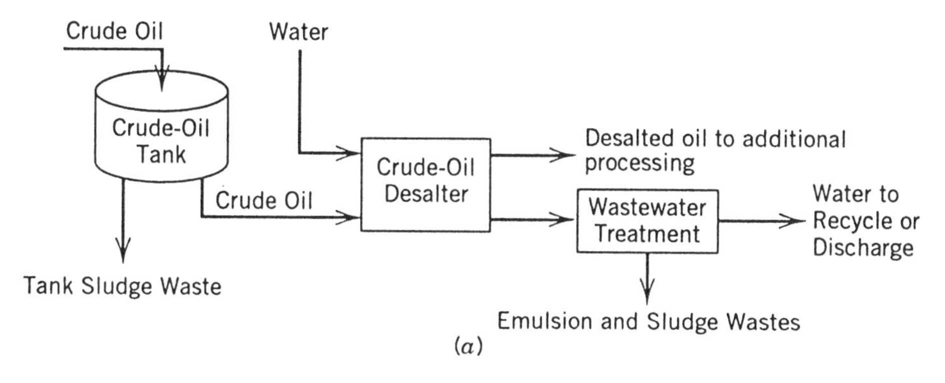

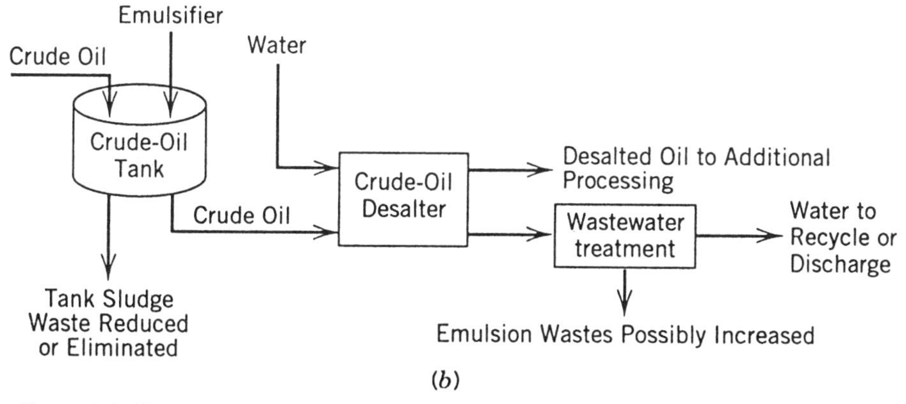

Figure 8-1. The emulsifying agent, added to the tank to reduce sludge, might result in more waste being formed when the crude oil is fed to the desalter: (*a*) before; (*b*) after.

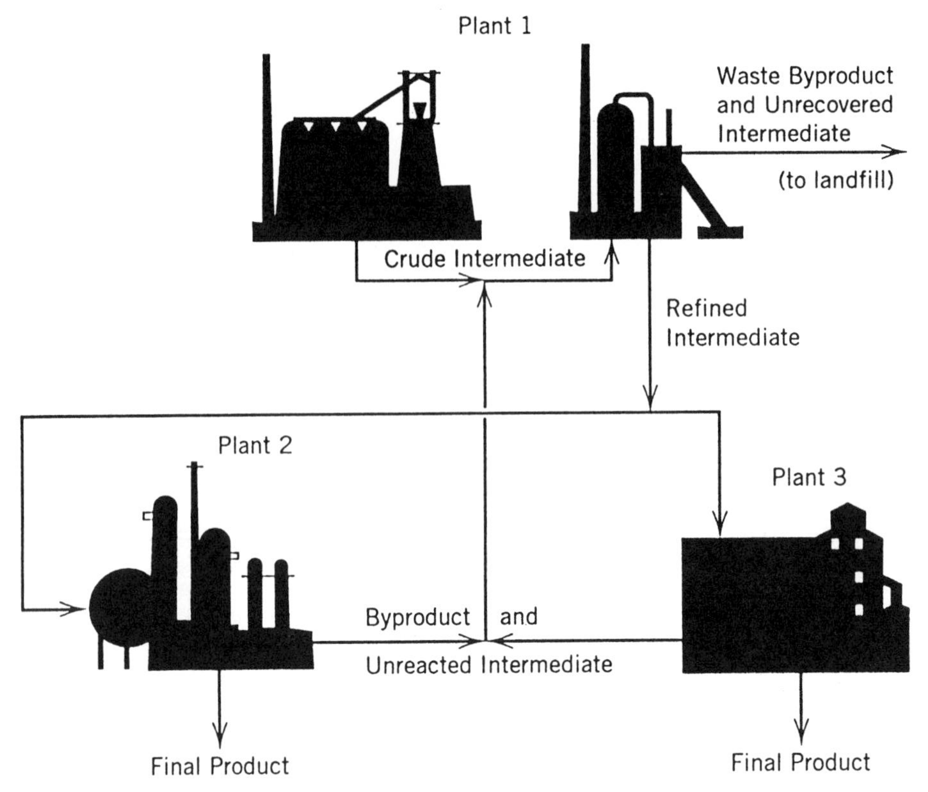

Figure 8-2. Before a proposed pollution prevention modification, plant 1 produces 17.5 million lb/yr of waste while plants 2 and 3 produce no waste (Tellus Institute, 1991).

consider the example of a large, integrated chemical manufacturing facility that operates plants 1, 2, and 3 (Tellus Institute, 1991). Plant 1 converts raw materials into an intermediate, which is then consumed by plants 2 and 3 to produce a final product. In the current configuration, shown in Figure 8-2, plant 1 produces 17.5 million lb/yr of wastes, while plants 2 and 3 produce no wastes. A proposed process modification, shown in Figure 8-3, would decrease the waste from plant 1 by 4.3 million lb/yr but would increase the wastes from plant 2 by 0.5 million lb/yr. Clearly, if the integrated operation of plants 1, 2, and 3 is considered as the flowsheet, then a simple mass balance reveals that this process modification would reduce wastes by 3.8 million lb/yr. On the other hand, if each plant is treated as an individual facility, responsible for the costs and management of its own wastes, then it would be very difficult to convince plant 2 to accept this process modification.

As yet another example of the importance of specifying system boundaries, consider a petroleum refinery that has agreed to accept used crankcase oil from consumers in a product stewardship campaign. By accepting this

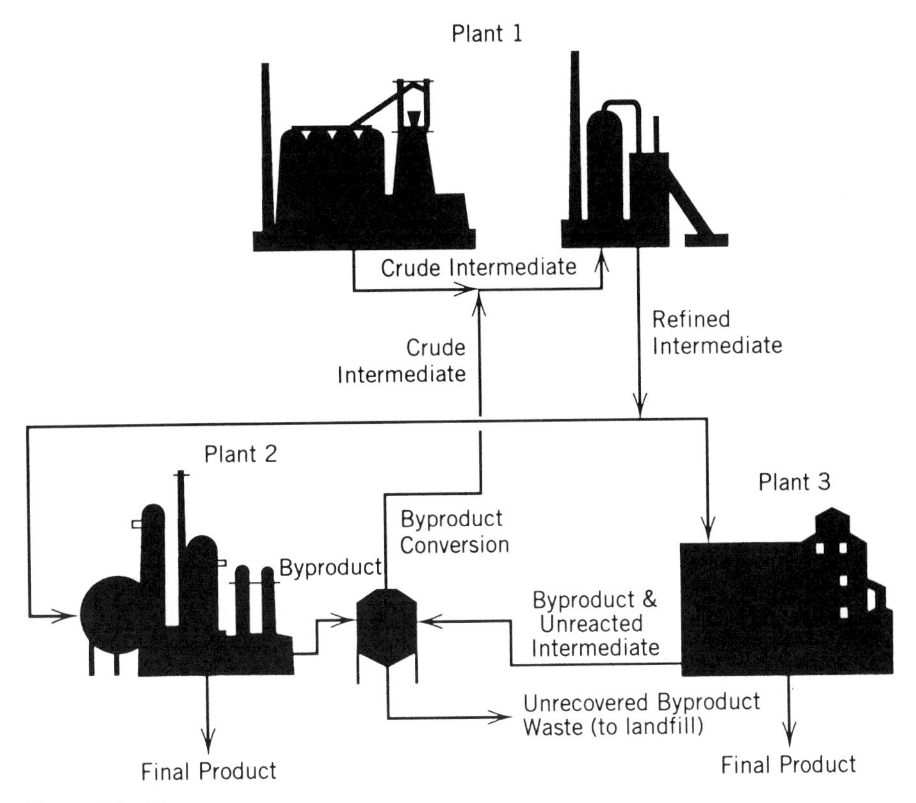

Figure 8-3. After a proposed pollution prevention modification at the manufacturing operation of Figure 8-2, plant 1 and plant 2 produce 13.2 and 0.5 million lb/yr of waste per year, respectively (Tellus Institute, 1991).

oil, which contains much higher levels of metals and other contaminants than most crude oils, the refiner increases both the quantity of waste generated in refining and the quantity of waste per unit of production. If the product stewardship campaign were viewed strictly from the standpoint of the refinery, the project would not proceed. If, on the other hand, one considers all of the wastes generated by the motor oil, from crude oil extraction to the handling of the used product, then it is clear that motor oil recycling prevents pollution. In this case, the desirability of a pollution prevention project depends on whether the mass flows are evaluated around the entire life cycle of a product, or around a single facility.

In all of these examples there is no correct or incorrect choice of system boundaries. The important points to note are that the system boundaries must be clearly defined, and that their specification can have a profound impact on the viability of a pollution prevention project.

8.1.1.2 *Mass Balances and Waste-Stream Measurements* After the system boundaries are selected, a second major decision that must be made in analyzing a process flowsheet is whether detailed mass-balance techniques yield significant information about waste flows. In some cases, mass-balance data are almost irrelevant for establishing waste flows. For example, in an API separator, the emissions to the atmosphere of benzene, toluene, ethylbenzene, and xylenes might be less than 0.01 kg/hr. If the flow rate of the oil from which these compounds were evaporating is over 10,000 kg/hr, estimating emissions using concentrations and mass flows into and out of the separator is unlikely to yield an accurate estimate of emissions, because the waste flows are over a million times lower than the process flows. The concentrations and flow rates would need to be accurate to six or more significant figures to provide acceptable data. This is typical in industries that produce bulk, commodity products (e.g., petroleum refining, chemical manufacturing, pulp and paper manufacturing, and mining/ore refining). In contrast, industries that produce high-value-added products, such as the pharmaceutical and electronics industries, may find mass balances an effective procedure for measuring emissions and wastes. For example, the most straightforward method for accounting for solvent losses from a parts cleaning operation may be to keep track of all solvent purchases and assume that most of the solvent is lost to evaporation or sludge formation.

8.1.2 Frameworks for Examining Flowsheets of Existing Processes for Pollution Prevention Opportunities

The design of a process for minimization of waste and the examination of an existing process for pollution prevention opportunities can both benefit from a qualitative analysis of the process flowsheet. In this section, procedures are discussed for analyzing existing processes.

After the boundaries of the process flowsheet have been established, and once waste flows and emissions have been estimated (whether by mass balances or by other methods), a flowsheet can be examined for pollution prevention opportunities. At this point the flowsheet analysis becomes analogous to flowsheet analysis for process safety, frequently referred to as HAZard and OPerability analysis (HAZ-OP). Well established procedures for HAZ-OP analysis exist (e.g., see Crowl and Louvar, 1990). The potential hazard associated with each process stream is evaluated qualitatively or quantitatively by systematically considering possible failure scenarios. A parallel analysis for pollution prevention would systematically consider each waste stream, posing questions such as

- Could the waste stream be eliminated with a change in raw materials?
- Could a change in process conditions (temperature, pressure, reactant concentrations) reduce waste generation?

- Could a change in process configuration reduce waste generation?
- Could the waste be a raw material for another process?

In addition, each unit operation in the flowsheet should be examined, and the pollution prevention options cited in previous chapters should be exploited.

To summarize, the principal steps for evaluating an existing process' flowsheet for pollution prevention opportunities are as follows:

1. Establish the bounds of the process.
2. Using mass balances or estimation procedures, perform a waste stream audit, recalling that as part of the audit, the composition of the waste stream and the mechanism of waste formation must be determined.
3. For each waste stream, systematically examine pollution prevention options.
4. For each unit operation, systematically examine pollution prevention options.
5. Evaluate the impact of each potential pollution prevention option on the entire process and on the finished product.

The following case studies illustrate how a systematic consideration of pollution prevention options can be accomplished. The case studies are drawn from a group of 15 cases prepared by DuPont and the U.S. Environmental Protection Agency (Report to the Government, 1993). Each case study begins with a description of the waste stream under consideration and the process that generates it. This description is followed by a listing of pollution prevention options and finally by an evaluation of these options.

8.1.2.1 ***Case Study of Qualitative Flowsheet Analysis for a Specialty Alcohols Wash Process*** A process flowsheet for a specialty alcohols wash process is shown in Figure 8-4. This figure shows that a batch of alcohol crude with acid residuals enters the wash kettle and is mixed with water, chemical scavenging agents, and isopropyl alcohol. The mixture is agitated and then allowed to settle. The specialty alcohol product separates from the wash and settles to the bottom of the kettle. The mixture is then drained from the bottom of the kettle through a sight glass monitored by an operator. The settled product leaves the kettle bottom first and is sent to an accumulator tank. When the operator sees that the product layer has drained and the aqueous wash has started to exit the kettle, the operator diverts the flow to a sump for disposal.

The product layer in the accumulator tank is then returned to the wash kettle for a second wash with water, scavengers, and isopropyl alcohol. Again, the mixture is agitated and then allowed to settle. The kettle is drained, with the specialty alcohol product going to the accumulator tank. Again, the operator diverts the aqueous wash to the wastewater sump for

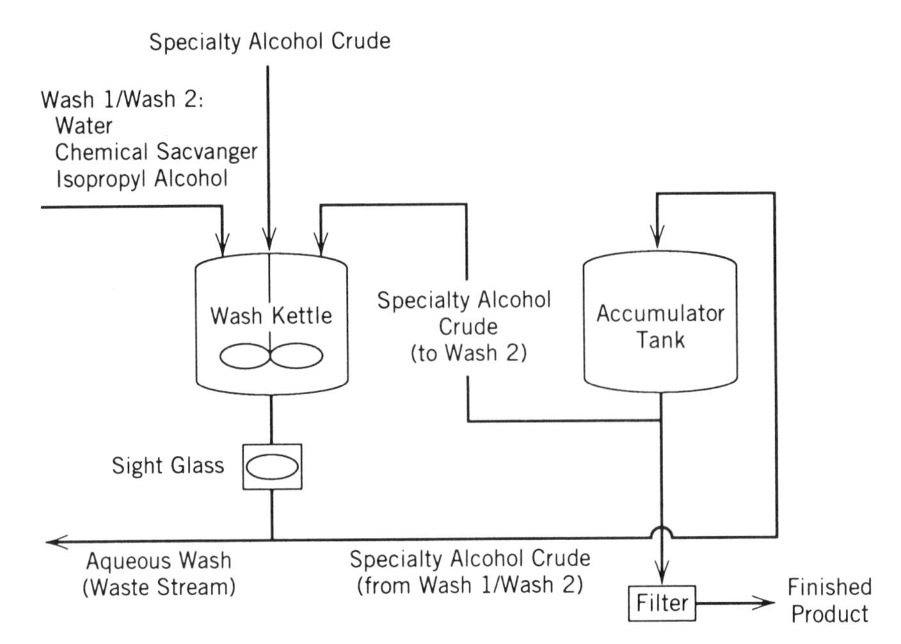

Figure 8-4. A specialty alcohols wash process (Report to the Government, 1993).

disposal. From the accumulator tank, the specialty alcohol is filtered and drummed for shipment as final product.

The waste stream from this process is typically composed of (by mass) 93.6% water, 5.1% isopropyl alcohol, 0.7% specialty alcohol, and 0.6% organic chemicals. Excluding water, isopropyl alcohol accounts for 80% of the waste stream and 94% of the total organic content. At present, the wash process produces 0.15 lb of total organic content in the waste stream for every pound of specialty alcohol produced. This waste stream is sent to the on-site wastewater treatment plant for disposal. Costs associated with the waste stream include the replacement cost of the isopropyl alcohol and the chemical scavengers, the yield loss represented by the specialty alcohol component, and the wastewater treatment cost.

A design team generated nine options for reducing waste in the specialty alcohols wash process in a brainstorming session, and recorded their ideas in a cause-and-effect chart as shown in Figure 8-5. Table 8-1 summarizes the team's informal evaluation of each option. Because acid neutralization offered a complete source reduction of waste with little capital cost, all options that prescribed variations on the present washing method were quickly eliminated. Options that prescribed changes in reaction chemistry or separation technology were poorly defined, and could not be implemented without significant research, capital expenditures, and long development times. This left acid neutralization as the only option worthy of an economic

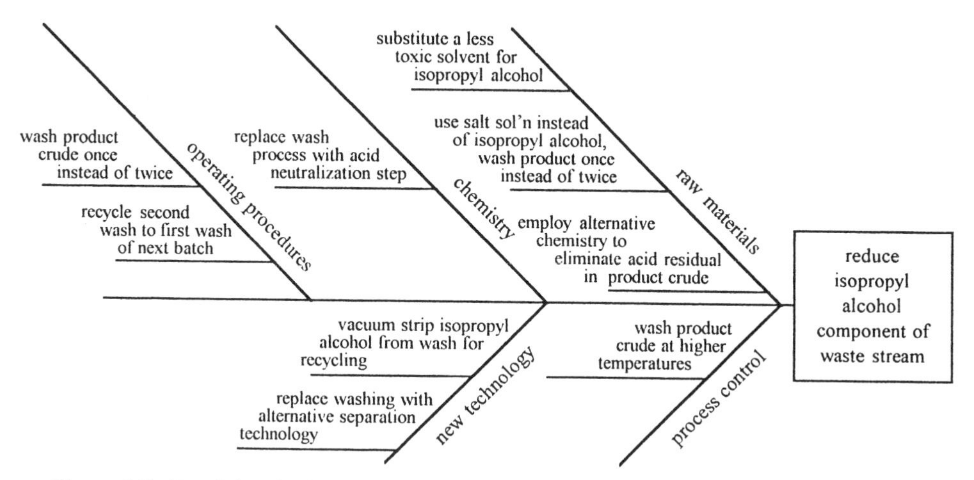

Figure 8-5. Specialty alcohols wash process waste minimization options (Report to the Government, 1993).

evaluation. It not only provided a complete source reduction of waste, but also met all the other process improvement goals.

Note how this case followed the general framework for identifying pollution prevention opportunities. First, the process boundary was defined and a waste flow and composition were specified. For this waste stream a series of pollution prevention alternatives were considered, including raw-material changes, changes in chemistry, and changes in technology. Then, for the existing unit operation, process control and operational changes were suggested. Each of these alternatives was then evaluated, completing steps 1 through 5 outlined in this section. A second case study, presented below, also follows this general methodology.

8.1.2.2 ***Case Study of Qualitative Flowsheet Analysis for an Organic Salt Process*** As shown in Figure 8-6, a series of reactions is used to produce a crude acid that is then processed into salable organic salt. One of the reactants is methanol, and excess amounts of it are required to force one of the reactions to completion. The crude acid from the reactors is fed to a neutralizer, where it is mixed with an alkaline compound and heated. From the neutralizer, the crude passes on to a crystallizer that cools the mixture, causing organic salt crystals to form. The crystal-bearing mixture passes from the crystallizer to a rotary vacuum filter, where the organic salt crystals collect in the outer part of the filter, forming a solid filter cake. The remaining liquid, or filtrate, consists of methanol, byproducts, and uncrystallized organic salt. After the acidic filtrate exits the rotary filter, it enters a neutralizing tank, where it is mixed with an alkaline compound to protect the downstream process equipment. From the neutralizing tank, the filtrate is sent to a

Table 8-1. Ranked summary of waste minimization options for specialty alcohols wash process

Option	Advantages	Disadvantages
Replace wash process with acid neutralization step	Elimination of the waste stream Attainment of other process improvement goals	
Substitute a less toxic solvent for isopropyl alcohol	Substitution of waste isopropyl alcohol with a less toxic substance	Alternative solvent is undefined Alternative solvent would pose disposal problems of its own No real reduction in the amount of waste produced
Use salt solution instead of isopropyl alcohol, and wash product once instead of twice	Reduction of total organic content load in waste stream by ~94%	Uncertain whether salt facilitates separation as well as isopropyl alcohol Would not attain other process improvement goals
Employ alternative chemistry to eliminate acid residual in product crude	Elimination of the waste stream	Alternative chemistry is undefined High research cost High capital cost Long implementation time

Wash product crude once instead of twice	Reduction of the isopropyl alcohol component of the waste stream by half	Would not attain other process improvement goals
Recycle second wash to first wash of next batch	Reduction of the isopropyl alcohol component of the waste stream by half	Would not attain other process improvement goals
Wash product crude at higher temperatures	Elimination or reduction of the isopropyl alcohol component of the waste stream	Uncertain chance of success Would not attain other process improvement goals
Vacuum-strip isopropyl alcohol from wash for recycling	Elimination of the isopropyl alcohol component of the waste stream	Not a source reduction High research cost High capital cost Long implementation time
Replace washing with alternative separation technology	Elimination of the waste stream	Uncertain chance of success High research cost High capital cost Long implementation time

Source: Report to the Government (1993).

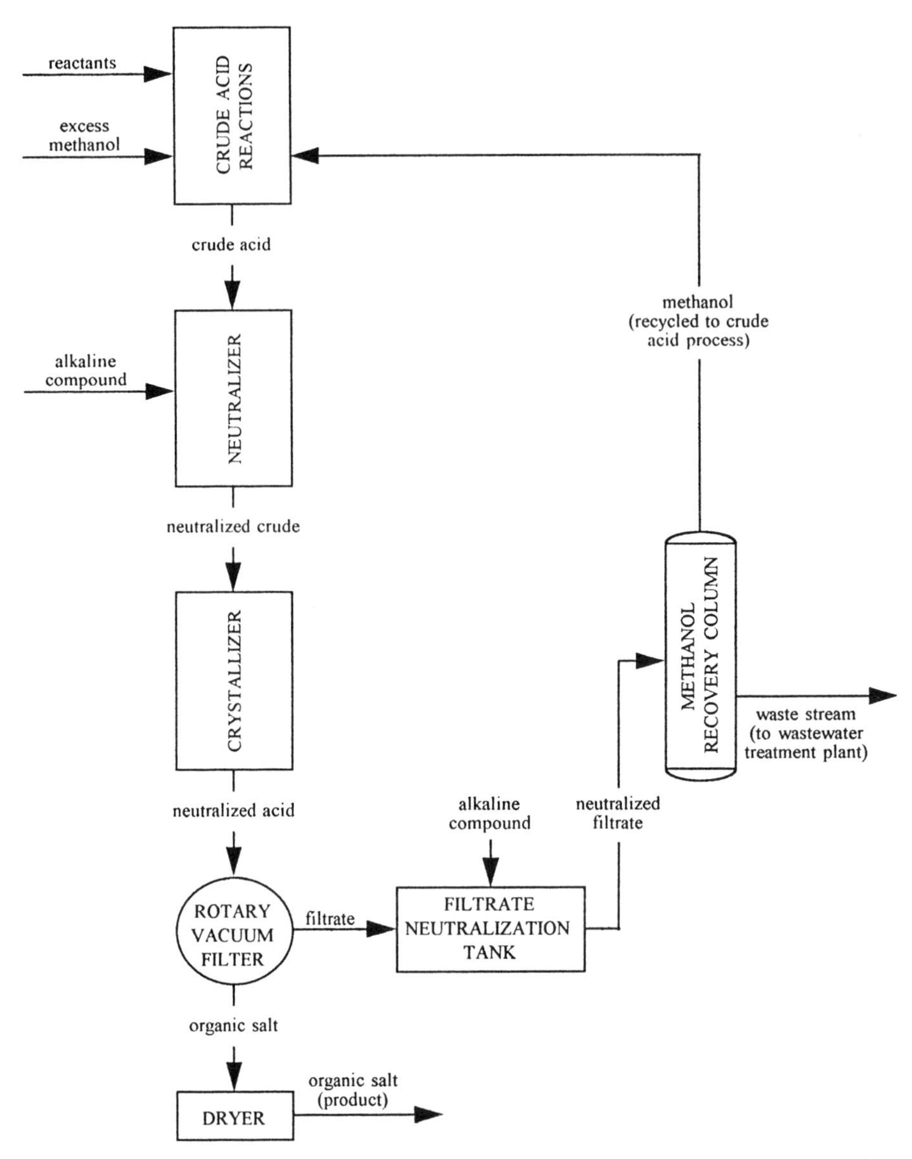

Figure 8-6. A process for producing organic salt (Report to the Government, 1993).

methanol-recovery column. There the filtrate is heated, causing the methanol to boil off and exit the process at the top of the column. The distilled methanol is collected and recycled back to the crude-acid process. The remaining liquid exits the column bottom and is sent to the on-site wastewater treatment plant for disposal.

The waste stream that is sent to disposal consists of (by mass) 88% water,

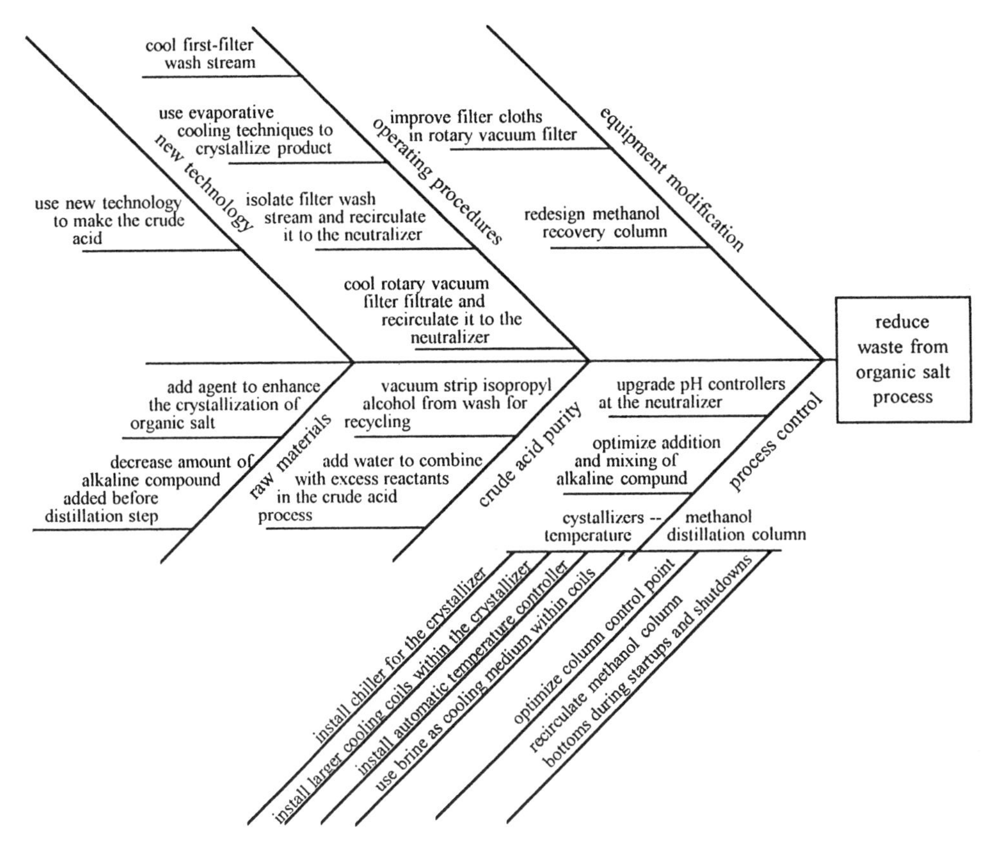

Figure 8-7. Options for minimizing waste from the organic salt process of Figure 8-6 (Report to the Government, 1993).

8.3% reaction byproducts, 2.2% organic salt, 1.0% alkaline compound, and 0.5% methanol. The amount of waste from this process is 0.14 lb of waste (dry basis) for every pound of organic salt produced. Costs associated with this waste stream include the yield loss represented by the unrecovered organic salt, the replacement cost of unrecovered methanol, and the costs of treating the stream at the wastewater treatment plant.

An assessment team generated 19 possible waste reduction options at a brainstorming session. These options were recorded in the cause-and-effect chart shown in Figure 8-7. Each of these options, which are summarized in Table 8-2, fall into one of three categories:

1. Source reduction of waste stream constituents (11 options)
2. Recovery and recycling of methanol or organic salt (five options)
3. Increased crude acid purity to reduce impurities that end up in the waste stream (three options)

Table 8-2. Ranked summary of waste minimization options for organic salt process

Option	Advantages	Disadvantages	Comments
Optimize reactant ratio in crude acid process (option 1)	Addresses a major cause of waste Low capital and operating costs		This option would reduce the amount of byproducts that form in the crude acid process
Use new technology to make the crude acid (option 2)	Very high source reduction potential	Very high operating costs	This option would ensure that virtually all of the reactants are consumed in the reaction that produces the crude acid
Add water to combine with excess reactants in the crude acid process (option 3)	High source reduction potential Low operating cost	Moderate capital cost Safety concerns	This option would prevent excess reactants in the crude acid process from forming byproducts
Recirculate methanol column bottoms during startups and shutdowns (option 4)	Moderate recycling potential Very good chance of success	Moderate capital cost	This option requires the installation of piping and control equipment
Install chiller for the crystallizer (option 5)	Moderate source reduction potential Very good chance of success	Moderate capital cost	Lower temperatures would cause more organic salt to crystallize, improving the product yield and reducing waste
Add agent to enhance the crystallization of organic salt	Low source reduction potential	Unknown chance of success	A chemical agent that would enhance crystallization has not yet been identified

Improve filter cloths in rotary vacuum filter	Low source reduction potential Low capital and operating costs	Low chance of success	At present, some organic salt is lost through the filter cloths; previous attempts to identify better filter cloths have failed
Optimize addition and mixing of alkaline compound	Presumptive source reduction	Low chance of success	The prevailing view is that the present method is already efficient
Optimize column control point	Small recycling potential	Low chance of success	This option involves finding a better location on the methanol recovery column for placing temperature and pressure controls
Install larger cooling coils within the crystallizer	Moderate source reduction potential Moderate chance of success	Very high capital cost	Larger cooling coils would cause more organic salt to crystallize, improving yield and reducing waste
Install automatic temperature controller	Moderate source reduction potential Moderate chance of success	Moderate capital cost	This option would provide better control of the rate of cooling within the crystallizer
Use brine as cooling medium within coils	Moderate source reduction potential Moderate chance of success	High capital cost High operation and maintenance costs	Replacing water with brine would permit cooler temperatures within the crystallizer
Decrease the amount of alkali added before distillation step	Low capital cost	Low waste minimization potential Low chance of success	The alkali protects process equipment from acidic crude; this option would reduce waste by the (small) amount of the alkali reduction

Table 8-2. (*Continued*)

Option	Advantages	Disadvantages	Comments
Use evaporative cooling technique to crystallize product	Moderate source reduction potential Moderate chance of success	Very high capital cost Very high operation and maintenance costs	This option would replace the present method of crystallizing the product
Upgrade pH controllers at neutralizer	Presumptive source reduction	Low chance of success	Assessment team achieved consensus that present pH control is not a problem
Isolate filter wash stream and recirculate it to the neutralizer		Low waste minimization potential Low chance of success	This option seeks to recover the small amount of organic salt that is lost in the filter wash stream
Cool filter wash stream		Low waste minimization potential Low chance of success	Water in the rotary filter washes impurities out of the filter cake; some crystals dissolve and wash away; cooling the washwater would reduce this (small) yield loss
Cool rotary vacuum filter filtrate and recirculate it to the neutralizer		Low waste minimization potential Low chance of success	This would recover more product, but requires a method for separating the salt from the rest of the filtrate before it recirculates to the neutralizer
Redesign methanol-recovery column		Low recycling potential Very high capital cost	This option hopes to achieve a more efficient distillation of the methanol from the waste stream

Source: Report to the Government (1993).

After weighing the advantages and disadvantages of each of the 19 options, the assessment team chose the following five options for a technical and economic feasibility analysis:

1. Optimize reactant ratio in the crude-acid process.
2. Use new technology to make the crude acid.
3. Add water to combine with the excess reactants in the crude-acid process.
4. Recirculate methanol column contents during startups and shutdowns.
5. Install a chiller for the crystallizer.

Optimizing the reactant ratio and adding water to the crude-acid process (options 1 and 3) both involve changes to the present process for making crude acid. The former option requires no capital investment, while the latter requires a moderate investment. Both options have high source reduction potentials, and both are technically feasible, although only plant trials can confirm their effectiveness. Using a new technology for making crude acid (option 2) has been successfully demonstrated in other applications. Although the waste minimization potential of this option is significant, the cost of implementing this option cannot be justified. Recirculating the methanol column bottoms during startups and shutdowns (option 4) is a recycling option that is easily implemented and requires very little capital investment. Installing a chiller for the crystallizer (option 5) is also easy to implement, but requires a moderate capital investment. Neither option 4 or 5 represents new technology, and both are technically feasible.

8.1.3 Hierarchical Design Procedures for Pollution Prevention

The previous section outlined methods for reviewing existing flowsheets. These procedures assume a baseline of the existing process and the waste and emissions it generates. Processes that are under initial development pose different challenges, because the design of a chemical process involves many decisions. Choices of raw materials, catalyst, reactor design, separator design, and level of energy integration can lead to a large number of possible flowsheets. A typical process may have 10^4 to 10^9 possible flowsheet configurations (Douglas, 1988), making it impossible to evaluate all the alternatives on the basis of waste generation or any other criterion. To systematically address flowsheet decisionmaking, a hierarchical approach has been developed (Douglas, 1988). Flowsheeting decisions are made sequentially at a variety of levels, as shown in Table 8-3. The number of possible flowsheets is reduced as the designer proceeds to higher and higher levels. This hierarchical design procedure allows the systematic incorporation of waste reduction alternatives at all stages of the design process. Shown in Table 8-4 are a

Table 8-3. Hierarchical decision levels for process synthesis

Level
1. Input information: type of problem
2. Input/output structure of the flowsheet
3. Recycle structure of the flowsheet
4. Specification of the separation system
4a. General structure: phase splits
4b. Vapor-recovery system
4c. Liquid-recovery system
4d. Solid-recovery system
5. Energy integration
6. Evaluation of alternatives
7. Flexibility and control
8. Safety

Source: Reprinted with permission from J. M. Douglas, "Process Synthesis for Waste Minimization," *Ind. Eng. Chem. Res.*, **31**(1), 238–243, Jan. 1992. Copyright © 1992 American Chemical Society.

number of waste reduction opportunities that could be considered at various design levels.

This framework offers a systematic approach to constructing a flowsheet that generates a minimum of waste. The advantage of this method is that it explicitly recognizes that waste reduction cannot be relegated to the end of the design process. Rather, it must be integrated into all decision levels.

This process can also be used to evaluate existing flowsheets. As an example, consider a case study drawn from the Amoco/US EPA Pollution Prevention Project at Amoco's Yorktown, Virginia refinery (Rossiter and Klee, 1995). In this example, the flowsheet of a fluidized-bed catalytic cracking unit, such as the one shown in Figure 8-8, is evaluated for pollution prevention options using the hierarchical framework. Because the system is already well defined, the analysis begins at level 2 of the hierarchy, which addresses the input/output structure of the flowsheet. At this level the following pollution prevention strategies were generated:

1. Improve quality of the feed to eliminate or reduce the need for the vapor line washing system.
2. Reduce steam consumption in the reactor so that there is less condensate to remove from the distillation system.
3. Within the regeneration system, the loss of fines is partly a function of the air input rate. A reduction in airflow (e.g., by using oxygen enrichment) is a possible means of reducing the discharge of fines.

Table 8-4. Possible waste reduction strategies to be evaluated at each design level

Level	Strategy
2.	Input/output structure of the flowsheet
	a. Problems caused by the reaction chemistry: change the chemistry
	b. Problems caused by air oxidation to NO_x: change to O_2 in oxidations
	c. Problems caused by spent catalysts: regenerate the catalyst
3.	Recycle structure of the flowsheet
	a. Problems caused by adding reactor diluents to shift the product distribution or to shift the equilibrium conversion: change the diluent
	b. Problems caused by adding heat carriers: change the heat carrier
	c. Problems caused by adding reactor solvents: change the solvent
4.	Specification of the separation system
	4a. General structure: phase splits
	4b. Vapor-recovery system
	a. Problems caused by absorber solvents: change the solvent
	b. Problems caused by regeneration of absorption beds: change the bed stripping agent
	c. Problems with removing spent absorbents: change to absorption or condensation
	d. Problems caused by the use of reactive absorbers to remove toxic materials
	4c. Liquid-recovery system
	a. Problems caused by stripping agents: change the agent
	b. Problems caused by extraction solvents: change the solvent
	c. Problems caused by crystallizer (recycle and) purge streams (almost pure water): reuse the purge water elsewhere in the process
	d. Problems caused by crystallizer purge streams (not almost pure water): remove the contaminants and recycle the water or look for a different separation system
	e. Problems caused by reactive crystallization byproducts: look for a different separation technique
	f. Problems caused by spent absorbents: regenerate the absorbent
	4d. Solid-recovery system
	a. Problems caused by cake washing: same as for crystallizer; filter mother liquor streams

Source: Reprinted with permission from J. M. Douglas, "Process Synthesis for Waste Minimization," *Ind. Eng. Chem. Res.*, **31**(1), 238–243, Jan. 1992. Copyright © 1992 American Chemical Society.

Two ideas were generated during review of the recycle structure (level 3):

1. The reactor uses 26,000 lb/hr of steam. This is provided from the utility steam system. If this could be replaced with steam generated from process water, the liquid effluent from the unit would be reduced. This would allow the condensate recovered within the distillation section to be revaporized and recycled to the reactor section. Volatile hydrocarbons contained in the

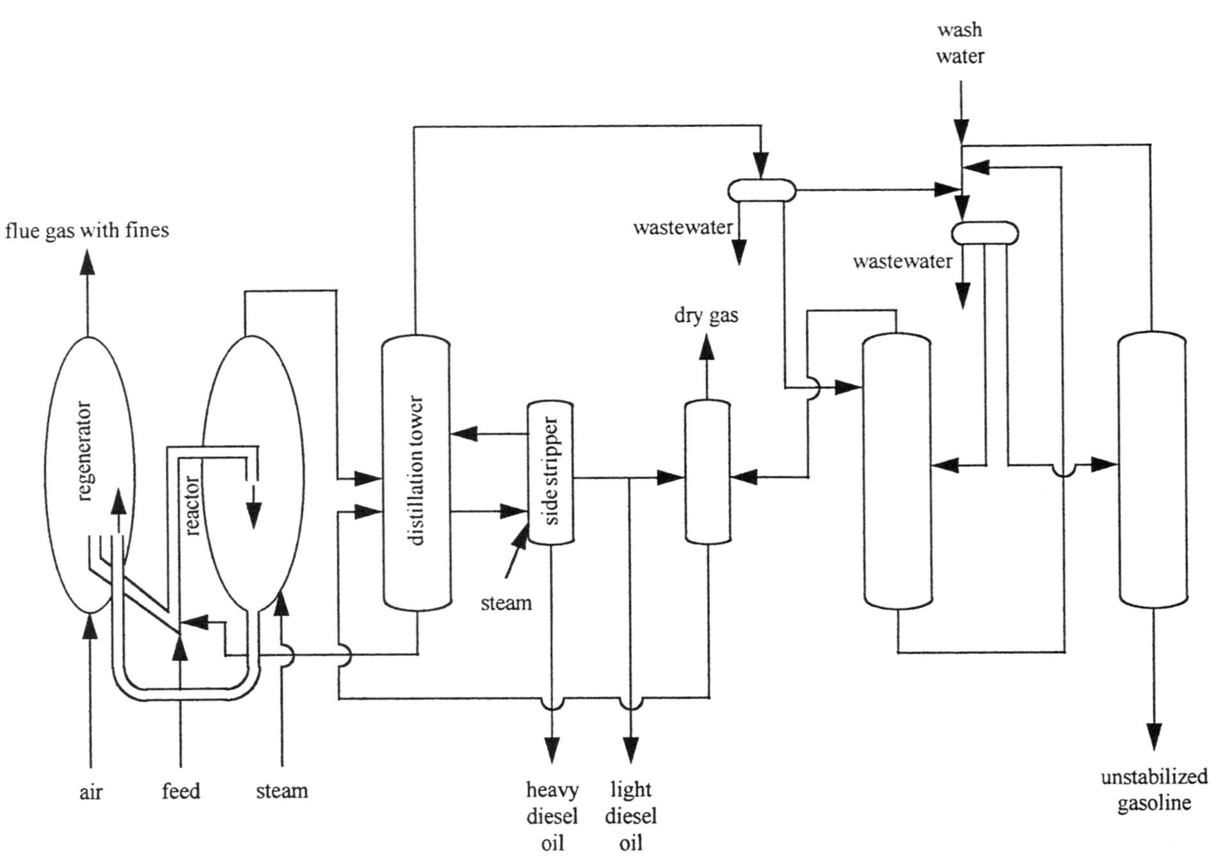

Figure 8-8. A fluidized-bed catalytic cracking unit at a refinery.

steam generated from the recovered condensate would be returned directly to the process. Regeneration consumes more than 11,000 lb/hr of steam. It may be possible to satisfy this duty with "dirty steam" as well, since the hydrocarbon content would be incinerated with the coke in the regenerator.

2. Used wash water is collected at several points and then purged from the process. If it could be recovered and recycled instead, or if recycled water from other sources could be used for washing in place of fresh water, fresh water usage and wastewater generation could both be reduced by about 10,500 lb/hr.

Three options were identified for separation systems (level 4):

1. Replace live steam stripping with reboilers.
2. Place additional oil–water separators downstream of existing condensate collection points and recover hydrocarbons.
3. Improve gas–solid separation downstream of the regenerator to eliminate loss of catalyst fines. This might simply require better cyclone and/or ductwork design, or electrostatic precipitation.

Energy efficiency was also considered (level 6). In steady state operation, the unit generates sufficient heat to be self-sustaining and requires no external heating utility. Much of the heat rejected by the unit is recovered in 150 psig (lb/in.^{2} gauge) of steam; the rest is rejected in coolers or leaves in hot gases. The primary objective of this level of analysis is to maximize the amount and the value of the recovered heat. Heat integration analysis, described in the next section, showed that it would be possible to

1. Increase heat recovery to 150 psig of steam by more than 10 million Btu/hr (nearly 31%).
2. Increase the steam pressure to nearly 600 psig and use the steam for power generation. After being used for power generation, the steam would be able to be used for other heating demands at the 150-psig level.

Some of the ideas generated in the hierarchical review of the fluid catalytic cracking unit at the Yorktown refinery are used as examples in describing pollution prevention management issues in Chapter 9 of this text. For now, it is sufficient to note that the analysis yields a systematic approach for incorporating pollution prevention into flowsheet design. When applied to existing processes such as the fluid catalytic cracking unit, this approach parallels some of the steps for flowsheet analysis described in the previous section.

8.2 QUANTITATIVE METHODS: MASS EXCHANGE NETWORK SYNTHESIS

Quantitative procedures for evaluating process flowsheets are becoming available and this section describes one of the more rigorous of these methods: mass-exchange network (MEN) synthesis. MEN synthesis is somewhat analogous to heat-exchange network (HEN) synthesis. In one case, energy efficiency is the goal; in the other case, mass efficiency is the goal. While this simple analogy has some flaws, a very brief discussion of HEN synthesis, which is conceptually simpler than MEN synthesis, serves to introduce the topic of MEN synthesis.

HEN synthesis is one of the most powerful flowsheeting tools available for maximizing the energy efficiency of a process. In HEN synthesis, all the heating and cooling requirements called for in a flowsheet are systematically examined to determine the extent to which streams that need to have their temperature raised can be heated by streams that need to be cooled. Heat transfer between streams, called *heat integration*, conserves fuel and reduces combustion-related emissions. A simple example of the potential for heat integration [taken from Douglas (1988)] is shown in Figure 8-9, where one stream needs to be heated from 50°C to 200°C and a second stream needs to be cooled from 200°C to 90°C. Both streams have the same flow rates and heat capacities. Instead of separately heating and cooling the two streams, the configuration shown in Figure 8-10 might be used. Detailed descriptions of HEN synthesis are available in most modern textbooks of chemical process design (e.g., Douglas, 1988).

MEN synthesis, originally developed by Vasilios Manousiouthakis and coworkers at UCLA, is the systematic generation of a cost-effective network of mass exchangers whose purpose is to preferentially transfer certain species from a set of rich (high concentration of pollutant) streams to a set of lean (low concentration of pollutant) streams. This method of synthesis can be conducted for any countercurrent, direct-contact mass-transfer operation, such as absorption, desorption, or leaching. As applied to pollution pre-

Figure 8-9. A simple heat exchange network synthesis problem: the hot stream needs to be cooled from 200°C to 90°C; the cold stream needs to be heated from 50°C to 200°C.

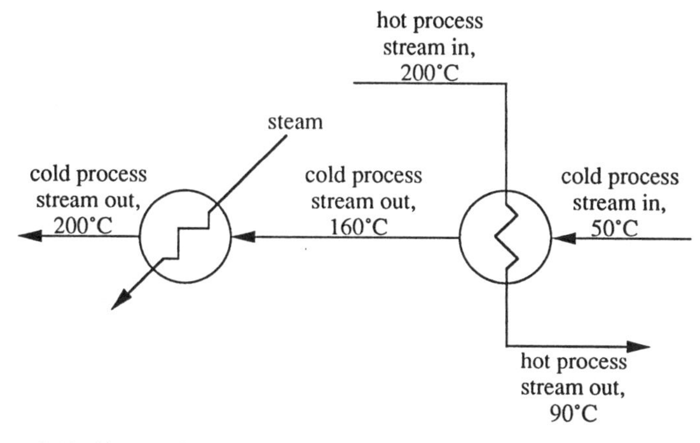

Figure 8-10. Heat-exchange network for the hot and cold streams of Figure 8-8.

vention, mass-exchange network theory determines the extent to which molecules that are potential pollutants can be transferred to streams in which they have a positive value. In refineries, for example, phenol is a pollutant in the water effluent of catalytic cracking units, desalter washwater, and spent sweetening waters. In other refinery streams, such as jet fuel, phenol can be a valuable additive. Implementation of a MEN might therefore prevent pollution in refinery wastewaters. A simplified schematic of a network for mass exchange of phenol in refineries is shown in Figure 8-11, where three phenol-rich streams R_i and three phenol-lean streams L_j are considered. Proper matching and exchange of phenol between the rich streams and lean streams within the process network is the key to optimization.

8.2.1 Thermodynamic Constraints for Mass Transfer

The amount of solute that can be transferred from a rich stream to a lean stream is limited by mass-balance constraints and equilibrium constraints, as follows: (1) the total mass transferred by the rich stream must be equal to that received by the lean stream, and (2) mass transfer is possible only if a positive driving force ϵ exists for all rich stream/lean stream matches. Methods for addressing these constraints in MEN synthesis are described below.

8.2.1.1 *Conservation of Mass* The equation describing the constraint for mass-balance is found by performing a material balance on the solute to be transferred from stream i to stream j of Figure 8-12, which results in the equation

$$R_i(y_{ki}^{\text{in}} - y_{ki}^{\text{out}}) = L_j(x_{kj}^{\text{out}} - x_{kj}^{\text{in}}), \tag{8.1}$$

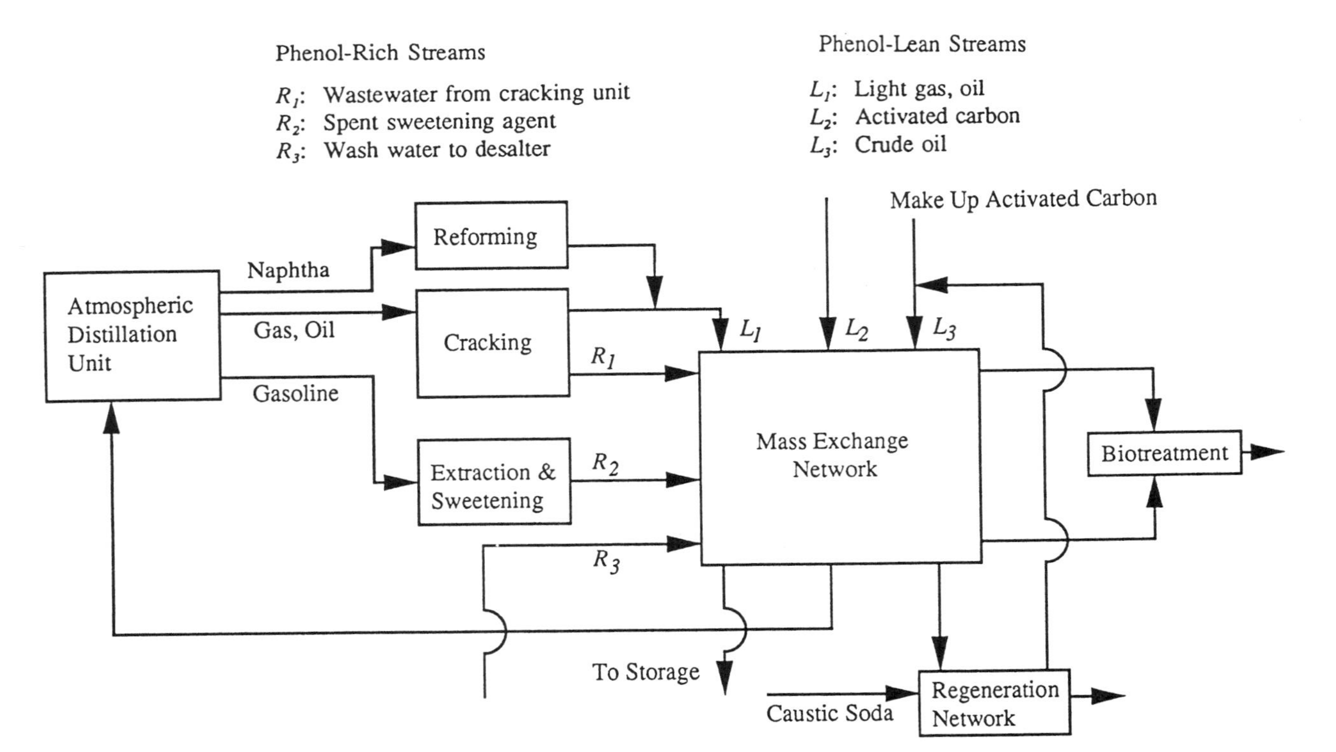

Figure 8-11. A mass-exchange network in an oil refinery (Allen et al., 1992).

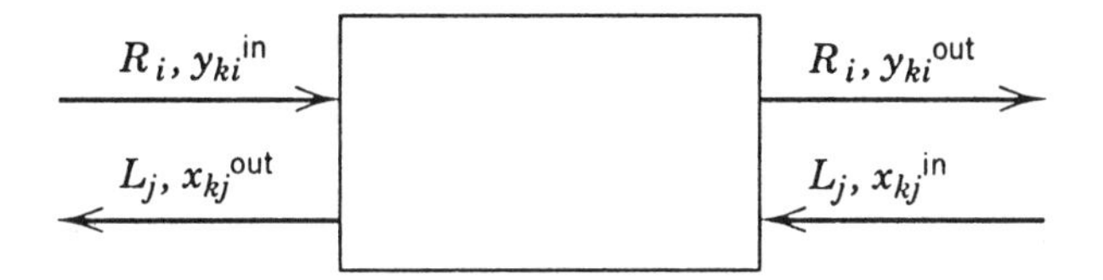

Figure 8-12. A mass exchanger.

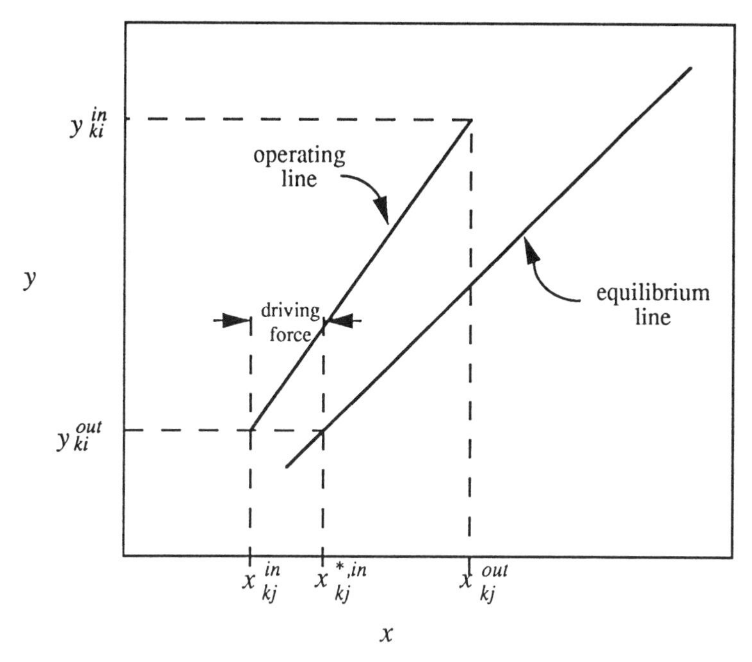

Figure 8-13. Equilibrium and mass-conservation constraints on mass-exchange network design.

where R_i is the flow rate of rich stream i, L_j is the flow rate of lean stream j, y_{ki} is the mass fraction of solute k in rich stream i, and x_{kj} is the mass fraction of the solute k in lean stream j. For the analysis in this chapter, the flow rates of the streams are assumed to be constant. While this is not strictly the case, it is a good approximation if the concentration of solute in the streams is low and little transfer of material other than solute occurs. Equation (8.1) represents the operating line for the mass transfer of solute from the rich stream to the lean stream. An operating line for contact between a single rich and a single lean stream is plotted in Figure 8-13. Note that the slope of the operating line is equal to L/R.

8.2.1.2 ***A Positive Driving Force*** Equilibrium between a rich stream and a lean stream can be represented by an equation of the form

$$y_{ki} = m_{kj} x_{kj}^* + b_{kj}, \tag{8.2}$$

where x_{kj}^* is the mass fraction of component k in stream j that is in equilibrium with the mass fraction y_{ki} in stream i. This linear type of equilibrium relationship is found in familiar expressions such as Raoult's law, Henry's law, and octanol/water partition coefficients. The constants of Equation (8.2) are thermodynamic properties and may be obtained through experimental data. A plot of an equilibrium line is shown in Figure 8-13. The positive driving-force constraint for mass transfer is satisfied when the equilibrium line lies to the right of the operating line. Equation (8.2) can be rewritten as

$$x_{kj}^* = \frac{y_{ki} - b_{kj}}{m_{kj}}. \tag{8.3}$$

8.2.2 The Tools of MEN Synthesis

Operating and equilibrium lines, discussed in the previous section to remind the reader of mass-exchange constraints, are familiar tools for separator design. However, in MEN synthesis, the design of individual separators is not initially considered. Instead, the streams to be contacted and the extent of transfer possible is determined. Equilibrium and operating lines are not useful for this task. The tools of MEN synthesis are composition interval diagrams and load lines.

8.2.2.1 ***Composition Interval Diagrams*** A composition interval diagram (CID) depicts the lean and rich streams under consideration in MEN synthesis. CIDs for the rich and lean streams of Table 8-5 are shown in Figure 8-14. To construct this diagram, an arrow is drawn for each stream with its tail at the entering mass fraction and its head at the exiting mass fraction. The following example provides further illustration of the process of mapping lean and rich streams onto a CID.

Table 8-5. Stream data for a scenario with two rich streams and one lean stream

Rich Streams				Lean Streams			
Stream	Flow Rate, kg/s	y^{in}	y^{out}	Stream	Flow Rate, kg/s	x^{in}	x^{out}
R_1	5	0.10	0.03	L	15	0.00	0.05
R_2	10	0.07	0.03	—	—	—	—

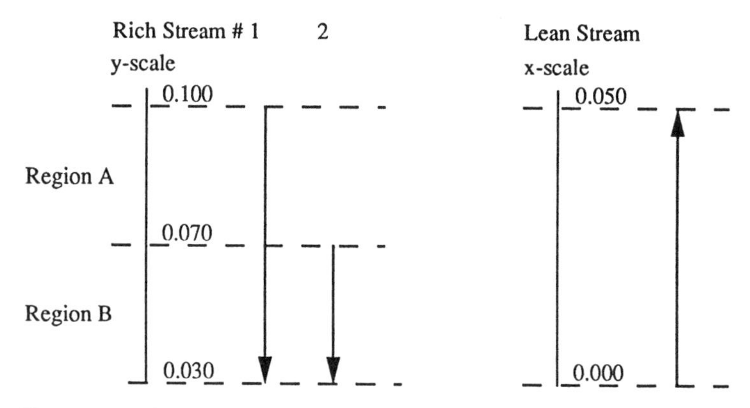

Figure 8-14. Composition interval diagrams for the streams of Table 8-5.

Table 8-6. Stream data for a scenario with three rich streams and one lean stream

Rich Streams				Lean Stream		
Stream	Flow rate, kg/s	y^{in}	y^{out}	Stream	x^{in}	x^{out}
R_1	5	0.10	0.03	L	0.00	0.14
R_2	10	0.07	0.03	—	—	—
R_3	5	0.08	0.01	—	—	—

Example 8-1 Constructing a Composition Interval Diagram Construct a CID for the rich streams of Table 8-6. With the aid of this CID, calculate the mass transferred out of the rich streams in kg/s within each region of the CID. The mass transferred from the rich streams within each region is equal to $(y^{out} - y^{in}) \times \Sigma R_i$, where y^{out} and y^{in} are the exiting and entering rich stream mass fractions, respectively, and ΣR_i is the sum of the rich stream flow rates in the region. Note that mass transferred is negative for the rich streams because they are losing mass.

Solution The rich streams are mapped from Table 8-6 to generate the CID shown in Figure 8-15. The mass transferred in each region is as follows:

$$\text{Region } 1 = (y^{out} - y^{in}) \times \Sigma R_i = (0.08 - 0.10)5 \text{ kg/s} = -0.10 \text{ kg/s}$$

$$\text{Region } 2 = (0.07 - 0.08)(5 + 5) \text{ kg/s} = -0.10 \text{ kg/s}$$

$$\text{Region } 3 = (0.03 - 0.07)(5 + 10 + 5) \text{ kg/s} = -0.80 \text{ kg/s}$$

$$\text{Region } 4 = (0.01 - 0.03)5 \text{ kg/s} = -0.10 \text{ kg/s}$$

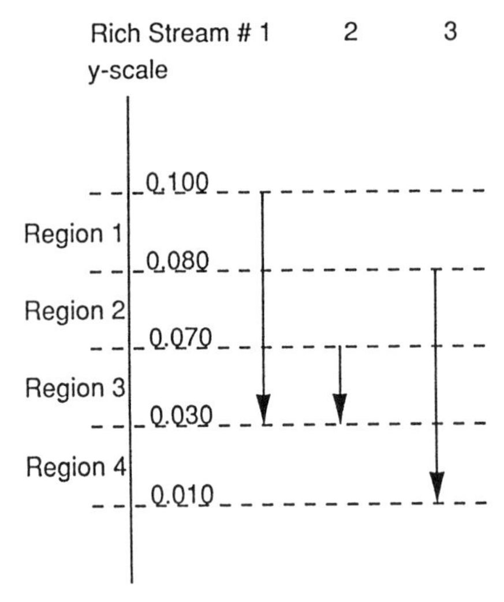

Figure 8-15. Composition interval diagram for the rich streams of Table 8-6.

In the CID of Figure 8-14, the compositions of the rich and lean streams are on separate axes. These axes can be combined by applying the equilibrium relationship. If the equilibrium relationship in the region of interest for the species considered in this problem is given by

$$y = 0.67x^*,$$

then a mass fraction of $y = 0.1$ in the rich stream is in equilibrium with a mass fraction of $x^* = 0.15$ in the lean stream. By converting the lean-stream compositions of Figure 8-14 to the rich-stream compositions with which they are in equilibrium, and vice versa, a combined CID with shared axes as shown in Figure 8-16 can be constructed.

Example 8-2 Constructing a Combined Composition Interval Diagram Construct a CID for the three rich streams and the lean stream of Table 8-6. The equilibrium relationship in the region of interest is

$$y = 0.67x^*.$$

Solution The combined CID is pictured in Figure 8-17.

8.2.2.2 *Load-Line Diagrams* Load lines depict the flow rate of solute transferred as a function of stream composition. A suitable set of axes for creating load lines is pictured in Figure 8-18. Constructing load lines for

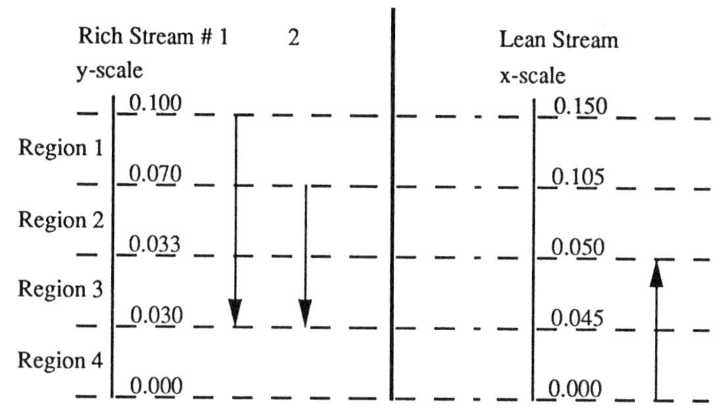

Figure 8-16. Combined composition interval diagram for the streams of Table 8-5.

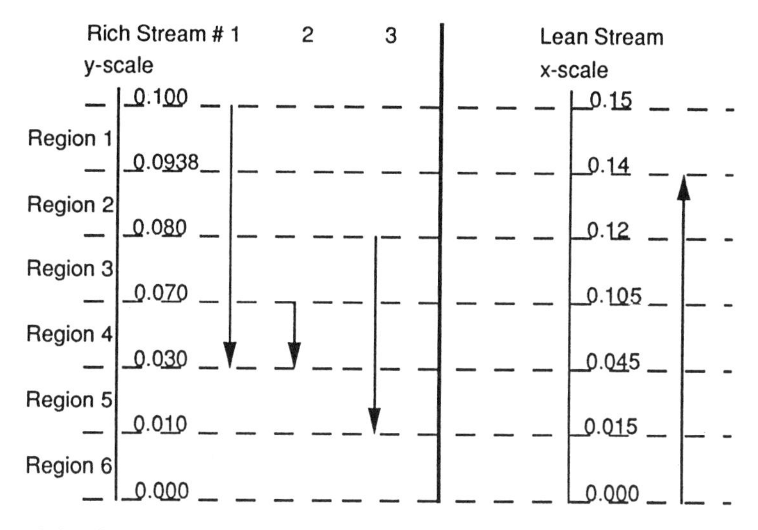

Figure 8-17. Combined composition interval diagram for the streams of Table 8-6.

single streams is simple. Take, for example, R_1 of Table 8-5. At its inlet, this stream has not begun exchange of solute, so one endpoint of its load line is $y^{in} = 0.10$, mass exchanged = 0 kg/s. At its outlet, this stream has exchanged 5 kg/s × (0.03 − 0.10) or −0.35 kg/s, so the coordinates of the other endpoint are (0.07, −0.35 kg/s). The amount of mass transferred is negative at this endpoint because mass is being transferred out of the stream. This load line is shown in Figure 8-19. Note that there is an arrow pointing down and to the left to indicate the direction of transfer. In the following example, you are asked to construct the load lines for the remaining streams of Table 8-5.

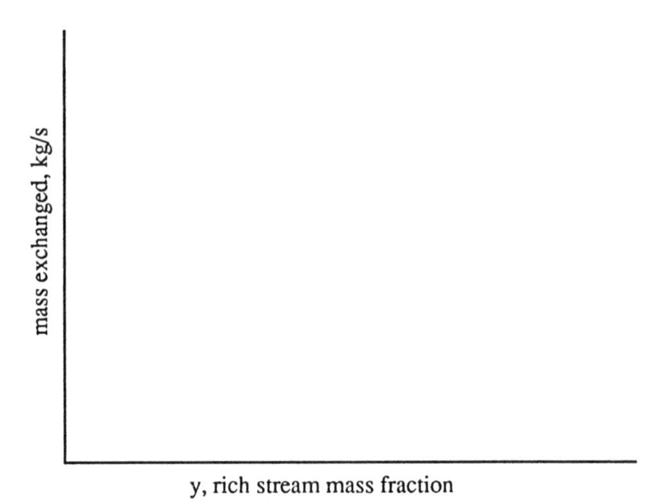

Figure 8-18. Axes for plotting load lines.

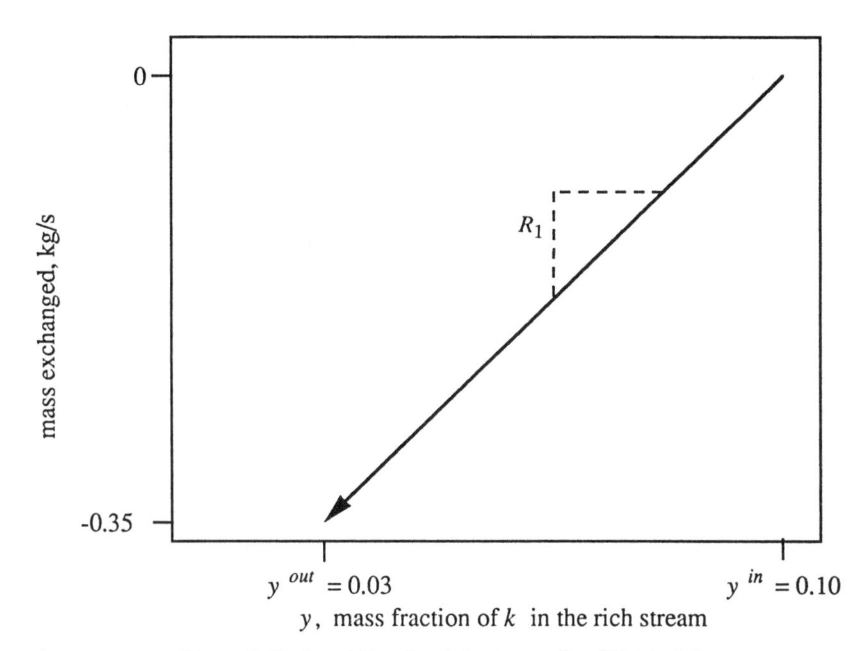

Figure 8-19. Load line for rich stream R_1 of Table 8-5.

Example 8-3 Constructing Load Lines On separate sets of axes, construct load lines for R_2 and L of Table 8-5. Remember that mass exchanged is equal to the flow rate multiplied by mass fraction out less mass fraction in.

Solution One endpoint of the rich stream is at y_{k2}^{in}, where no mass has been

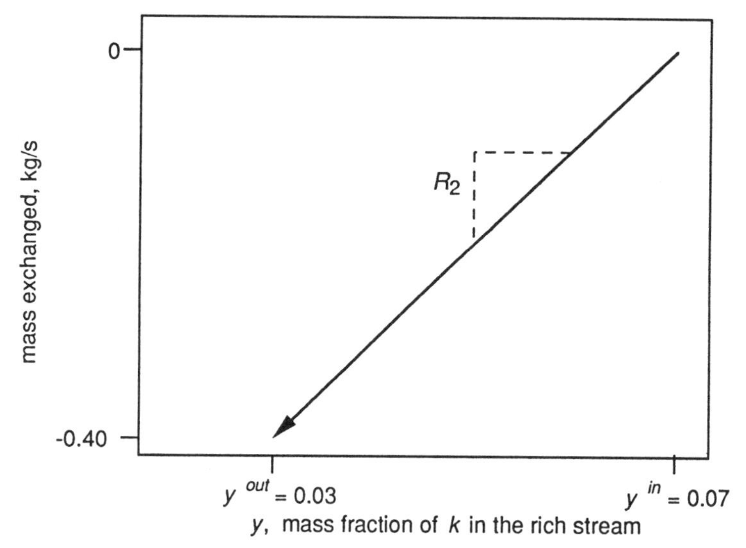

Figure 8-20. Load line for rich streams R_2 of Table 8-5.

exchanged. Therefore, at kg/s = 0, $y = 0.07$. The other endpoint is (0.03, 10 kg/s × (0.03 − 0.07) = − 0.4 kg/s). The endpoints of the lean stream load line are 0 kg/s, $x = 0$ and 0.75 kg/s, $x = 0.05$. Plotting these points on the given axes, the graphic relationship between mass transferred (kg/s) and mass fraction shown in Figures 8-20 and 8-21 is obtained. Note that the slope of the load line for the rich stream is (0.4 − 0)/(0.07 − 0.03) or 10 kg/s, which is the same as its flow rate. The same is true of the lean-stream load line.

The load lines of Example 8-3 are for a single rich and a single lean stream. When there is more than one rich stream or more than one lean stream to consider, a line representative of the multiple streams must be constructed. This line, called a *composite load line*, is the sum of the individual load lines and is developed with the aid of a CID.

As an illustration, the composite load line for the rich streams of Table 8-5 is plotted in Figure 8-22. This composite line is the sum of the load lines of Figures 8-19 and 8-20. It consists of two segments, corresponding to the regions shown in the CID of Figure 8-14. The rich streams with mass fractions of <0.1 and >0.07 ($0.1 \leq y \leq 0.07$) are in region A. Only R_1 falls into this category, so the total flow rate of the rich streams in this composition range is 5 kg/s. The starting point for the load line is $y = 0.1$ and 0.0 kg/s transferred. Recall that the mass transferred in each CID region is equal to the mass fraction exiting the region minus the mass fraction entering the region, multiplied by the sum of the flow rates in the region. Therefore, at $y = 0.07$, 5 kg/s × (0.07 − 0.1) or −0.15 kg/s have been transferred, and the endpoint of the load line in this region is (0.07, −0.15 kg/s). As before, the slope of the load line equals the mass flow rate of the stream. The rich streams with

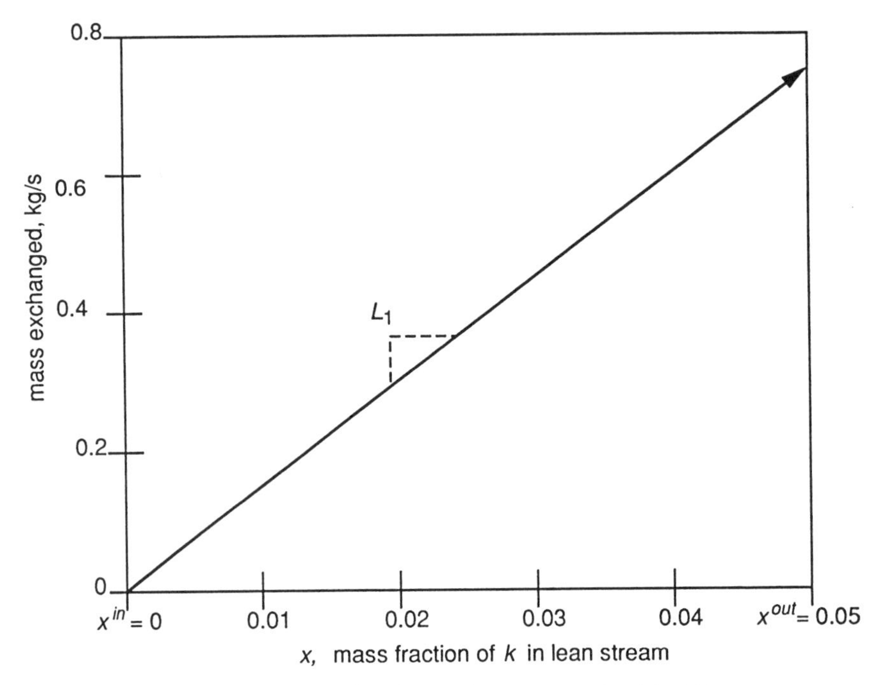

Figure 8-21. Load line for the lean stream of Table 8-5.

mole fractions less than 0.07 and greater than 0.03 are in Region B. Both rich streams fall into this region. When the load line is plotted, it has a slope equal to the sum of the flow rates of all streams in this region. The starting point for this segment of the load line is the termination point of the previous segment.

Example 8-4 Constructing a Composite Load Line Plot the composite load line for the rich streams of Table 8-6. On the basis of the total amount of mass transferred from the rich streams, calculate the minimum flow rate required for the lean stream. In the example that follows, the validity of using this simple overall mass-balance technique for determining the minimum lean stream flow rate is examined.

Solution The endpoints for the segments of the composite load line for the rich streams are: $(0.1, 0)$, $(0.08, -0.10\,\text{kg/s})$, $(0.07, -0.20\,\text{kg/s})$, $(0.03, -1.0)$ and $(0.01, -1.1)$. See Figure 8-23. Note that to plot a continuous load line the cumulative mass transferred is used; the mass transferred in each interval is added to the mass transferred in all previous intervals. If the mass transferred by rich streams equals the mass gained by the lean stream, and (from before) total mass transferred from the rich streams is 1.10 kg/s, then from

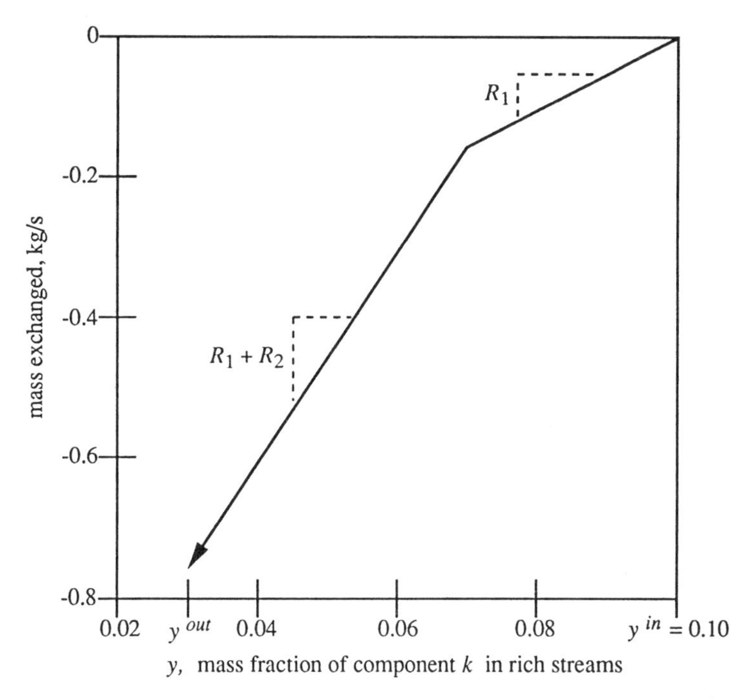

Figure 8-22. Composite rich-stream load line for the rich streams of Table 8-5.

Table 8-6, mass gained by the lean stream is equal to $L(0.14 - 0.0)$, where L is the flow rate of the lean stream. Thus

$$L(0.14 - 0.0) = 1.10$$

or

$$L = 7.86 \text{ kg/s}.$$

The next step in constructing load-line diagrams is to plot the lean and rich streams on the same axes. As with combined CIDs, load-line diagrams can be combined by making use of the equilibrium relationship, which for this specific numerical example is assumed to be

$$y = 0.67x^*$$

in the region of interest. A combined figure can be made by following several different conventions, each of which gives the same final results. The convention used here for constructing the combined figure for the streams of Table 8-5 is to first plot the load line of the lean stream as shown in Figure 8-21. The rich-stream composite load line is added to the figure after

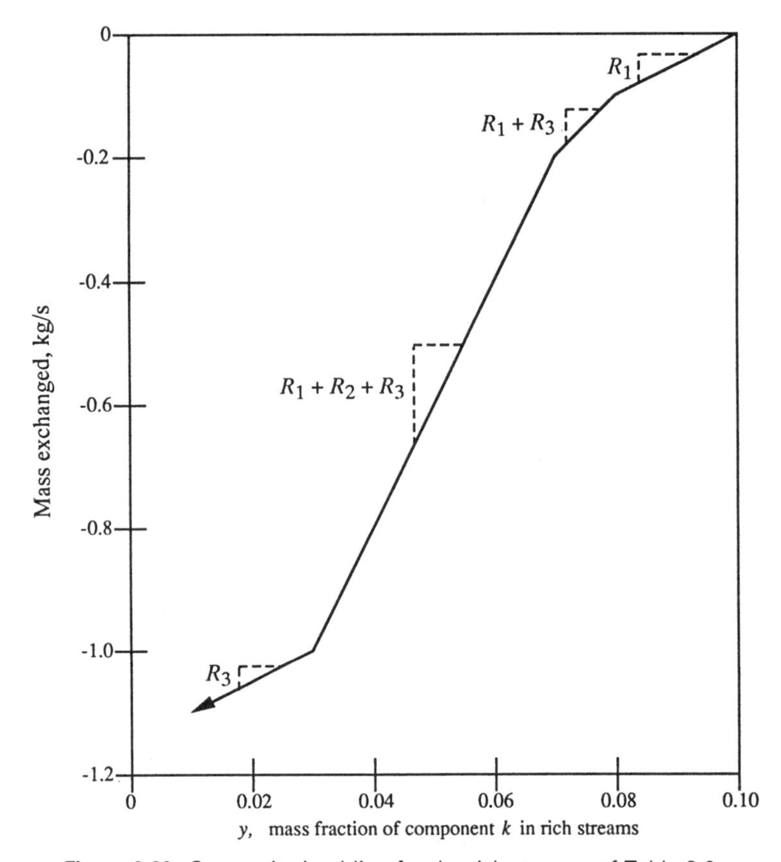

Figure 8-23. Composite load line for the rich streams of Table 8-6.

converting the rich-stream mass fractions into the lean-stream mass fractions with which they are in equilibrium. These conversions were made in order to construct the CID of Figure 8-16. The rich-stream composite load line is free to move vertically; its placement determines the contact between the lean and rich streams. The rich-stream load line has this freedom to move vertically because the values for mass exchanged on the y axis are not absolute; they are useful only in relative terms, that is, in terms of the differences in mass transferred between points. Therefore, the composite rich-stream load line begins at x^* (the x axis) = lean stream mass fraction with which the rich stream is in equilibrium = 0.10/0.67 = 0.15 and continues downward and to the left with a slope of $0.67R_1$. The next point falls at $x = 0.07/0.67 = 0.10$, *where the slope changes to* $0.67(R_1 + R_2)$. The load line ends where $x = 0.03/0.67 = 0.045$. The load lines for the lean and rich streams of Table 8-5 are plotted together in Figure 8-24. As stated before,

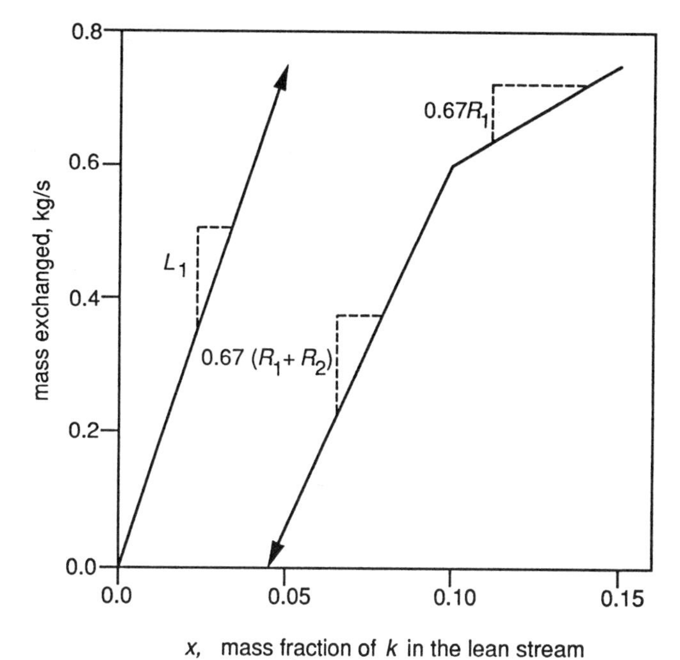

Figure 8-24. Combined lean/rich-stream load lines for the streams in Table 8-5.

the composite rich-stream load line could have been located in any number of different vertical positions.

In Figure 8-24, the lean-stream load line falls to the left of the composite rich-stream load line at every point. This indicates that the desired mass exchange is thermodynamically feasible, and that the transfer could be accomplished using exchangers of finite size with no need for mass exchange into or out of any other streams. In Figure 8-25, the lean-stream load line lies to the left of the rich-stream load line, except at a point, called the *pinch point*, where the lines meet. Mass exchange in a case such as this is thermodynamically feasible, but would require an infinitely large mass exchanger (e.g., an infinite number of trays or stages). Therefore, there is a practical requirement that conditions be manipulated so that a positive horizontal ϵ exists between the load lines. This ϵ is the driving force for mass transfer. If at any point the lean stream load line lies to the right of the rich-stream load line, as shown in Figure 8-26, mass exchange in the desired direction is not thermodynamically feasible. In fact, if streams with such characteristics are contacted, mass exchange from the lean stream to the rich stream occurs. This infeasible situation could be made feasible by moving the rich-stream load line down, but as discussed later, there is a utility cost associated with moving the rich stream down in a case such as this.

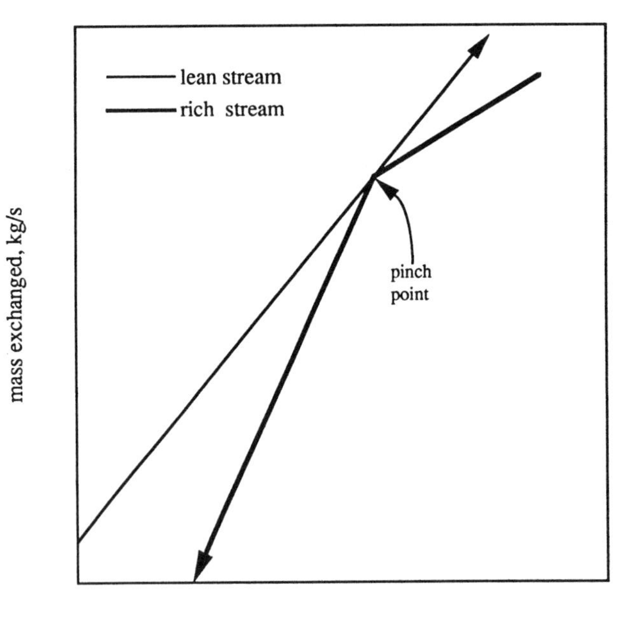

x, mass fraction of k in the lean stream

Figure 8-25. Load lines forming a pinch.

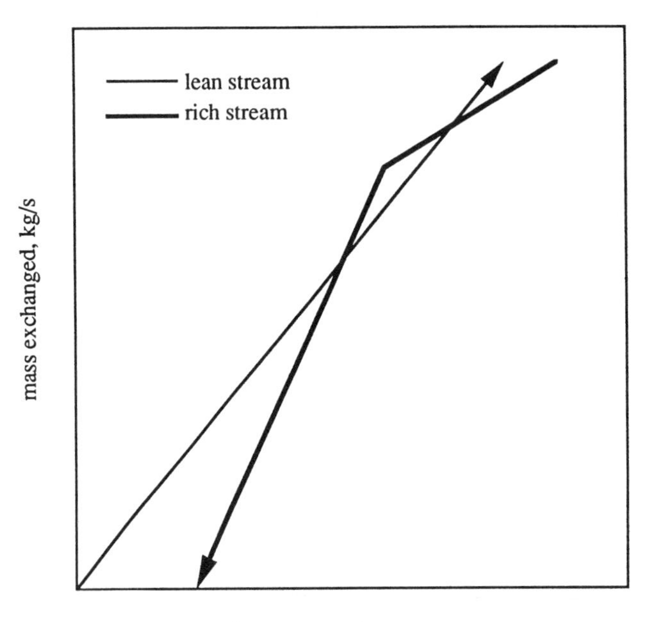

x, mass fraction of k in the lean stream

Figure 8-26. Load lines depicting thermodynamic feasibility.

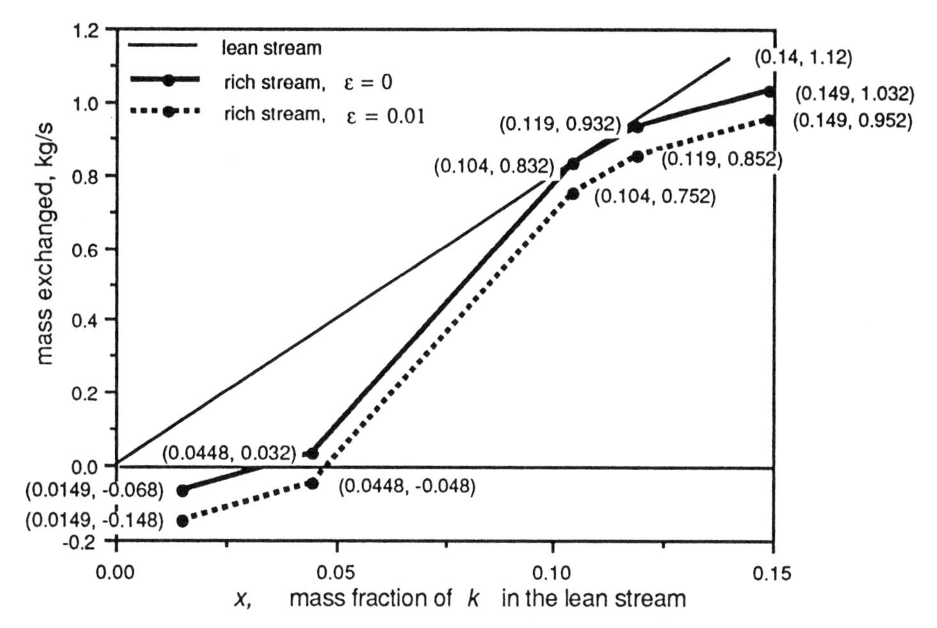

Figure 8-27. Combined lean and composite rich-stream load lines for the streams of Table 8-6.

Example 8-5 Plotting Load Lines on a Single Set of Axes Using the CID you generated in Example 8-2, plot the load lines of the rich and lean streams described in Table 8-6 on the same diagram so that a pinch exists. Assign the lean stream a flow rate of 8 kg/s. Can the specified target concentrations be achieved solely through contact between the lean and rich streams? What does this indicate about the solution in Example 8-4 for the minimum lean-stream flow rate required? Is there a lean-stream flow rate at which all the desired mass exchange can occur solely through contact between the streams? If yes, what is that flow rate?

Solution The easiest way to add the composite rich-stream load line to the lean stream diagram is to determine where the pinch point occurs, plot it, and work outward. From examination of Figure 8-23, it appears that the point where $x = 0.07/0.67 = 0.104$ is the pinch point. The value for mass exchanged at this point is found using the equation for the lean stream load line, which is

$$\text{Mass exchanged} = (8\text{ kg/s})x.$$

Therefore, the first point to plot for the composite rich-stream load line is [0.104, 8(0.104)kg/s = 0.832 kg/s]. See Figure 8-27 for the remaining points on the composite rich-stream load line. Target concentrations cannot be

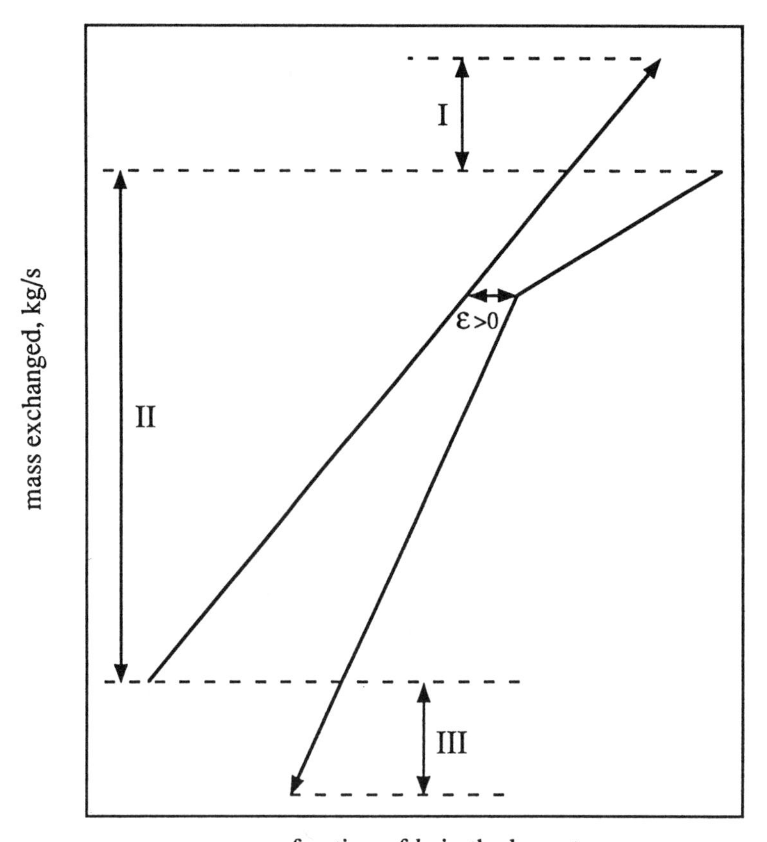

Figure 8-28. The three regions of a mass-exchange network.

achieved solely through contact between the streams, because the target concentration of the rich stream is lower than its concentration after contact with the lean stream (bottom left of Figure 8-27) and because the target concentration of the lean stream is higher than its concentration after contact with the rich stream (top right of Figure 8-27). The lean-stream flow rate here (8 kg/s) is greater than the minimum lean-stream flow rate calculated in Example 8-4, but it is insufficient to effect all the necessary mass exchange. There is no lean-stream flow rate at which all the desired mass exchange can be achieved solely through contact between the streams.

Diagrams that combine the rich- and lean-stream load lines also show the amount of excess mass-transfer capacity available from the lean stream and the amount of excess mass-transfer capacity available from the rich stream. These regions were deliberately, if unrealistically, omitted in Figure 8-24 for the sake of illustrating thermodynamic feasibility. In Figure 8-28, three regions—labeled I, II, and III—are identified. In region I, the lean stream has the capacity to exchange more mass and become richer, but there is a "shortage" of rich stream. The lean stream must be brought up to its specified concentration in a manner other than through mass exchange with the rich stream. For instance, solute k may be added to it. In region II, mass exchange can occur through contact between the rich and lean streams. In region III, the rich stream is capable of mass exchange, but there is a "shortage" of lean stream. An external lean stream mass separating agent is required to achieve the rich stream's target concentration. For example, an adsorbent such as activated carbon might be used to take up the excess k, which is a pollutant in the rich stream. For the lean and rich streams depicted in Figure 8-28, the least amount of k and external lean-stream or mass-separating agent is required when the rich-stream load line is manipulated to form a pinch point ($\epsilon = 0$). As ϵ increases, operating costs (the cost of k and the cost of the mass separating agent) increase and capital costs (the cost of the network) decrease. As ϵ decreases, operating costs decrease and capital costs increase. It is possible to find the ϵ at which total annualized costs are minimized.

Example 8-6 Finding External Mass-Separating Agent Requirements Use the graph you created in Example 8-5 to find the minimum amount of k (k_{req}) the lean stream needs to have added to it after contact with the rich stream and the minimum amount of k (k_{rem}) a mass separating agent must remove from the rich stream after contact with the lean stream. At the values you find, the mass-exchange network has an infinitely high capital cost.

Solution The amount of k required is found by noting that the top right y-axis coordinate of the lean stream is 1.12 kg/s and the top right y-axis coordinate of the rich stream is 1.032 kg/s. Therefore,

$$k_{req} = (1.12 - 1.032)\text{kg/s} = 0.088\ \text{kg/s}.$$

To find k_{rem}, find the bottom left y-axis coordinate of the lean stream (0) and the bottom left y-axis coordinate of the rich stream (−0.068 kg/s). A mass-separating agent therefore needs to remove k at a rate of

$$k_{rem} = (0 - (-0.068))\text{kg/s} = 0.068\ \text{kg/s}.$$

If the optimum ϵ for the streams of Table 8-6 is determined from an economic analysis to be equal to 0.01, the composite rich-stream load line must be moved downward until the point with a value of $x = 0.104$ (the former pinch point) is 0.01 to the right of the lean-stream load line. As before, the equation of the lean-stream load line is

$$\text{Mass exchanged} = (8\ \text{kg/s})x.$$

Therefore, the y-axis value for the composite rich-stream load line at $x =$ 0.104 is

$$\text{Mass exchanged} = 8\ \text{kg/s}(0.104 - 0.01) = 0.752\ \text{kg/s},$$

and the first point to plot on the composite rich-stream load line is (0.104, 0.752 kg/s). The remaining composite rich-stream load line points are shown in Figure 8-27. Values for k'_{req} and k'_{rem} are calculated as in Example 8-6:

$$k'_{\text{req}} = (1.12 - 0.952)\text{kg/s} = 0.168\ \text{kg/s}$$

and

$$k'_{\text{rem}} = 0 - (-0.148\ \text{kg/s}) = 0.148\ \text{kg/s}.$$

8.2.3 Determining Which Streams Contact Each Other in a Mass-Exchange Network

MEN synthesis is powerful because of its graphic determination of the pinch point and its ability to show whether mass exchange is thermodynamically feasible. The culmination of all the previous steps is the synthesis of a mass-exchange network. Combined load-line diagrams show where the lean and rich streams contact each other in a mass-exchange network. For example, inspection of Figure 8-24 reveals that rich stream 1 contacts the lean stream as the lean stream exits the mass-exchange network. Then, where the lean-stream mass fraction is $0.60/L_1 = 0.04$, the lean stream is split and one-third contacts rich stream 1 while the remaining two-thirds contacts rich stream 2. The lean stream enters this mass-exchange network at the same point where both rich streams exit.

Example 8-7 Pairing Streams in a Mass-Exchange Network Describe how the streams contact each other in the optimal mass-exchange network of Figure 8-27. (This is the network where the value of ϵ is equal to 0.01.) To begin, rich stream 1 contacts the lean stream when the lean stream exits the mass-exchange network. Further down the mass-exchange network, the lean stream is split into two parts, with one third contacting rich stream 1 and two-thirds contacting rich stream 2. Complete this description for the mass-exchange network, and give the lean-stream and rich-stream mass fractions at which the splits/junctions occur.

Solution As stated in the problem statement, rich stream 1 contacts the lean stream where the lean stream exits the mass-exchange network. At this point the mass fraction of k in the lean stream is $(0.952\ \text{kg/s})/(8\ \text{kg/s}) = 0.119$ (found from the equation of the lean-stream load line) and the rich-stream mass fraction is $0.149 \times 0.67 = 0.10$ (read directly from the composition interval diagram). Where the lean-stream mass fraction is $0.852/8 = 0.107$ and the rich-stream mass fraction is $0.119 \times 0.67 = 0.08$, the lean stream is split, with one-third contacting rich stream 1 and two-thirds contacting rich stream 2. The lean stream is split three ways when its mass fraction is $0.752/8 = 0.094$ and the rich-stream mass fraction is $0.104 \times 0.67 = 0.07$, with one-fourth contacting rich stream 1, one-half contacting rich stream 2, and the remaining fourth contacting rich stream 3. To find the mass fraction of k in the lean stream when the rich streams exit the mass-exchange network, the equation of the corresponding portion of the composite rich-stream load line must first be found. It is known that the slope of this line is

$$m = 0.67(R_1 + R_2 + R_3) = 13.4\ \text{kg/s}.$$

The intercept is found by using one of the known points as follows:

$$0.752\ \text{kg/s} = 13.4\ \text{kg/s}(0.07/0.67) + b.$$

Therefore, $b = -0.648$ kg/s. The mass fraction at which this line crosses the x axis can be found from

$$0 = (13.4\ \text{kg/s})x - 0.648\ \text{kg/s}.$$

This means that the rich streams exit the mass-exchange network when the mass fraction of k in them is $0.67 \times 0.648/13.4 = 0.0324$. The lean-stream mass fraction at this point is 0. See Figure 8-29 for a flowsheet of this network.

Stream matching around the pinch point is particularly complex (Douglas, 1988).

8.2.4 MEN Synthesis Examples

The practical application of MEN synthesis can be made clearer through examination of solved problems. The following example illustrates the usefulness of MEN analysis in determining whether mass exchange between a rich and a lean stream is a feasible pollution prevention strategy. All that is needed is information about the flow rates and target and supply concentrations of the lean and rich streams, and an appropriate equilibrium relationship. This example also illustrates the use of a nonlinear equilibrium relationship.

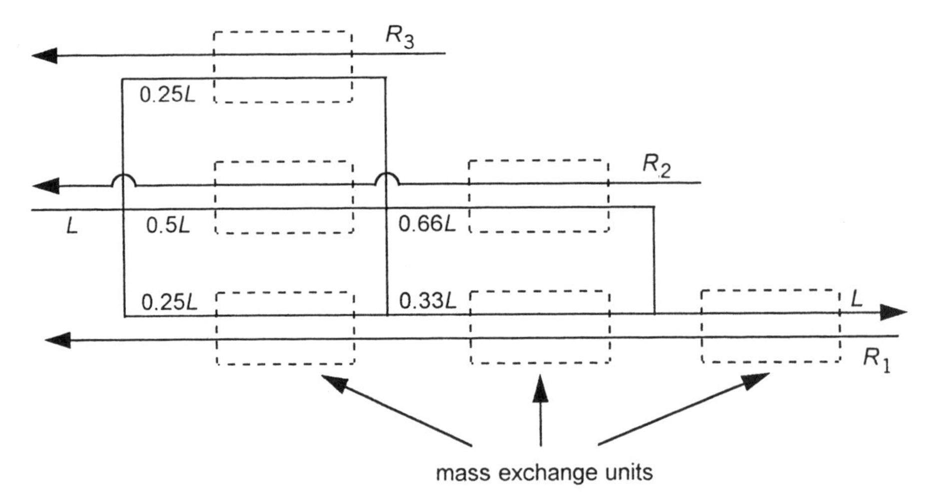

Figure 8-29. Contact in a mass-exchange network between the lean and rich streams of Table 8-6.

Example 8-8 Preventing Waste at a Pesticide Formulation Plant The active ingredients in pesticides are generally too potent to be used or marketed in pure form. To formulate pesticides, the active ingredients are sometimes diluted with water or a solvent; in other cases they may be mixed with a solid such as sand or clay. One product at a large pesticide formulation plant is created by mixing a liquid active ingredient with sand in a ratio of 9 g of active ingredient per kilogram of sand. Between runs, the mixer is generally cleaned with sand that becomes final product, but when the mixer requires maintenance, it must be cleaned with high-pressure steam. Steam cleaning generates 20,000 L/yr of wastewater with an active ingredient concentration of 3 g/L. This wastewater stream is currently mixed with wastewater streams containing other active ingredients and treated, generating solid waste with high disposal costs. The objective in this example is to determine whether wastewater from steam-cleaning the mixers could be purified to the extent desired by contacting it with sand in a continuous cross-current adsorber. The sand from the adsorber would become final product, eliminating waste due to steam-cleaning the mixers. The Freundlich equation,

$$S = bC^m,$$

where S is the concentration of the adsorbate loading on the solid and C is the concentration of the adsorbate in the liquid, provides a good fit for adsorption of liquids onto solids (McCabe et al., 1985). To be conservative, it can be assumed that 9 g/kg is the maximum loading for the active ingredient on sand. Another conservative approximation is to set the exponent m equal to 0.7, which is its value for Aldicarb in sand (Lafrance et al., 1988). Aldicarb

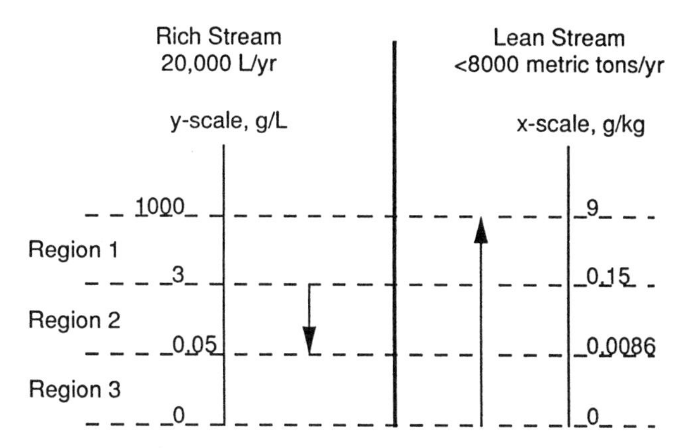

Figure 8-30. Combined composition interval diagram for pesticide-contaminated wastewater.

is a pesticide known for its high mobility through soil. These approximations result in the equilibrium expression

$$S = 0.07C^{0.7},$$

where S is in g/kg and C is in g/L. Since this equation probably represents a less favorable relationship than actually exists between the active ingredient of this example and sand, it is suitable for a preliminary analysis. What is the minimum flow rate of sand required in the adsorber if the target concentration of the wastewater stream is 0.05 g/L? Is this pollution prevention strategy feasible if 8000 metric tons of the formulation of this active ingredient with sand are produced at the pesticide formulation plant each year?

Solution The combined composition interval diagram is pictured in Figure 8-30. In this case, the flow rate of the lean stream is unknown so the rich-stream load line is plotted first on the load-line diagram. Because of the nonlinear equilibrium expression, the solution is best obtained by plotting the rich-stream load line against the lean-stream equilibrium concentrations, so that the lean-stream load line is linear. The lowest possible slope for the lean-stream load line (minimum flow rate of sand) is obtained when the lean-stream load line begins to the left of the end of the rich-stream load line, and ends at a pinch where the rich-stream load line begins. The combined load-line diagram is given in Figure 8-31. It can be seen from the slope of the lean-stream load line that the minimum flow rate of sand required to purify the wastewater to the extent desired is [0 − (−59,000 g/yr)]/(0.15 g/kg − 0) = 400 metric tons/yr, which is more than an order of magnitude less than the sand used to formulate this particular pesticide. The next step in this analysis would be to determine the economic viability of such an adsorber.

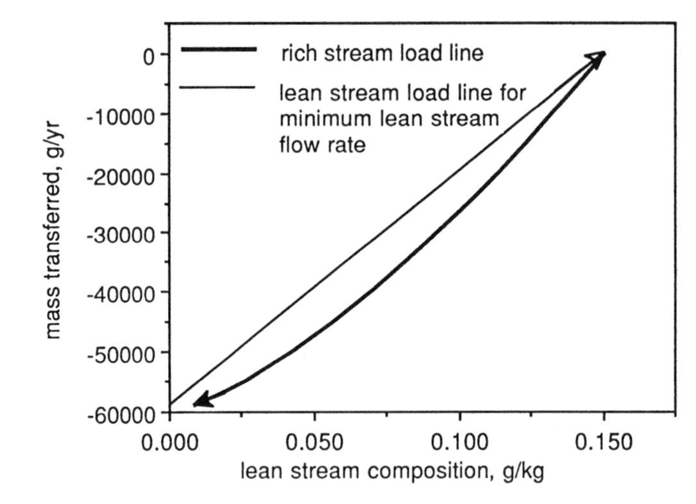

Figure 8-31. Load-line diagram for pesticide-contaminated wastewater.

In the next example, the objective is to determine which of two lean streams removes the most metal from a solvent and a flux created during printed-circuit-board (PCB) manufacture.

Example 8-9 Recovery of Metal from Waste Streams in Printed Circuit Board Manufacture PCB manufacture begins with plastic or fiberglass boards laminated with metal foil. Holes are drilled in these boards in order to attach components or to provide conductive pathways. A layer of material called photoresist is added, then a photolithographic process is used to create an image of the desired circuitry on the board, usually using UV light. Next, the photoresist that became soluble during the photolithographic process is removed and the exposed metal is etched away. Sometimes further solder plating and reflowing of the remaining pattern is necessary. In this example, a PCB manufacturer is currently using an ammonia-based etchant to remove the metal plating. Because metal recovery from ammoniacal solutions is problematic (Foecke, 1988), this large facility is converting to a new etchant that allows for metal recovery with a simple drop in temperature. Two such etchants are available, both with identical cost and quality characteristics. It has been proposed that the new etchant be used in a countercurrent extraction operation to reduce the metal content of (1) the halogenated solvent used to develop the photoresist and (2) the reflow flux. Therefore, the etchant becomes a lean stream because it is desirable to transfer the metal into it so that it can be recovered, as shown in Figure 8-32. The halogenated solvent and flux are rich streams. Concentrations and flow rates of the four streams are given in Table 8-7. The inlet and outlet concentrations of the etchant

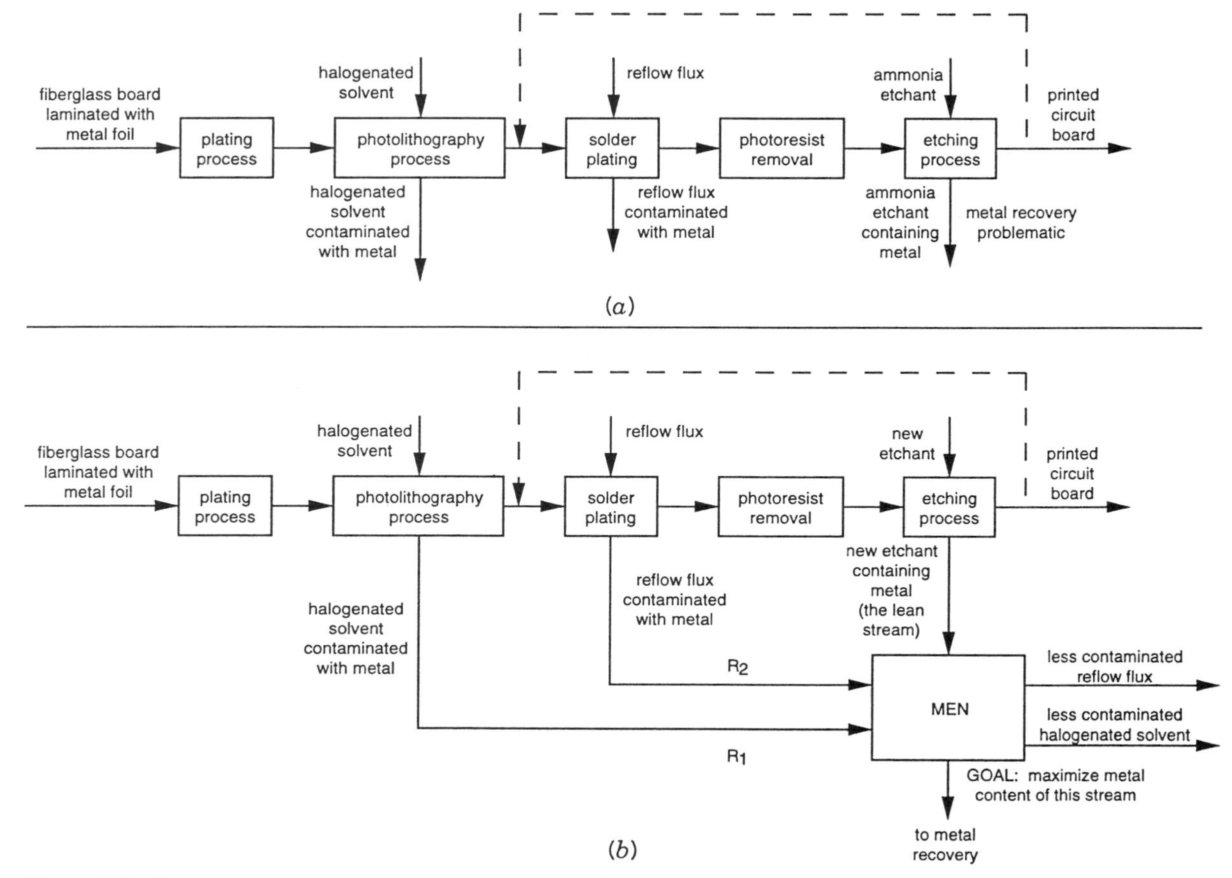

Figure 8-32. Process diagram for the proposed modification of Example 8-9: (*a*) current process; (*b*) proposed process.

Table 8-7. Lean and rich streams in PCB manufacture

			Concentration, g/L	
Stream Description	Stream Label	Flow Rate, L/hr	Inlet	Outlet
New etchant 1	L_1	3.79	40	80
New etchant 2	L_2	3.79	35	60
Halogenated solvent	R_1	0.379	50	—
Reflow flux	R_2	0.379	20	—

streams enclose the region in which they are most effective at removing the metal. Equilibrium relationships are as follows:

$$L_1 - R_1: \quad y_1 = 1.1x_1,$$

$$L_1 - R_2: \quad y_2 = 0.4x_1 + 1,$$

$$L_2 - R_1: \quad y_1 = 0.9x_2,$$

$$L_2 - R_2: \quad y_2 = 0.5x_2 + 2,$$

where x is the concentration in g/L in the lean stream and y is the concentration in g/L in the rich stream. Because the two etchants have the same characteristics, the facility will choose the one that removes the most metal from the rich streams, and therefore allows for the greatest metal recovery. Which etchant is this?

Solution Outlet concentrations for the rich streams are not given in the problem statement. However, it is desirable to remove as much of the metal as possible from the rich streams, so the CIDs can be created by setting their target concentrations at zero. Two CIDs are necessary for solving this problem: one for L_1 and the other for L_2. The CID for L_1 is shown in Figure 8-33. To create the load-line diagram, the lean stream is plotted first, as shown in Figure 8-34. The pinch can be located by recognizing that the slope of the rich-stream load line is always less than the slope of the lean-stream load line. (In fact, once this is recognized, the solution can be obtained without creating a load-line diagram, but for the sake of generality, load-line diagrams are constructed in this solution.) Therefore, the pinch is at point (40,0). Points on the rich-stream load line below and to the left of the pinch point do not provide information useful in solving this problem. The slope of the rich-stream load line at the pinch point, obtained from the equilibrium relationships for L_1 and the rich streams, is

$$1.1R_1 + 0.4R_2 = 0.5685 \text{ L/hr},$$

and the next point on the composite rich-stream load line is at $x_1 = 45.45$ g/L

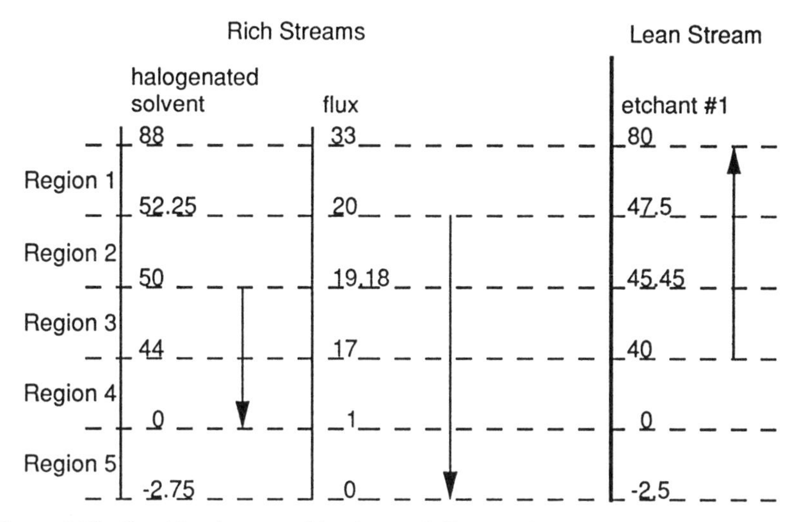

Figure 8-33. Combined composition interval diagram for etchant 1 of Example 8-9.

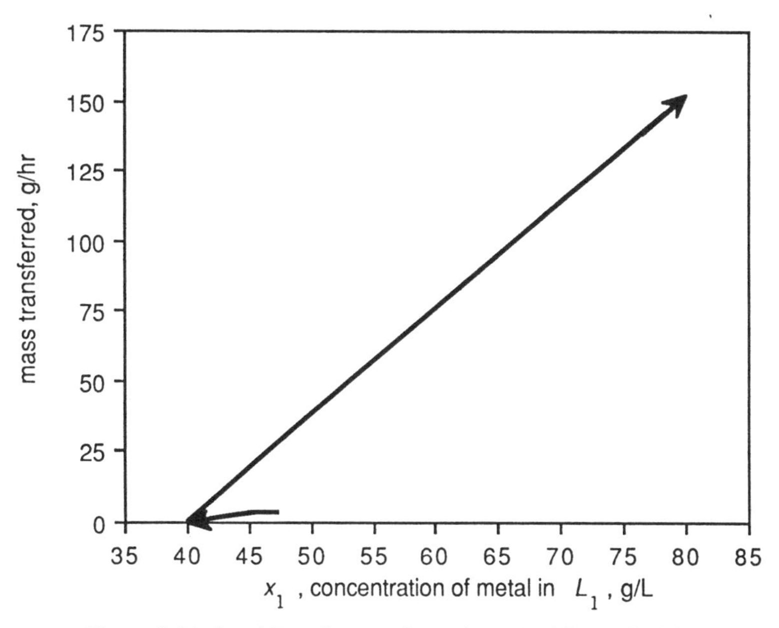

Figure 8-34. Load-line diagram for etchant 1 of Example 8-9.

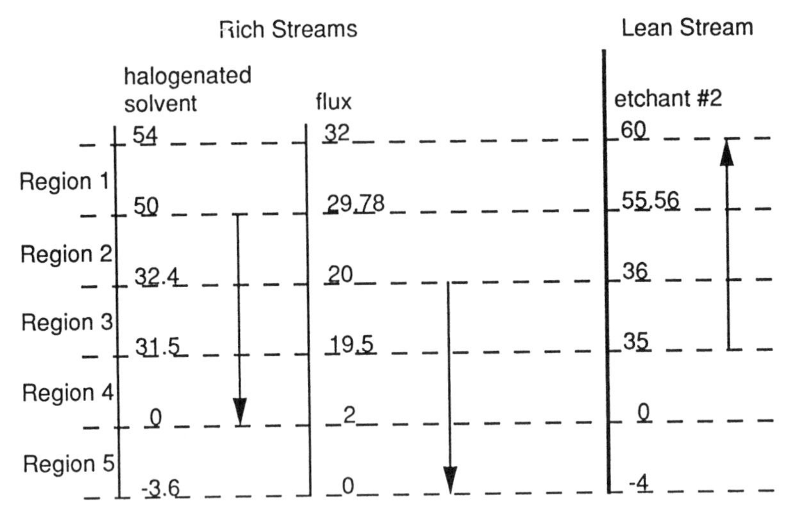

Figure 8-35. Combined composition interval diagram for etchant 2 of Example 8-9.

(from Figure 8-33). The value for mass exchanged at this point, found by using the definition of the slope of a line, is

$$0.5685\ \text{L/hr} \times (45.45\ \text{g/L} - 40\ \text{g/L}) = 3.098\ \text{g/hr}.$$

The endpoint of the rich-stream load line occurs at $x_1 = 47.5$ g/L, and the slope to this point is $0.4R_2 = 0.1516$ L/hr. Therefore, this point falls at

$$3.098\ \text{g/hr} + 0.1516\ \text{L/hr} \times (47.5\ \text{g/L} - 45.45\ \text{g/L}) = 3.4\ \text{g/hr}.$$

This is the theoretical maximum amount of metal recoverable from the solvent and the flux when etchant 1 is used. The CID and load-line diagram for L_2, constructed similarly, are shown in Figures 8-35 and 8-36. The theoretical maximum amount of metal recoverable when the second etchant is used, 7.2 g/hr, is over twice that of the metal recovered when the first etchant is used.

The best solution for Example 8-9 may be to use both etchants. The optimization might then consist of varying the lean-stream flow rates while keeping the number of exchangers to a minimum. Optimizing when there is a choice of lean streams and lean-stream flow rates can be done using linear programming techniques, which have been developed. Analytical methods for determining the pinch point in a MEN problem and problems solved using linear programming techniques can be found in works published by

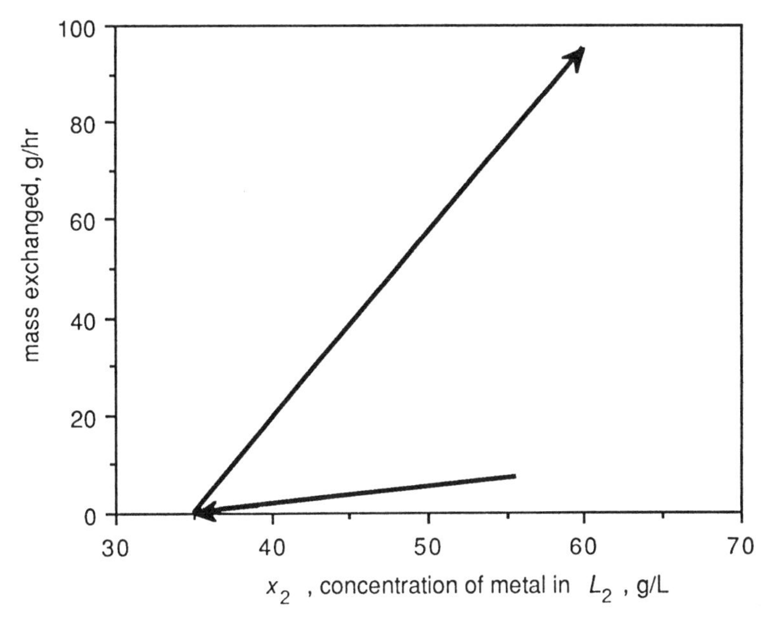

Figure 8-36. Load-line diagram for etchant 2 of Example 8-9.

the group that developed MEN synthesis (e.g., in El-Halwagi and Manousiouthakis, 1989, 1990).

8.3 SUMMARY

Flowsheet analysis, either qualitative or quantitative, is a powerful tool for pollution prevention. In material flow analysis, careful consideration must be given to system boundaries because their placement can have a profound effect on the viability of a pollution prevention option. Material flow analysis in the form of mass balances is in some cases a straightforward method for obtaining information about waste flows. Once system boundaries have been determined and the waste flows and emissions have been estimated, the waste streams and unit operations in the flowsheet can be examined in a systematic manner for pollution prevention opportunities.

Mass-exchange network synthesis provides a systematic framework of analysis for determining the extent to which a pollutant in a waste stream can be transferred to a stream in which it has a positive value. Such transfer could potentially prevent harmful materials from entering the environment.

PROBLEM STATEMENTS

1. The oil recycling plant shown in Figure 8-37 generates two phenol-containing waste streams, R_1 and R_2. The characteristics of these streams are listed in Table 8-8. Phenol is to be removed from these waste streams using the lean streams described in Table 8-9. The equilibrium data for the transfer of phenol to the jth lean stream is given by $y = m_j x_j$, where the values of m_j are 2.00 and 1.53 for S_1 and S_2, respectively. Throughout this problem, a minimum allowable composition difference of $\epsilon_j = 0.001$ kg phenol/kg rich stream is to be used.
   a. Plot the composition interval diagram for the rich and lean streams.
   b. Plot the composite load lines for the rich and lean streams on a diagram with mass exchanged on the y-axis and rich-stream mass fraction y_i on the x axis. Use the maximum lean-stream flow rates. Where is the pinch point?
   c. How much phenol can be transferred from the rich streams to the lean streams?

OPEN-ENDED PROBLEM STATEMENTS

1. Using the framework described in Section 8.1, identify pollution prevention options for a nitroaromatics process. In the process, aromatic compounds and nitric acid are first mixed in a reactor where they form nitroaromatic compounds and reaction byproducts, as shown in Figure 8-38. The crude leaving the reactor proceeds to a distillation step, where nitroaromatic compounds are removed from the crude and purified. In the distillation step, a detarring column is used to boil off the nitroaromatic product, leaving the reaction byproducts, unrecovered product, and distillation tars to accumulate within the reboiler. A pump removes these wastes continuously. The flow from the reboiler must be carefully measured and controlled to prevent a buildup of reaction byproducts to unsafe levels. This flow control is accomplished by means of a flow meter and valve. The flow rate is so low that a $\frac{1}{8}$-in. flow tube is used to obtain accurate measurements. At extremely low flows the valve and flow meter become plugged, forcing a process shutdown. To prevent this occurrence, the operation runs at a higher than optimal flow rate. The waste stream leaving the aromatics column, which is sent to incineration, consists of 76% unrecovered product, 20% byproducts, and 4% distillation tars (by mass). For every pound of nitroaromatic produced, 0.014 lb of waste is created. Costs associated with this waste stream include the yield loss represented by unrecovered product and the cost of incinerating the waste stream.
2. Using the framework described in Section 8.1, identify pollution prevention options for a process for making diphenol ether. In this process, a solid organic salt is slurried with solvent within a slurry tank and

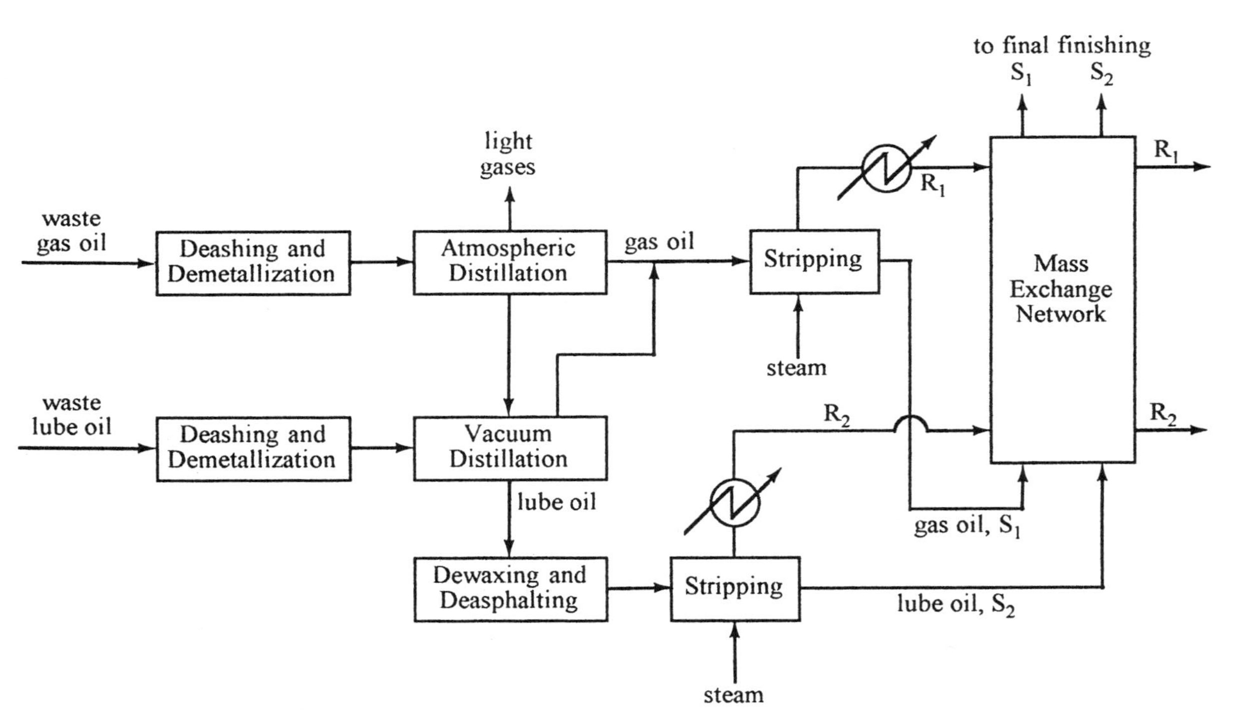

Figure 8-37. Flowsheet of an oil-recycling plant with phenolic wastes.

Table 8-8. Data for the waste streams of Problem Statement 1

Stream	Description	Flow Rate G_i, kg/s	Supply Composition y_i^s	Target Composition y_i^t
R_1	Condensate from first stripper	2	0.050	0.010
R_2	Condensate from second stripper	1	0.030	0.006

Table 8-9. Data for the process lean streams of Problem Statement 1

Stream	Description	Upper Bound of Flow Rate L_j, kg/s	Supply Composition x_j^s	Target Composition x_j^t
S_1	Gas oil	5	0.005	0.015
S_2	Lube oil	3	0.010	0.030

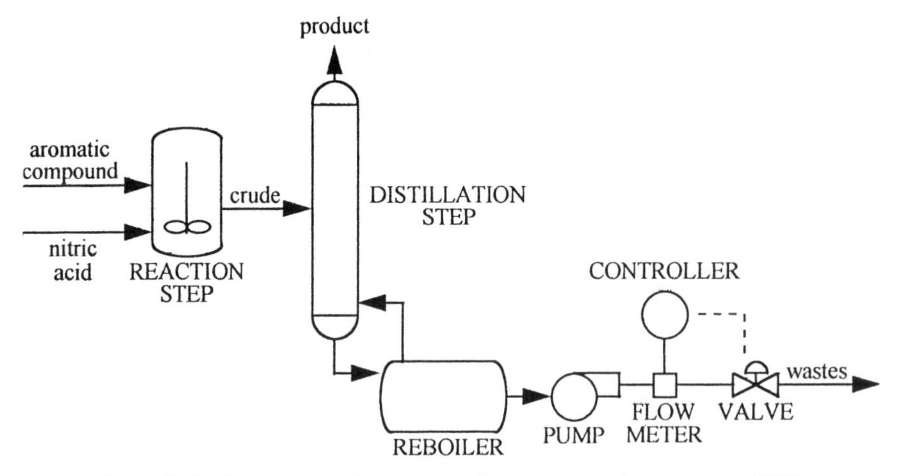

Figure 8-38. A nitroaromatics process (Report to the Government, 1993).

then introduced to the reaction vessel, as shown in Figure 8-39. In the reaction vessel, the slurry is mixed with chlorinated phenol and heated. The resulting reaction produces diphenol ether and a small amount of byproducts. When the reaction is completed, a vacuum is applied to the equipment to lower the boiling point of the solvent. The solvent then boils up the distillation column and collects in the solvent receiving tank for eventual reuse. About half of the solvent in the reaction vessel is recovered in this way, enough to nearly fill the receiving tank. After the solvent-recovery step, the reaction mass is dumped into a drowning tank full of water. It is then washed with a large, continuous stream of water. The diphenol ether settles to the bottom of the tank, while

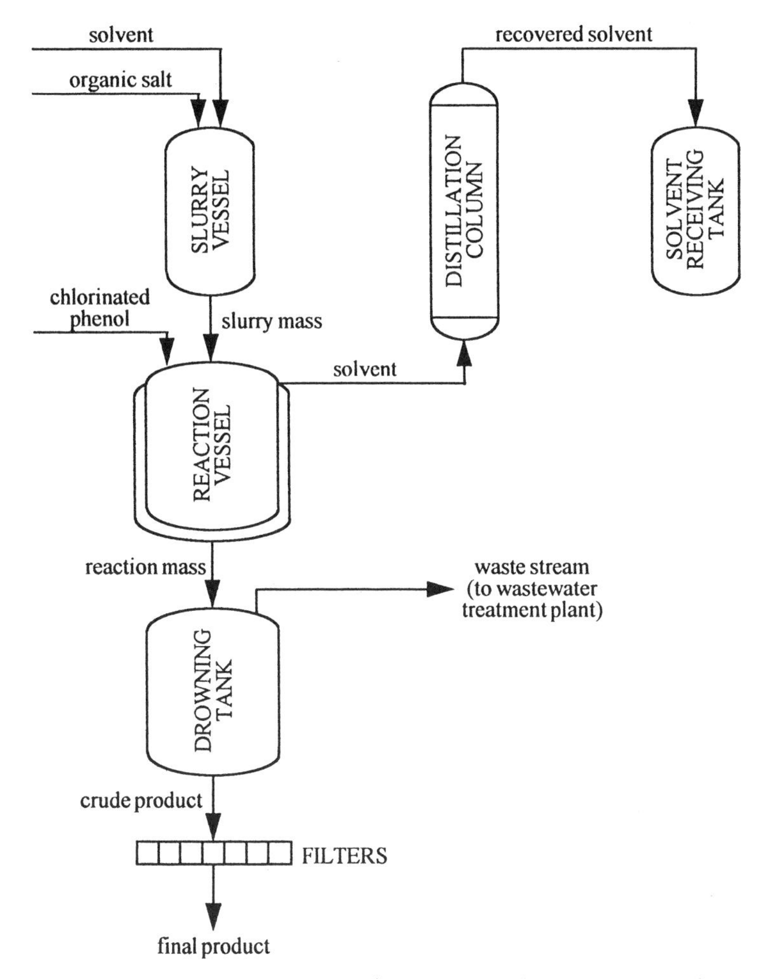

Figure 8-39. Diphenol ether process (Report to the Government, 1993).

the remaining solvent and reaction byproducts dissolve in the aqueous wash. The waste-bearing water is siphoned off continuously from the top of the tank and sent to the on-site wastewater treatment plant for disposal. After the wash step, the free liquid is filtered off. The diphenol ether is then sent to another facility on-site for further processing. Water comprises almost 99% of the waste from the drowning tank. On a dry basis, this waste stream would typically be 50% inorganic salt, 36% solvent, 11% unreacted raw material, and 3% reaction byproducts by mass. For every pound of diphenol ether produced, 0.24 lb of waste (dry basis) is produced. Major costs associated with the wastewater stream are the costs of wastewater treatment and the raw material costs represented by the unrecovered solvent. Yield losses from this process are extremely low.

REFERENCES

American Petroleum Institute (API), "Waste Minimization in the Petroleum Industry," API Publication 849–00020, Washington, DC, 1991.

Crowl, D. A. and J. F. Louvar, *Chemical Process Safety: Fundamentals with Applications*, Prentice-Hall, Englewood Cliffs, NJ, 1990.

Douglas, J. M., *Conceptual Design of Chemical Processes*, McGraw-Hill, New York, 1988.

Douglas, J. M., " Process Synthesis for Waste Minimization," *Ind. Eng. Chem. Res.*, **31**(1), 238–243, Jan. 1992.

El-Halwagi, M. M. and V. Manousiouthakis, "Synthesis of Mass Exchange Networks," *AIChE J.*, **35**(8), 1233–1244, Aug. 1989.

El-Halwagi, M. M. and V. Manousiouthakis, "Automatic Synthesis of Mass Exchange Networks with Single Component Targets," *Chem. Eng. Sci.*, **45**(9), 2813–2832, 1990.

Foecke, T. L., "Waste Minimization in the Electronics Products Industries," *J. Air Pollution Control Assoc.*, **38**(3), 283–291, March 1988.

Lafrance, P., L. Ait-ssi, O. Banton, P. G. C. Campbell, and J. P. Villeneuve, *Water Pollution Research J. Can.*, **23**(2), 253–269, 1988.

McCabe, W. L., J. C. Smith, and P. Harriott, *Unit Operations of Chemical Engineering*, McGraw-Hill, New York, 1985.

Report to the U.S. Government in satisfaction of consent decree in *U.S. v. DuPont*, Docket No. 91cv768(JFG) "DuPont Chambers Works Waste Minimization Project: Chambers Works—Deepwater, New Jersey," May 1993.

Rossiter, A. P. and H. Klee, "Hierarchical Process Review for Waste Minimization," in *Waste Minimization through Process Design*, A.P. Rossiter, ed., McGraw-Hill, New York, 1995.

Tellus Institute, "Alternative Approaches to the Financial Evaluation of Industrial Pollution Prevention Investments," prepared for the New Jersey Department of Environmental Protection, Project P32250, 1991.

9

MANAGEMENT OF POLLUTION PREVENTION ACTIVITIES AT INDUSTRIAL FACILITIES

The previous four chapters of this text have described processes for identifying and characterizing waste streams and proposing pollution prevention projects to minimize the impacts of those streams. Once the options for pollution prevention have been identified, they must themselves be evaluated. The first portion of this chapter describes considerations that are particularly important in the economic analysis of pollution prevention projects. Typical economic evaluations of projects are far from comprehensive in their treatment of the hidden, deferred, and qualitative costs of waste generation, and methods for estimating or evaluating these costs are described in this chapter. The second portion of this chapter is devoted to descriptions of procedures for ranking the pollution prevention options available to a facility. Time and budget limitations make it impossible for facilities to pursue all the pollution prevention options available to them. However, it is not necessarily obvious which of the possible strategies available to a facility are superior. In such cases, some type of systematic prioritization scheme to rank the available pollution prevention options should be employed, as described in this chapter.

9.1 ECONOMIC EVALUATION OF POLLUTION PREVENTION PROJECTS

The economic evaluation of engineering projects typically involves estimation of equipment, installation, raw-material, energy, and maintenance costs. Disposal and pollution control costs are often factored into these calculations in determining economic rates of return, but other regulatory and social costs are not. In this section, total cost assessment of waste management alternatives is described, and the hidden costs, future liabilities, and less tangible costs associated with waste generation are discussed. The section closes with a summary of commercially available packages for economic evaluation of pollution prevention projects.

9.1.1 Total Cost Assessment of Pollution Control and Prevention Strategies

Traditional economic measures of engineering projects evaluate equipment, raw-material, energy, operating, and maintenance costs. These evaluations generally overlook some of the costs of waste generation. A more complete accounting of environmental costs is referred to as *total cost assessment*. One approach to total cost assessment is illustrated in Figure 9-1. Four types of

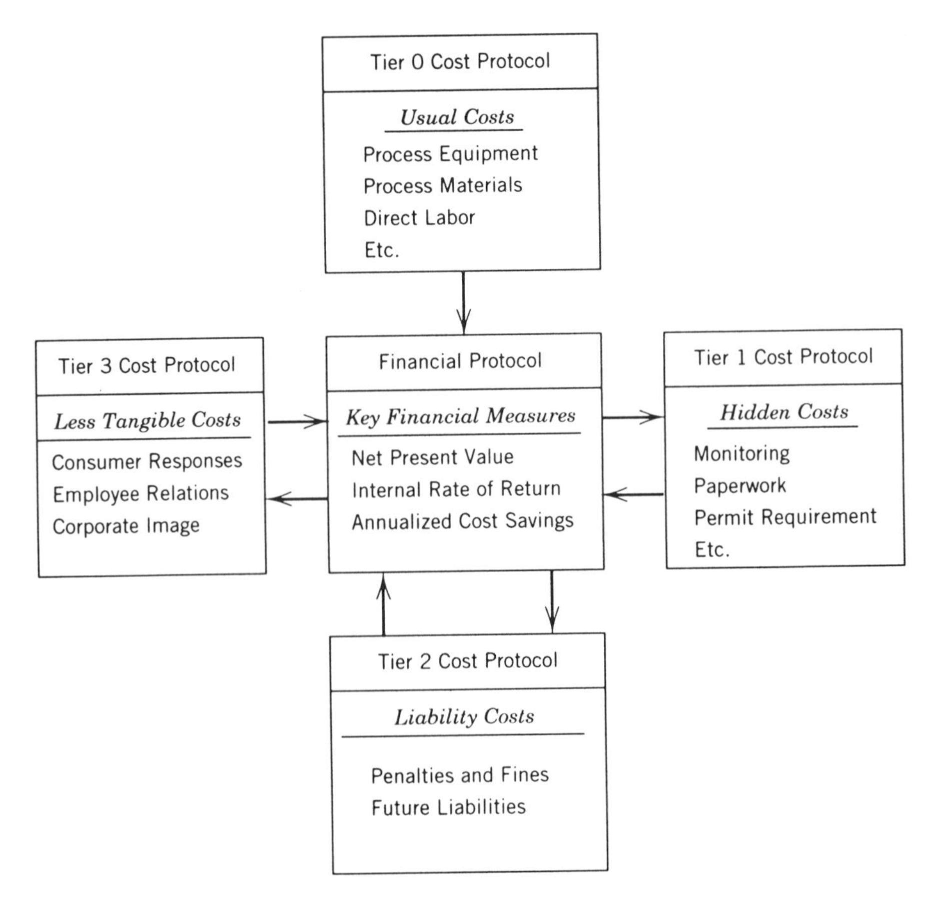

Figure 9-1. Schematic of EPA methodology for performing a total cost assessment (US EPA, 1989).

costs are identified, labeled tier 0, tier 1, tier 2, and tier 3. Tier 0 costs are the "usual" costs that are included in a conventional analysis of a project. Tier 1 costs include permitting, reporting, monitoring, manifesting, and insurance costs and are often referred to as "hidden" costs because they are usually treated as overhead costs and are not directly charged to a project. Waste disposal costs are sometimes treated as overhead costs as well. Tier 2 costs include future liabilities, which are extremely difficult to accurately evaluate. Even more difficult to evaluate are tier 3 costs, which include consumer responses, employee relations, and public image. All four tiers of costs provide information for financial analysis methods, where measures such as rate of return and payback periods are evaluated.

A thorough analysis of tier 0 costs is the first step in performing a total cost assessment. These costs are described in a number of textbooks on engineering economics (e.g., Valle-Riestra, 1983) and are not the subject of this section. Instead, approximate procedures for quantitatively estimating hidden costs such as permitting, reporting, monitoring, manifesting, insurance, and waste management costs are provided along with qualitative and quantitative methods for evaluating future liabilities. Tier 3 costs are briefly described as well.

9.1.1.1 ***Evaluating Hidden Costs*** As mentioned above, tier 1 costs are often charged to overhead accounts, and as a consequence are regarded as "hidden." As a result, very simple opportunities for waste reduction, and therefore cost avoidance, are sometimes not exploited. For example, replacing an environmentally hazardous material with a more benign alternative may be difficult to justify if there are no direct cost savings. However, if the substitute material eliminates the need for reporting in the Toxic Chemical Release Inventory (TRI) as well as the need for manifesting residues that are regulated under the Resource Conservation and Recovery Act (RCRA), the actual cost savings associated with the alternate material may be substantial.

Evaluation of some of the hidden costs associated with waste generation and emissions is straightforward. For example, waste taxes and fees, as well as monitoring and analysis costs, can be quantitatively evaluated. Reporting and paperwork costs can be much more challenging to estimate, particularly for large organizations where these efforts are centralized. Reporting and paperwork costs for RCRA are used in this section to illustrate a mechanism for estimating such costs.

The level of reporting required under RCRA depends on whether the facility is a large- or small-quantity generator, whether the facility transports waste, and whether the facility is considered to be a treatment, storage, and disposal site. Costs associated with RCRA reporting might include notification, reporting, recordkeeping, manifesting, labeling, monitoring/testing, planning/studies/modeling, training, inspections, preparedness/protective equipment, closure/post closure assurance, and/or insurance and special taxes. The U.S. EPA has provided data for estimating the costs of the first five of these categories. These data are summarized in Tables 9-1 to 9-5. Costs are estimated by multiplying the projected frequency of a reporting occurrence by the sum of the nonlabor and labor costs per occurrence. As shown in the tables, the costs estimated by applying this method can vary by orders of magnitude from facility to facility, depending on factors such as the number of waste streams that are generated or managed, whether the management takes place on-site, and the salary level of the responsible individual. Nevertheless, the general procedures are reasonable and can be tailored to fit individual facilities.

As shown in Tables 9-1 to 9-5, the costs associated with notification, reporting, recordkeeping, manifesting, and labeling for RCRA reporting can be substantial. The potential importance of these costs in an economic analysis of alternatives is illustrated in the following example.

Example 9-1 Consideration of Hidden Costs during Material Substitution Evaluation A small facility must make a decision on whether to substitute compound A for compound B in their parts-cleaning operations. The facility would require the same quantity of compound A as it does of compound B, and compounds A and B have roughly the same volumetric cost. However, expenditures other than material costs set the two compounds apart. For instance, compound B is listed as a hazardous chemical and compound A is not, so the facility has to pay higher waste transport and disposal costs for compound B than it would for compound A. On the other hand, compound B has qualities that make it superior to compound A for parts cleaning, so that operating expenses are projected to be lower if compound B is used. When these two factors are considered, projected expenses for the compound A scenario are $1000 more annually than for the compound B scenario. However, it is suggested that if a tier 1 cost estimate for compound B is made and hidden costs are included in the analysis, compound A will be proven to be the more economically desirable alternative. Using the information in Tables 9-1 to 9-5, perform a tier 1 analysis to determine whether the facility should substitute compound A for compound B in its parts-cleaning operations. The facility is currently a small-quantity generator of compound B and has chosen to hire outside consultants to perform their reporting, recordkeeping, and manifesting tasks. The hourly rate for the outside consultants is $100. The employees who label compound B wastes cost the facility an average of $30/hr. No exception reports on compound B have ever been, or are expected to be, filed. Recordkeeping for compound B has been minimal, at six occurrences per year. Compound B is temporarily stored at the site and transported biweekly to a treatment, storage, and disposal facility, so labeling and manifesting expenses occur 26 times a year. Manifesting is routine and takes no more than 15 min for each occurrence.

Solution Use of compound B incurs no notification costs because the facility

Table 9-1. Method for estimating RCRA notification costs

$$\underset{(\$/\text{yr})}{\text{Annual cost}} = \underset{(\text{No./yr})}{\text{Frequency of occurence}} \times \left(\text{Nonlabor cost per occurrence} + \text{Labor costs per occurrence} \right)$$

RCRA Category	Notification Provision and Citation for Relevant Legislation	Frequency of Notification (Occ[a]/yr)	Nonlabor Cost per Notification ($/Occ)	Time Required per Notification (hr)	Average Wage Rate of Person Completing Notification ($/hr)	Approximate Annual Cost ($/yr)[b]
Exporter of hazardous waste	Exportation of Hazardous Waste Notification 40CFR §262.53	1	2	2–3	25–100	50–300
Treatment storage or disposal facility	RCRA Foreign Source Notification 40CFR §164.12(a), 40CFR §165.12(a)	0–5	1	2	25–100	0–1,000
	RCRA Permit Confirmation 40CFR §164.12(b)	1–4	1	2	25–100	50–800
	Local Notification of Operations 40CFR §264.37 40CFR §265.37	1	3	40	25–100	100–4,000
	Manifest Discrepancy Notification 40CFR §264.72 40CFR §265.72	0–1.25	1	2	25–100	0–20,000

[a]Occurence(s).
[b]Costs given to one significant figure.
Source: US EPA (1989).

Table 9-2. Method for estimating RCRA reporting costs

$$\text{Annual cost (\$/yr)} = \text{Frequency of occurence (No./yr)} \times \left(\text{Nonlabor cost per occurrence} + \text{Labor costs per occurrence}\right)$$

RCRA Category	Reporting Provision and Citation for Relevant Legislation	Frequency of Reporting (Occ/yr)	Nonlabor Cost per Report ($/Occ)	Time Required per Report (hr)	Average Wage Rate of Person Completing Report ($/hr)	Approximate Annual Cost ($/yr)[a]
RCRA waste generator (large- or small-quantity)	RCRA Biennial Report 40CFR §262.41	0.5	5	8	25–100	100–400
Large-quantity generator (LQG)	LQG Exception Report 40CFR §262.42(a)	0.1–1.5	1	2	25–100	6–300
Small-quantity generator (SQG)	SQG Exception Report 40CFR §262.42(b)	0–0.1	1	0.25	25–100	0–4

Waste exporter	Primary Exporter's Annual and Exception Reports 40CFR §262.55 40CFR §262.56	1–2.5	2	2.5	25–100	60–600
Treatment, storage, or disposal facility (TSDF)	TSDF Biennial Report 40CFR §264.35 40CFR §265.75	0.5	5	8–40	25–100	100–2,000
TSDF	Unmanifested Waste Report 40CFR §264.76 40CFR §264.75	0–1.25	1	1	25–100	0–1,200
TSDF	Release, Fire, Explosion and Closure Reporting	2	2	5	25–100	250–1,000

[a]Costs given to one significant figure.

Source: US EPA (1989).

Table 9-3. Method for estimating RCRA recordkeeping costs

$$\underset{(\$/yr)}{\text{Annual cost}} = \underset{(No./yr)}{\text{Frequency of occurence}} \times \left(\text{Nonlabor cost per occurrence} + \text{Labor costs per occurrence} \right)$$

RCRA Category	Recordkeeping Provision and Citation for Relevant Legislation	Frequency of Recordkeeping (Occ/yr)	Nonlabor Cost per Record ($/Occ)	Time Required per Record (hr)	Average Wage Rate of Person Completing Record ($/hr)	Approximate Annual Cost ($/yr)[a]
RCRA waste generator	Reports, Test Results, Waste Analysis Records 40CFR §262.40	5–100	1	0.25	10–100	10–2,500
RCRA waste exporter	Exporters Records and Notification Records 40CFR §262.57	5	1	0.25	10–100	10–100
Hazardous waste transporter	Manifesting Records 40CFR §263.22	0–200	1	0.25	10–100	0–5,000
TSDF	Operating Record 40CFR §264.73 40CFR §165.73	250	1	0.25	10–100	600–6,000

[a]Costs given to one significant figure.

Source: US EPA (1989).

Table 9-4. Method for estimating RCRA manifesting costs

$$\text{Annual cost (\$/yr)} = \text{Frequency of occurence (No./yr)} \times \left(\text{Nonlabor cost per occurrence} + \text{Labor costs per occurrence}\right)$$

RCRA Category	Manifesting Provision and Citation for Relevant Legislation	Frequency of Manifesting (Occ/yr)	Nonlabor Cost per Manifest ($/Occ)	Time Required per Manifest (hr)	Average Wage Rate of Person Completing Manifest ($/hr)	Approximate Annual Cost ($/yr)[a]
RCRA waste generator	Off-Site Transport Manifesting 40CFR §262, Subpart B	4–100	0.5	0.25–1	25–100	20–10,000
TSDF	Manifesting 40CFR §264.71 40CFR §265.71	4–500	0.5	0.25–1	25–100	20–50,000

[a]Costs given to one significant figure.

Source: US EPA (1989).

Table 9-5. Method for estimating RCRA labeling costs

$$\underset{(\$/\text{yr})}{\text{Annual cost}} = \underset{(\text{No./yr})}{\text{Frequency of occurence}} = \left(\text{Nonlabor cost per occurrence} + \text{Labor costs per occurrence}\right)$$

RCRA Category	Labeling Provision and Citation for Relevant Legislation	Frequency of Labeling (Occ/yr)	Nonlabor Cost per Label ($/Occ)	Time Required per Label (hr)	Average Wage Rate of Person Completing Label ($/hr)	Approximate Annual Cost ($/yr)[a]
RCRA waste generator	Package Marking and Transportation Labeling 40CFR §262.31 40CFR §262.32 40CFR §262.33	4–500	20	0.75	15–50	100–30,000

[a]Costs given to one significant figure.

Source: US EPA (1989).

is neither an exporter of hazardous waste nor a treatment, storage, and disposal facility. From Table 9-2, it can be seen that the RCRA Biennial Report generally takes 8 hr to complete, so annual reporting costs are

$$0.5/\text{yr} \times (\$5 + 8\text{ hr} \times \$100/\text{hr}) = \$400/\text{yr}.$$

Annual recordkeeping costs, from Table 9-3, are

$$6/\text{yr} \times (\$1 + 0.25\text{ hr} \times \$100/\text{hr}) = \$200/\text{yr}.$$

Annual manifesting costs, from Table 9-4, are

$$26/\text{yr} \times (\$0.5 + 0.25\text{ hr} \times \$100/\text{hr}) = \$700/\text{yr}.$$

Finally, annual labeling costs are taken from Table 9-5 and are

$$26/\text{yr} \times (\$20 + 0.75\text{ hr} \times \$30/\text{hr}) = \$1000/\text{yr}.$$

Thus, the total hidden costs of using compound B are $2000/yr (to one significant figure), making compound A the most economical parts-cleaning solvent for the facility.

Additional hidden costs are incurred by requirements of other federal regulations such as the Clean Air Act, the Clean Water Act, Superfund, and the Occupational Health and Safety Act. Tables similar to Tables 9-1 through 9-5 are available for some of these regulations (US EPA, 1989). State and local reporting requirements can also be significant. Although the techniques presented in this section are perhaps imprecise, particularly for large organizations, any effort to systematically measure hidden costs will likely lead to better economic decisionmaking.

9.1.1.2 ***Evaluating Future Liabilities*** Tier 2 costs are the costs of future liabilities. Such liabilities can take the form of either penalties and fines or personal injury and property damage settlements. Penalties and fines are due to regulatory noncompliance. Personal injury and property damage settlements may be the result of legislation like the Comprehensive Environmental Response, Compensation, and Liability Act (CERCLA, also known as "Superfund"). Liability costs can be staggering: some large corporations set aside hundreds of millions of dollars annually for such costs. However, evaluating liability costs and incorporating them into economic evaluations of projects can be very difficult. One source of difficulty is the uncertainty of the costs. It is impossible to predict at the time when a project is evaluated whether and when the waste streams associated with it will generate liability under Superfund. Both the liability costs and the time at which these liabilities occur can dramatically impact a project's economic viability, as shown in the following example.

Example 9-2 The Effect of Liability on the Viability of a Project During 1979 and 1980, company A legally disposed of 104 drums of hazardous waste in a landfill. The landfill subsequently leaked, and the landfill operator is now bankrupt. Although its wastes were disposed of legally, company A is now liable for $100,000 in cleanup costs under the joint and several liability provisions of Superfund legislation. When the wastes were generated, at the rate of one drum per week in 1979 and 1980, disposal costs were $10 per drum. In 1978 the company evaluated a process modification that would have eliminated this waste stream. Capital costs for the project were estimated to be $2000. Operating costs were estimated to be $5 per drum of waste avoided. What would the present value for the pollution prevention project have been at the start of the project (October 1, 1978), assuming that the capital equipment was purchased on October 1, 1978 and that the unit began operation on January 1, 1979? Also assume that the process generating the wastes shut down on December 31, 1980. At first, ignore liability costs and use a 10% annual interest rate, compounded weekly. Repeat this calculation incorporating the liability costs. Assume that the liability occurred on October 1, 1990. The present value of operating costs can be determined using the formula

$$\mathrm{PV} = \frac{(\mathrm{OC}/p)[[1 + (i/p)]^{pn} - 1]}{(i/p)[1 + (i/p)]^{pn}},$$

where PV is present value, i is effective annual interest, p is the number of periods per year, n is the number of years, and OC/p represents the operating costs per period. The following equation is used to convert a future worth to its present value:

$$\mathrm{PV} = \frac{F}{[1 + (i/p)]^{pn}},$$

where F is future worth and PV, i, p, and n are as before.

Solution The present value of the operating costs is determined using the given formula where PV is present value evaluated on 1/1/79, the date when operations started; i is interest (10% per year); p is the number of periods per year ($p = 52$); n is the number of years ($n = 2$); and OC/p is the operating costs per period ($\mathrm{OC}/p = \$(10 - 5) = \$5$, a $5 savings per drum). Substituting in values yields

$$\mathrm{PV} = \$471.$$

The relationship for adjusting this present value to the value on October 1, 1978 is

$$PV = \frac{\$471}{[1 + (i/p)]^{pn}},$$

where $n = \frac{1}{4}$ and PV = \$459. Results of these calculations and the net present value of the project on October 1, 1978 are as follows:

Category	Present Value on 10/1/78
Capital costs	−\$2000
\$5 net savings per drum, 104 drums, 1 drum per week beginning 1/1/79	\$459
Net present value on 10/1/78	−\$1541

To repeat the calculation taking liability into account, determine the present value on October 1, 1978 for a \$100,000 liability expense that occurred on October 1, 1990, according to the equation

$$PV = \frac{\$100{,}000}{[1 + (i/p)]^{pn}},$$

where $n = 12$ and PV = \$30,154. Using this estimate, the net present value of the pollution prevention project, including avoided liability costs, is \$28,613. This is much more favorable than the net present value without the liability.

The previous example clearly illustrated the potential for future liability costs to dramatically impact a project's economic viability. Unfortunately, it is impossible to know, with certainty, whether a particular waste stream will result in liability and when that liability will occur. Estimates can be based on experiences with landfill failure. For example, a company's records might indicate the percentage of landfills that failed within a given time period. Estimates might also be based on the average expected lifetime of a landfill, which is approximately 20 years. Reconsider the previous example, this time using an estimated potential for landfill failure.

Example 9-3 Estimating Future Liability In 1978, company A evaluated a process modification that would eliminate a waste stream that was being generated at a rate of one drum per week and that was expected to be generated for a 2-year period. Capital costs for the modification were estimated to be $2000, and operating costs were estimated to be $5 per drum of waste treated. Disposal costs for the waste stream were $10 per drum. In making the evaluation, the company took into account that 10% of the landfill operators it deals with have created a liability, and that the liability occurred an average of 10 years after the company finished disposing of its wastes. The present value of cleanup costs at failed landfills in 1978 was anticipated to be $290 per drum of waste (the present value of the $100,000 for 100 drums used in the previous example), and company A expects these costs to rise at the same rate as the interest rate. Calculate what the present value for the pollution prevention project was at the start of the project (October 1, 1978), assuming that the capital equipment was purchased on October 1, 1978 and that the unit began operation on January 1, 1979. Include expected liability costs (10% failure at 10 years) and use a 10% annual interest rate, compounded weekly.

Solution The present values of capital cost and net savings per drum were calculated in Example 9-2 and are −$2000 and $459, respectively. The present value of projected liability costs is

$$0.1(104\text{ drums})(\$290/\text{drum}) = \$3016.$$

Therefore, the net present value of the pollution prevention project is

$$-\$2000 + \$459 + \$3016 = \$1475.$$

This example shows that there can be a great deal of difference between knowing with certainty that a waste stream will result in a liability, and knowing that, on average, a certain fraction of waste streams result in a liability. Further uncertainty is introduced by the magnitude of the liability. Cleanup costs vary according to the stipulated requirements and the techniques involved. Also, if a generator is the only principal responsible party at a Superfund site, it will be responsible for the costs of the entire cleanup. If there are multiple principal responsible parties, the cleanup costs are divided among them. Consider yet another example.

Example 9-4 The Effect of Shared Liability Company A is one of 10 companies equally responsible for $100,000 in cleanup costs from a failed landfill. In 1979 and 1980, when company A generated the 104 drums of waste that went to the landfill, disposal costs were $10 per drum. In 1978 the company evaluated a process modification that would have eliminated the waste stream. Capital costs for the modification were estimated to be \$2000, and operating costs were estimated to be \$5 per drum of waste avoided. Calculate what the present value for the pollution prevention project would have been at the start of the project (October 1, 1978), assuming that the capital equipment was purchased on October 1, 1978 and that the unit began operation on January 1, 1979. Incorporate the liability cost mentioned above, and assume that the liability occurred on October 1, 1990. Use a 10% interest rate, compounded weekly.

Solution The present values of capital cost and net savings per drum were calculated in Example 9-2 and are -\$2000 and \$459, respectively. The present value of a \$10,000 liability expense occurring on October 1, 1990 is

$$\mathrm{PV} = \$10{,}000/(1 + i/p)^{pn},$$

where the value for i is 10%, for p is 52, and for n is 12. Therefore, the present value of the liability is \$3015, so the net present value of the pollution prevention project is \$1474.

Partly because of the tremendous uncertainty associated with estimates of future liabilities, they are rarely included in economic evaluations of projects. There is another factor that impedes the inclusion of these liabilities. According to the Tellus Institute (1991), "[f]or publicly traded companies, liability estimation is controversial because the Securities and Exchange Commission requires firms to report liabilities to stockholders and to accrue assets to cover these future costs." This level of responsibility for cost estimates that might be inaccurate by orders of magnitude drives publicly traded companies to either avoid quantitative evaluation of liability costs or to exercise extreme caution in estimating potential liability.

As shown in the next section on prioritizing pollution prevention options, future liabilities are sometimes evaluated qualitatively instead of quantitatively. For example, a particular pollution prevention alternative might be assumed to either reduce or increase liability. It is possible to include a qualitative evaluation such as this in the economic evaluation of projects.

9.1.1.3 *Less Tangible Costs* The least tangible of the economic factors that might be considered in evaluating pollution prevention projects are grouped into the third tier of costs. Consumer responses to improved product quality or improved corporate image, employee responses to improved environmental stewardship, and potential improvements in worker health and safety due to pollution prevention could all be considered. Such factors are even more difficult to quantitatively evaluate than tier 1 and tier 2 costs, and this section merely notes their existence.

9.1.2 Comprehensive Methods of Economic Analysis

Several well-documented methodologies for economic evaluation of pollution prevention projects are publicly available. Some that have been critically reviewed (Tellus, 1991) include

- A method developed with General Electric by ICF, Incorporated (in 1987)
- A method developed for the U.S. Environmental Protection Agency by ICF, Incorporated (in 1989)
- Software developed by the George Beetle Company for the U.S. Environmental Protection Agency

The Tellus Institute has also developed its own software package, called P2/FINANCE. Table 9-6 provides contact information for ordering these comprehensive packages.

Table 9-6. Comprehensive economic evaluation packages

	Financial Analysis of Waste Management Alternativies (GE Methods)
Prepared for:	General Electric, Corporate Environmental Programs Richard W. MacLean, Manager
Prepared by:	General Electric and ICF Incorporated
Publication date:	1987
Contents:	Workbook, worksheets, and financial calculation software developed with Lotus 1–2-3, Version 2.01.
	Pollution Prevention Benefits Manual (EPA Method)
Prepared for:	Office of Solid Waste/Office of Policy, Planning and Evaluation, U.S. Environmental Protection Agency, Dr. Ron McHugh
Prepared by:	ICF Incorporated, 9300 Lee Highway, Fairfax, VA 22031
Printing date:	October 1990
Contents:	Manual and worksheets
	Precosis
Prepared for:	U.S. Environmental Protection Agency, Center for Environmental Research Information, Cincinnati, OH
Prepared by:	George Beetle, George Beetle Company, 533 Arbutus Street, Philadelphia, PA 19119
Publication date:	1989
Contents:	Manual and software

Source: Tellus Institute (1991).

9.2 RANKING POLLUTION PREVENTION PROCESS MODIFICATIONS

There are two basic steps to prioritizing pollution prevention activities at a large industrial facility. In the first step, waste streams and emissions are evaluated. On the basis of the evaluations, a set of these streams are selected for further investigation. The process of ranking waste streams was described at the end of Chapter 5. In the second step, process modifications that reduce or eliminate the streams chosen in the first step are identified and the process options are ranked. Two techniques for ranking pollution prevention modifications are illustrated in this section, using examples from the petroleum refining industry. The first technique is to assign dimensionless scores to criteria, as was described in the section on ranking waste streams. The second technique, called the *analytic hierarchy process*, is a systematic method for ranking alternatives when there is a diverse set of decisionmaking criteria to consider.

It is difficult to ascertain which pollution prevention options among a set are superior for the same reasons that it is difficult to prioritize waste streams for pollution prevention projects; there are many criteria that determine the priority of the options, and these criteria do not have the same units of measurement. Nevertheless, these variables must somehow be aggregated if the "best" pollution prevention projects are to be selected.

Criteria for ranking options might include

1. Where the project falls in the waste management hierarchy, with source reduction favored over recycling favored over waste treatment
2. The reduction of waste volume or disposal and treatment costs
3. Ease of implementation
4. Proven performance
5. Safety and health risks
6. Quantifiable results

These criteria were used to screen 250 pollution prevention modifications at a large refinery (Balik and Koraido, 1991). Another set of criteria used to rank pollution prevention opportunities along with the bases of evaluations for the criteria are given in Table 9-7. The criteria in this table were used during the Amoco-US EPA Pollution Prevention Project at Amoco's Yorktown refinery (Amoco/US EPA, 1992).

9.2.1 Screening of Pollution Prevention Process Modifications

The simplest way to overcome the difficulty associated with criteria of different units of measure is to assign a dimensionless score to the criteria, as in the case for the waste stream prioritization scheme described in Chapter 5. If possible, the dimensionless score should have an absolute basis. For

Table 9-7. Bases for evaluation of criteria in Yorktown project

Criteria	Basis for Evaluation
Risk reduction	Determine the relative impacts on human risk each option would have if implemented
Technical characteristics	
Resource utilization	Find the number of decreases or increases in raw material and utilities for each option
Timeliness	Determined as either short- (<3 years), medium- (3–6 years) or long-term (>7 years); shorter-term options considered more desirable
Release reduction	Quantitative estimate of release reduction in tons/yr, and categorization of the reduction mode into one of four categories (listed in descending order of preference): source reduction, recycling, treatment, and disposal
Transferability	Evaluated in terms of whether or not the technology required to implement the option could be transferred to other facilities near the refinery, other refineries, or other industries; highest priority given to options transferable to other refineries, followed by transferability to the community and other industries, in that order
Cost factors	
Operation and maintenance	Quantitative estimate in $/yr
Capital	Quantitative estimate in $
Liability	Determined qualitatively for three types of liability: remedial (for cleanup of current disposal sites in the future), catastrophic, and product; options determined to cause either a marginal decrease, no change, a marginal increase, or a significant increase in these types of liability

Source: Amoco/US EPA (1992).

example, if scores from 0 to 5 are assigned, with five indicating most desirable, a score of 5 might be assigned to the capital cost criteria for those projects with no capital costs. It is clear in this case that the absolute best a project can do is to have no capital costs. It would be more difficult to decide which level of capital cost should receive a score of 0. For a very small facility, a capital cost of $25,000 might put a project out of reach, while at a large facility any project with capital costs under $200,000 might be considered to have very low capital costs. Also, scores should not be assigned merely on the basis of the range of values for the criterion among the projects being considered. All the options, for example, may have very low capital costs, and it would not be proper to assign the project with the highest capital costs a score of 0 in such a case.

Fourteen of the proposed pollution prevention options for the Yorktown project are listed and briefly described in Table 9-8. The following example illustrates how these pollution prevention options might be screened.

Example 9-5 Ranking of Pollution Prevention Options Using Unweighted Criteria Evaluations In Table 9-9, ratings of 0–5 points for each criterion used in the Yorktown project and each pollution prevention option listed in Table 9-8 are given. (Note, however, that while the point assignments in this table are reasonable, they are not the ones that were used in the project.) It is also indicated in this table whether a modification is required for compliance with pending regulations. Obtain an overall ranking for the 14 options by adding together the points for each of the eight criteria.

Solution Sample calculation for rerouting the desalter washwater:

$$0 + 5 + 5 + 0 + 5 + 5 + 5 + 3 = 28 \text{ total points.}$$

Results for all the options are given in Table 9-10. Rerouting the desalter washwater, reducing the barge loading emissions, and improving the sour-water system are top-priority options while replacing the cyclones, installing an electrostatic precipitator, upgrading the drainage system, and upgrading the treatment plant are the lowest-priority options.

As in the case for prioritizing waste streams, the criteria used to rank pollution prevention options might be weighted so that criteria perceived of as being the most important affect the results of the ranking the most, as illustrated in the following example.

Example 9-6 Ranking of Pollution Prevention Options Using Weighted Criteria Repeat the ranking of Example 9-5, this time assuming that management places a higher value on the safety and well-being of its employees and the community than it places on any other criterion. Do this by giving the risk reduction criterion twice the relative weight of all the other criteria. How is the new ranking different from that of Example 9-5?

Solution Sample calculation for rerouting the desalter washwater:

$$0 \times 2 + 5 + 5 + 0 + 5 + 5 + 5 + 3 = 28 \text{ total points.}$$

Table 9-8. Waste reduction options studied at Amoco's Yorktown refinery

Option No.	Option Label	Brief Description
1	Reroute desalter water	Reduce secondary emissions from the refinery wastewater system by installing a line to carry hot desalter effluent water directly to the wastewater treatment units instead of draining it into the process water sewer, which is ventilated to the atmosphere
2	Replace cyclones	Replace the current particle collection equipment (cyclones) that capture fine catalyst particles generated by a petroleum cracking unit
3	Install ESP	Install an electrostatic precipitator to replace the cyclones described in option 2
4	Eliminate coker blowdown pond	This pond's fugitive emissions can be eliminated by changing procedures to eliminate the pond
5	Install double seals on tanks	Install double seals on all storage tanks with external floating roofs and add internal floaters to tanks with fixed roofs to reduce fugitive emissions
6	Keep soils out of drain	Keep soils out of the sewer by sweeping the roadways and concrete areas and installing sewer boxes designed to reduce soil movement into the sewer system; soil in the sewer system not only increases the mass of solids in the wastewater sludge but also makes hydrocarbons that adsorb onto the soils more difficult to treat
7	Blowdown system upgrade	Reduce hydrocarbon emissions by replacing existing atmospheric blowdown stacks, which discharge directly to the atmosphere, with flares
8	Drainage system upgrade	Reduce secondary emissions by installing above-grade pressurized sewers and segregating storm water and process water
9	Treatment plant upgrade	Replace the API separator, a part of the wastewater treatment system that is currently uncovered, with a covered gravity separator and air flotation system to reduce secondary emissions and capture hydrocarbon vapors

Table 9-8. *(Continued)*

Option No.	Option Label	Brief Description
10	Modify sampling systems	Current sampling lines are open-ended; when flushed prior to sampling, they increase sewer loadings; replace existing sampling stations with flowthrough sampling stations to reduce the oil loading in the sewer and oil drained to the deck
11	Reduce barge loading emissions	Barges are used to transport some refinery products; loading the barges generates fugitive emissions; install a marine vapor-loss-control system to reduce barge loading emissions
12	Sour-water system improvements	Refinery sour waters typically are wastewaters containing NH_3 and H_2S—these gases are typically removed from the water and purified in a sour water stripper; the H_2S can cause odor problems, so, this improvement would upgrade the sour-water stripper to reduce H_2S emissions
13,14	Quarterly or annual LDAR[a] program	Institute a LDAR program, either quarterly or annually, to reduce fugitive emissions

[a]Leak detection and repair.

Results for all the options are given in Table 9-11. Rerouting the desalter washwater, reducing barge loading emissions, and instituting an LDAR program are the highest-priority options; replacing the cyclones, installing an electrostatic precipitator, upgrading the drainage system, and upgrading the treatment plant are the lowest-priority options. The ranking is similar to the ranking produced in Example 9-5.

It is possible to choose one criterion, such as risk to human health, for ranking pollution prevention options. However, the rankings obtained will depend heavily on the choice of criterion, as shown in the following example.

Example 9-7 Ranking of Pollution Prevention Options Using a Single Criterion Rank the projects of Example 9-5 again, on the basis of a single criterion. List the highest-rated projects, assuming that risk reduction is the sole criterion, capital cost is the sole criterion, release reduction is the sole criterion, and regulatory compliance is the sole criterion. Compare your results to the results of the previous examples.

Table 9-9. Point assignments for prioritizing the waste reduction options[a]

		Ranking Criteria								
			Technical Characteristics				Cost Factors			
	Waste Reduction Option	Risk Reduction	Resource Utilization	Timeliness	Release Reduction	Transferability	Operation and Maintenance	Capital	Liability	Required for Future Regulatory Compliance
1.	Reroute desalter water	0	5	5	0	5	5	5	3	No
2.	Replace cyclones	0	4	2	0	5	3	3	0	No
3.	Install ESP	0	4	2	0	5	1	3	2	No
4.	Eliminate coker blowdown pond	0	0	5	0	2	5	5	3	Yes
5.	Install double seals on tanks	1	4	0	1	0	5	5	5	Yes
6.	Keep soils out of drain	0	4	2	1	5	5	5	5	No
7.	Blowdown system upgrade	1	2	2	5	0	4	4	2	Yes
8.	Drainage system upgrade	1	2	5	0	0	1	0	3	Yes
9.	Treatment plant upgrade	1	2	5	0	0	0	1	1	Yes
10.	Modify sampling systems	0	5	2	0	0	5	5	5	Yes
11.	Reduce barge loading emissions	5	1	5	2	5	4	4	2	Yes
12.	Sour-water system improvements	0	4	5	0	5	5	5	3	No
13,14.	Quarterly/annual LDAR program	1	5	5	1	0	5	5	4	Yes

Key: 5 = high priority, 0 = low priority.

Table 9-10. Ranking based on equal weighting factors

Option	Total Points	Overall Option Ranking (1 = high)
Reroute desalter water	28	1
Replace cyclones	17	10
Install ESP	17	10
Eliminate coker blowdown pond	20	8
Install double seals on tanks	21	7
Keep soils out of drain	27	3
Blowdown system upgrade	20	8
Drainage system upgrade	12	12
Treatment plant upgrade	10	13
Modify sampling systems	22	6
Reduce barge loading emissions	28	1
Sour-water system improvements	27	3
Quarterly/annual LDAR program	27	3

Table 9-11. Ranking with risk reduction considerations given weight of 2

Option	Total Points	Overall Option Ranking (1 = high)
Reroute desalter water	28	2
Replace cyclones	17	10
Install ESP	17	10
Eliminate coker blowdown pond	20	8
Install double seals on tanks	22	6
Keep soils out of drain	27	4
Blowdown system upgrade	20	8
Drainage system upgrade	13	12
Treatment plant upgrade	7	13
Modify sampling systems	22	6
Reduce barge loading emissions	38	1
Sour-water system improvements	27	4
Quarterly/annual LDAR program	28	2

Solution The highest ranking projects for each of the single criteria are given in Table 9-12. Note that there is no overlap between the highest-priority options for the first three criteria; that is, none of the projects appears as a highest ranked option more than once. Most of the options have high priority when ranked by regulatory compliance. Rerouting the desalter wash-water was an option given high priority in both of the previous examples, but it is given low priority if the options are ranked according to regulatory compliance.

Table 9-12. Highest rated projects for ranking based on a single criterion

Criterion		Highest-Rated Projects
Risk reduction	11.	Reduce barge loading emissions
Capital costs	1.	Reroute desalter water
	4.	Eliminate coker blowdown pond
	5.	Install double seals on tanks
	6.	Keep soils out of drain
	10.	Modify sampling systems
	12.	Sour-water system improvements
	13,14.	LDAR program
Release reduction	7.	Blowdown system upgrade
Regulatory compliance	4.	Eliminate the coker blowdown pond
	5.	Install double seals on tanks
	7.	Blowdown system upgrade
	8.	Drainage system upgrade
	9.	Treatment plant upgrade
	10.	Modify sampling systems
	11.	Reduce barge loading emissions
	13,14.	LDAR program

These examples show that more robust rankings, relatively insensitive to modest change, are obtained when a set of criteria are considered. Stated slightly differently, if multiple ranking criteria are used, there is far less variability between different ranking schemes than if a single criterion is used.

9.2.1.1 ***Rankings Based on Regulatory Compliance*** Projects required for regulatory compliance are not necessarily rated highly using other criteria. Even members of regulatory agencies, when charged with establishing a rating system that includes multiple criteria, create ranking systems that do not necessarily put compliance projects at the top of the list. This finding can be interpreted in a number of ways. Industry lobbyists would argue that this means that environmental regulations do not make sense. A more optimistic view is that new regulations will be promulgated that encourage the options with the highest overall rating, once the basis for the rankings is understood by policymakers. In practice, the lesson to be learned is that relying only on compliance to identify pollution prevention activities is probably not an effective long-term strategy.

9.2.2 Hierarchical Evaluation of Pollution Prevention Process Modifications

The method presented for ranking pollution prevention options in this section is a more detailed hierarchical ranking that includes a systematic procedure for determining criteria weights. In the Yorktown project, the *analytic hierarchy process* (AHP) was adopted for ranking the options. This method, among others, is described by Saaty (1982). The AHP was chosen because it is useful for making decisions involving a large number of diverse criteria. There are five steps in the AHP, the first of which is to determine a goal and identify the important decisionmaking criteria. The second step is to organize the criteria into a hierarchical structure that gives the relationship of each criterion to the goal. Each level in the hierarchy can be influenced only by the level above it and can influence only the level below it. The next step is to quantify the relative significance of each criterion by making pairwise comparisons between the criteria at each level in the hierarchy. Then each option is evaluated for each criterion. The final step in the AHP is to make adjustments to the previous steps because of new information gathered during the ranking process. Adjustments may also be necessary if the ranking produced is sensitive to changes in criteria weights and option evaluation scores.

To understand the basis for a hierarchical ranking, consider liability costs, just one of the eight ranking criteria of Table 9-7. In developing a score for the liability cost criterion, the costs associated with possible future remediation actions, catastrophic releases, and product liability could be considered. Each of these three potential liability costs could be thought of as a new ranking criterion, but logically, they belong under the general heading of liability costs.

The example of the liability cost criterion shows that some of the criteria in Table 9-7 in the previous section can be expanded into multiple subcriteria. Careful examination of the list of criteria reveals that some of the criteria really are the subcriteria of a larger objective. For example, operation and maintenance expenditures, capital costs, and liability costs are all cost factors.

Taking this view, the criteria can be broadly grouped into technical characteristics, cost factors, and risk reduction. The technical characteristics criterion includes four subcategories. The cost criterion consists of three subcategories, and risk reduction has no subcategories. At the lowest category level, evaluations are made. For example, the timeliness subcategory is evaluated as either short-, medium-, or long-term.

Given this hierarchical structure, how is an overall score for a project found? The score is evaluated by starting with the lowest point in the hierarchy. The tree structure in Figure 9-2 may be helpful in following these calculations. As an example, consider the factors making up the liability costs. First, relative weights are assigned to remedial, catastrophic, and product liabilities by comparing them pairwise with each other. If the weights

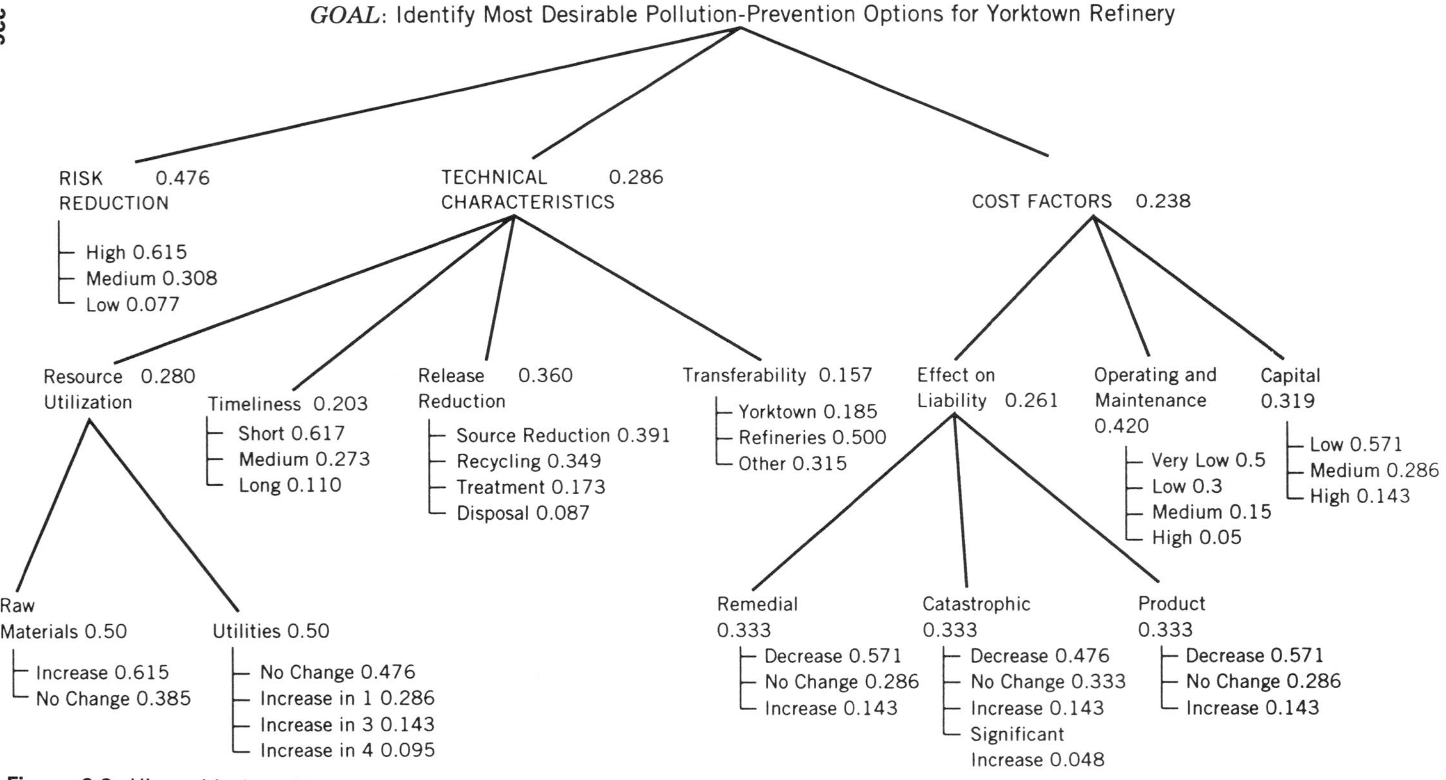

Figure 9-2. Hierarchical ranking system and weighting system developed by Amoco Corporation and the United States Environmental Protection Agency (Amoco/USEPA, 1993).

are equal, then each factor makes up 33.3% of the liability subcategory. Next, an evaluation is assigned to each factor. If the remedial liability evaluation for a project is that it would result in a decrease in liability (score = .571), the catastrophic liability evaluation is that it would cause an increase in liability (score = .143), and the product liability evaluation is that it would cause an increase in liability (score = .143), then the net score for the liability subcategory is

$$.29 = .333 \times .571 + .333 \times .143 + .333 \times .143.$$

The operation and maintenance and the capital cost subcategories are not broken into additional factors. By making pairwise comparisons, the Yorktown project workgroup developed weights of .420, .319, and .261 (Amoco/US EPA, 1993) for the operation and maintenance, capital, and liability costs, respectively. If the operation and maintenance cost evaluation revealed these costs to be low and the capital cost evaluation revealed them to be high, then the score for the cost criterion would be

$$.25 = .420 \times .3 + .319 \times .143 + .261 \times .29.$$

To evaluate the total project score, relative weighting factors for the three main branches of the hierarchy must be established. The Yorktown project workgroup assigned weights of .476, .286, and .238 to the risk reduction, technical characteristics, and cost factors, respectively. If risk reduction were evaluated to be high and the combined technical characteristics evaluations led to a score of .35, then the overall project score would be

$$.45 = .476 \times .615 + .286 \times .35 + .238 \times .25.$$

The criteria weights and option evaluations used to produce a ranking can be somewhat arbitrary and in some cases based on order-of-magnitude estimates. A sensitivity analysis can be performed to determine whether a ranking is unstable because of the uncertainties inherent in the weight and evaluation assignments by perturbing the variables and finding the new ranking. If the ranking produced from the perturbed variables is similar to the original ranking, the original ranking is considered to be robust. This approach was used in the Yorktown project. A group of Amoco representatives established weights for ranking criteria, and a group of regulators and policymakers from the state of Virginia and the U.S. Environmental Protection Agency also established weights for ranking criteria. The ranges

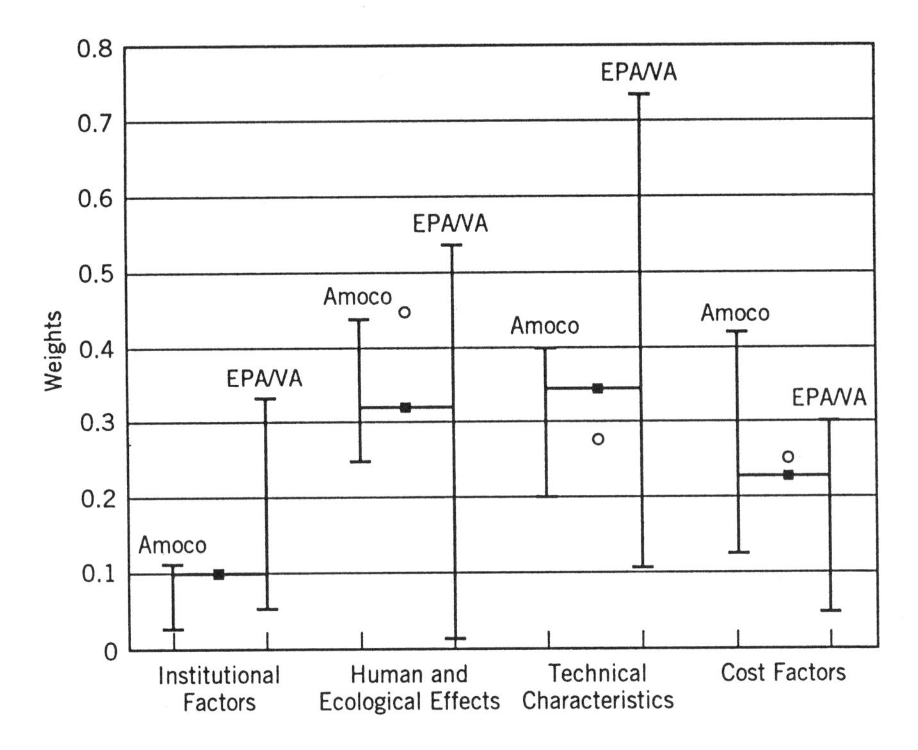

Figure 9-3. Comparison of weights assigned to ranking criteria by Amoco representatives and by employees of the US EPA and the state of Virginia. The bars represent the range of weights proposed by inidviduals in each group (Amoco/USA EPA, 1992).

of weighting factors recommended by these two groups are shown in Figure 9-3.

The EPA/Virginia group weighted institutional factors higher than did the Amoco group. Otherwise the range of assigned weights was quite comparable. Table 9-13 shows the scores for the pollution prevention options of Table 9-8 based on criteria weights from the Amoco representatives, the EPA/Virginia group, and the project workgroup. Comparison of the rankings reveals that most of the rankings remain the same, although the absolute numerical scores differed. The only significant difference in the rankings was that the Amoco representative weights gave the blowdown system project a lower ranking than did the other two groups.

The advantage that a hierarchical ranking system has over a simple listing of criteria is its capability to handle criteria at a number of levels of detail. This is particularly important in the relatively arbitrary process of establishing weights and assigning values to evaluations. Use of the hierarchial ranking system requires the establishment of relative weights between criteria that

Table 9-13. Comparison of option rankings based on Amoco representatives, EPA/Virginia representatives, and project workgroup weighting factors

Option	Score Using Project Workgroup Weights	Score Using Amoco Representative Weights	Score Using EPA/VA Representative Weights
Reduce barge loading emissions	100	100	100
Install double seals on tanks	74	93	87
Blowdown system upgrade	63	68	66
Quarterly LDAR program	60	91	67
Annual LDAR program	60	91	67
Modify sampling systems	53	81	59
Reroute desalter water	52	76	60
Sour-water system improvements	47	68	51
Keep soils out of drain	46	69	59
Eliminate coker blowdown pond	45	49	49
Drainage system upgrade	37	49	39
Replace cyclones	34	45	42
Treatment plant upgrade	32	40	34
Install ESP	32	42	39

Source: Amoco/USEPA (1993).

are similar to each other in level of detail, such as risk, technical characteristics, and cost factors. It does not require the establishment of a relative ranking between liability costs and risk reduction, two criteria that are different in their level of detail.

Different companies and even different individuals within the same company will choose different criteria for ranking their pollution prevention options. It should be noted that the intent in this chapter is not to identify the single, correct set of criteria and weights for ranking pollution prevention options, for there is no single, correct set of criteria and weights.

9.3 SUMMARY

In this chapter, some of the potential economic benefits of pollution prevention projects that are not normally considered in project evaluation were identified. These potential benefits were presented with the goal of providing some general ideas about estimating the total costs of waste generation and to guide the reader to more detailed descriptions of pollution prevention economics. Following EPA terminology, the potential nontraditional economic benefits of pollution prevention can be grouped into three tiers, with tier 1 consisting of the economic benefits that are easiest to quantify and tier 3 consisting of the economic benefits that are most difficult to quantify. Methods and examples presented in this chapter for estimating hidden costs and future liabilities clearly demonstrate that the potential economic benefits of pollution prevention vary greatly from facility to facility. Furthermore, for some of the benefits, economic valuations are highly uncertain.

Two procedures for ranking pollution prevention options at industrial facilities were also described in this chapter. The first method is semiquantitative and involves assigning dimensionless scores to each option. A second related scoring system based on a hierarchical approach was also presented. The purpose of this chapter was to illustrate generalized methodologies and to demonstrate how the choice of ranking criteria can influence ranking results, and the most important conclusions to be retained from the material are

1. When multiple ranking criteria are used, different groups often come up with ranking schemes that generate similar lists of highly rated projects.
2. When a single ranking criterion is used, projects selected are highly dependent on the choice of criterion.

In short, ranking strategies based on multiple criteria are robust.

QUESTIONS FOR DISCUSSION

1. One sensitivity analysis for the Yorktown project ranking was done by performing the ranking using weights developed by diverse interest groups. Suggest additional sensitivity analyses that might be performed to test the robustness of the ranking.
2. What are the advantages of assigning dimensionless scores to the evaluations for a set of criteria and adding the scores to rank pollution prevention options instead of applying the hierarchical process described in this chapter? Could these two procedures be hybridized or combined to provide the best of both worlds?
3. Were the Yorktown project pollution prevention options that were required for regulatory compliance ranked highly by the project workgroup?

PROBLEM STATEMENTS

1. Repeat the calculations of Example 9-1 using the lowest wage rates given in Tables 9-1 through 9-5. Which compound is economically

Table 9-14. Selected criteria evaluations for the Yorktown project

Option	Net Release Reduction, tons/yr	Capital Costs, $1000	Operating and Maintenance Costs, $1000	Timing, years
Reroute desalter water	52.4	1,000	163	<3
Replace cyclones	245	8,300	4,743	4–7
Install ESP	442	9,100	4,774	4–7
Eliminate coker blowdown pond	130	2,000	175	<3
Install double seals on tanks	592.2	1,682	105	>7
Keep soils out of drain	530	213	275	4–7
Blowdown system upgrade	5,096	4,521	912	4–7
Drainage system upgrade	112.5	188,000	3,146	<3
Treatment plant upgrade	58	22,500	5,184	<3
Modify sampling systems	63	87	−45	4–7
Reduce barge loading emissions	768	4,700	338	<3
Sour-water system improvements	18	605	142	<3
Annual LDAR program	319.5	5	−11	<3
Quarterly LDAR program	510.5	5	31	<3

Source: Amoco/USEPA (1993).

favorable when hidden costs are included, assuming these low wage rates?

2. Repeat the calculations of Example 9-3, assuming that the average lifetime of an existing landfill is expected to be 20 years. How different is your result from the result of Example 9-3?

OPEN-ENDED PROBLEM STATEMENTS

1. Put the first set of project ranking criteria presented in Section 9.2 [the criteria reported by Balik and Koraido (1991)] into a hierarchical structure and describe the steps you would take to evaluate them. Also, assign weights to these criteria.
2. How would you assign dimensionless scores to the evaluations of the Yorktown project that are shown in Table 9-14? Explain your rationale for determining what evaluation should lead to the lowest possible score and what evaluation should lead to the highest possible score.

REFERENCES

Amoco/United States Environmental Protection Agency (Amoco/US EPA), "Amoco-US EPA Pollution Prevention Project, Yorktown, Virginia, Project Summary," available through NTIS as PB92–228527, June 1992.

Amoco/United States Environmental Protection Agency (Amoco/US EPA), "Amoco-US EPA Pollution Prevention Project, Yorktown, Virginia, Projects, Evaluations, and Ranking," available through NTIS as PB92–228626, July 1993.

Balik, J.A. and S.M. Koraido, "Identifying Pollution Prevention Options for a Petroleum Refinery," *Pollution Prevention Review*, **1**, 273–293, Summer 1991.

Saaty, T. L., *Decision Making for Leaders: The Analytical Hierarchy Process for Decisions in a Complex World*, Lifetime Learning Publications, Belmont, CA, 1982.

Tellus Institute, "Alternative Approaches to the Financial Evaluation of Industrial Pollution Prevention Investments," prepared for the New Jersey Department of Environmental Protection, Project P32250, 1991.

United States Environmental Protection Agency (US EPA), "Waste Minimization Opportunity Assessment Manual," Publication EPA/625/7–88/003, July 1988.

U.S. Environmental Protection Agency (US EPA), "Pollution Prevention Benefits Manual Volume I: The Manual," Office of Policy Planning and Evaluation and Office of Solid Waste (report never officially released; however, copies are available on request), 1989.

Valle-Riestra, J. F., *Project Evaluation in the Chemical Process Industries*, McGraw-Hill, New York, 1983.

10

POLLUTION PREVENTION CASE STUDY PROBLEM MODULES

This chapter contains case study problem modules that draw on the mesoscale pollution prevention tools presented in Chapters 5–9. The case studies are structured so that problems can either be worked as the material in Chapters 5–9 is covered or as a capstone design problem performed after the completion of all the material in the mesoscale section of this text. The modules are comparable in difficulty.

10.1 SULFUR RECOVERY AT REFINERIES

As mentioned in Chapter 2, there are six pollutants, referred to as *criteria pollutants*, for which National Ambient Air Quality Standards (NAAQS) were established through the Clean Air Act: (1) particulate matter <10 μm in diameter (PM_{10}), (2) sulfur dioxide (SO_2), (3) nitrogen oxides (NO_x), (4) carbon monoxide (CO), (5) ozone (O_3), and (6) lead (Pb). NAAQSs are time-averaged concentrations that cannot be exceeded in the ambient air more than a specified number of times in a year. Attainment areas are those that are in compliance with the NAAQSs; nonattainment areas are those that are not. In nonattainment areas, permitting for a new source or for a major modification of an existing source requires that lowest achievable emission rate (LAER) technology be used. Also, any increase in emissions in a nonattainment area must be offset by decreases in emissions from other sources in the same region. Existing sources in nonattainment areas are also subject to emission control measures. Although volatile organic compounds (VOCs) are not on the list of criteria pollutants, facilities emitting VOCs can be impacted by the NAAQS for ozone because of the role VOCs play in the creation of ozone in the lower atmosphere. The same holds true for

NO_x: in a region that is an attainment area for NO_x but a nonattainment area for ozone there may be strict controls placed on NO_x emissions.

All fossil fuels, including crude oil, contain organic sulfur compounds. When the fuels are burned, sulfur oxides (SO_x) are emitted. Sulfur oxides can directly cause respiratory problems, but emissions of these compounds are a concern primarily because they react in the atmosphere, forming acids. Over time, deposition of these acids through precipitation and diffusion can increase the acidity of natural bodies of water and change local ecosystems.

In this section, the impact of process selection on waste generation is examined, focusing on the fate of sulfur around the fluidized-bed catalytic cracking unit (FCCU) in a refinery. Figure 10-1 is a simplified refinery flowsheet showing a feed stream to an FCCU. Crude oil is distilled first at atmospheric pressure, producing atmospheric distillate products: naphtha, which is further refined to gasoline; kerosine, which is further refined to become jet fuel; and diesel or No. 2 fuel oil for home heating. The bottoms from atmospheric distillation are further distilled under vacuum to produce vacuum distillate. This distillate is a heavy oil, too heavy to be blended directly into "premium" products. It can be fed to an FCCU where the large hydrocarbon molecules, containing 20–40 carbon atoms, are "cracked" or broken into smaller molecules with less than 12 carbon atoms that can be blended into gasoline.

The catalyst in an FCCU becomes deactivated very rapidly during contact with the oil feed. This is primarily because coke, a class of highly dehydrogenated hydrocarbons, is formed on the surface of the catalyst. To maintain the activity of the FCCU catalyst, it is continually circulated in a fluidized state between the riser, where the cracking reactions take place, and a regenerator. In the regenerator, air is contacted with the hot catalyst, burning the coke off of the catalyst. This produces a flue gas stream as shown in Figure 10-1. Because crude oil and all the products distilled from crude oil—including the vacuum distillate fed to the FCCU—contain organic sulfur compounds, the coke formed on the FCCU catalyst contains sulfur. When the coke is burned in the regenerator, the sulfur in the coke forms sulfur oxides.

In conventional FCCUs, a small amount of the fuel-bound sulfur in the feed is converted to H_2S in the riser. Therefore, sulfur in the feed can be either converted to H_2S in the riser, converted to SO_2 in the regenerator, or incorporated into the products leaving the FCCU (e.g., in gasoline or diesel fuel). If the sulfur remains in the products, SO_x is formed when the products are used as fuel. The Clean Air Act Amendments of 1990 include regulations for sulfur levels in gasoline because of this. If the sulfur is converted to H_2S in the riser, it can be fed to a Claus plant for sulfur recovery and the production of salable sulfur. The balance between H_2S formation, SO_2 formation, and the sulfur remaining in the products depends on the catalyst used in the cracker. New catalysts have been developed that favor the formation of H_2S in the riser in order to take advantage of this effective alternative for recovering the sulfur as a salable product.

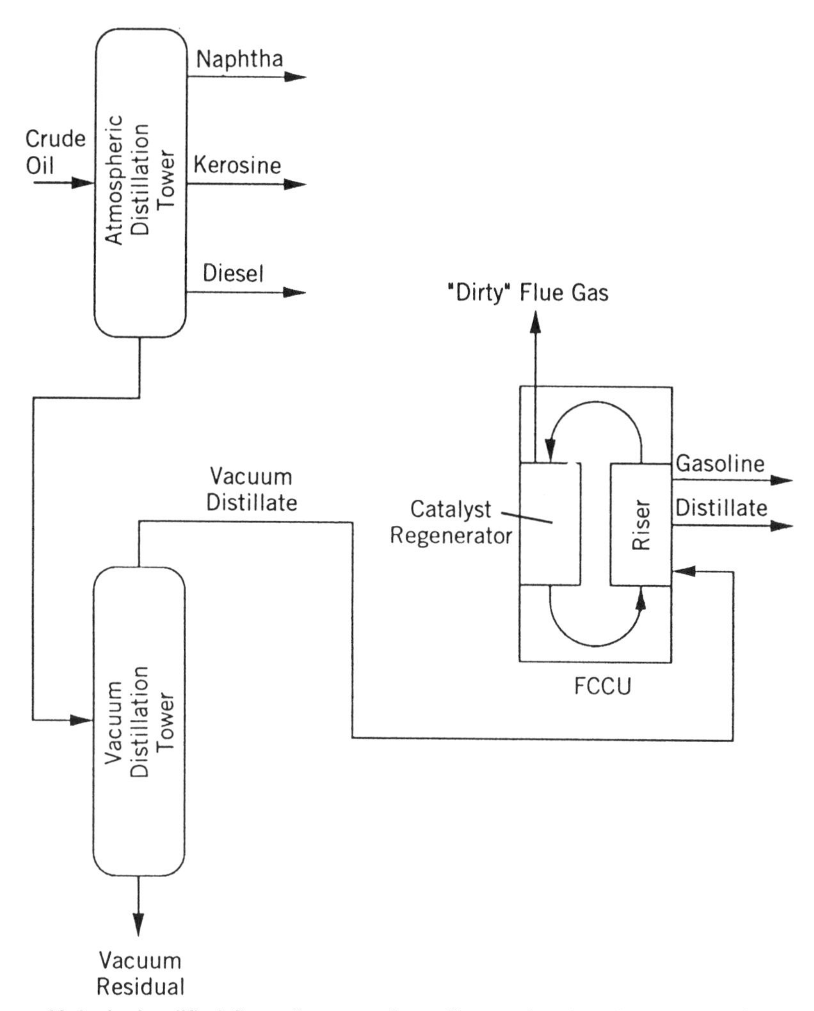

Figure 10-1. A simplified flow diagram of a refinery, showing the source of the feed stream to the fluidized-bed catalytic cracking unit (Allen and Rosselot, 1994).

10.1.1 Reducing SO_x Emissions from Fluidized-Bed Catalytic Cracking Units

In various parts of the United States, air regulations require refiners to install equipment to reduce the SO_x emitted from their FCCUs. There are a number of approaches to such reduction. Two approaches are considered in this module: the addition of flue gas scrubbing equipment (an end-of-pipe control) and the removal of sulfur from the FCCU feed by hydrotreating.

10.1.1.1 ***(Chapter 5) SO_x Emissions at Refineries*** SO_2 emissions from the nation's largest 122 refinery emitters total 420,000 tons/yr and comprise

Table 10-1. Refinery emissions of SO_2 by EPA region

EPA Region (Headquarters)	SO_2 Emissions, tons/yr
1 (Boston)	—
2 (New York)	4,700
3 (Philadelphia)	8,600
4 (Atlanta)	15,000
5 (Chicago)	130,000
6 (Dallas)	170,000
7 (Kansas City, KS)	22,000
8 (Denver)	41,000
9 (San Francisco)	22,000
10 (Seattle)	11,000
Total	420,000

Source: EPA (1994b).

approximately 3% of the nation's total SO_2 emissions, according to the National Air Data Branch of the United States Environmental Protection Agency's Office of Air Quality Planning and Standards (EPA, 1994a). Sulfur dioxide emissions from refineries are given by EPA region in Table 10-1. In Table 10-2, the SO_x emission profiles for two large refineries in California are given. This table shows that refineries emit SO_x from a number of processes, but the primary sources are process heaters and boilers, where fossil fuels are burned, and FCCUs. The data also illustrate the fact that SO_x emissions can vary dramatically from refinery to refinery, depending on the type of crude oil refined and the methods of sulfur removal that are employed.

Problem Statement

10-1. Calculate the distribution (by percentage) of SO_x emissions by process type for each of the refineries in Table 10-2. Compare the distribution of SO_x emissions from the two refineries and postulate some of the reasons behind the differences in the profiles.

10-2. Using the data for refinery 1 in Table 10-2, rank the processes (first column) by SO_x emissions per individual unit. If the costs per unit of reducing SO_x emissions by a given fraction were roughly the same for all types of processes, which units would be the first to target for SO_x reduction? Which would be the last to target?

10.1.1.2 ***(Chapter 6) Preventing FCCU SO_x Emissions*** Figure 10-2 depicts the addition of a flue-gas scrubbing process to the refinery shown in Figure 10-1. With this process, the basic FCCU remains unchanged. The scrubber removes SO_x from the flue gas before it is emitted to the atmosphere

Table 10-2. SO_x emissions by process type for two large refineries

Process Type	Refinery 1		Refinery 2	
	No. of Units	SO_x Emissions, tons/yr	No. of Units	SO_x Emissions, tons/yr
Industrial external combustion boiler				
No. 6 residual-oil-fired	0	—	5	1.5
No. 1 and No. 2 distillate-oil-fired	4	11	0	—
Process-gas-fired	10	93	8	95
Modified Claus plant (for sulfur recovery)	3	62	3	81
Fugitive emissions from inorganic chemical manufacturing	0	—	1	15
Process heaters				
Oil-fired	1	0.1	0	—
Natural-gas-fired	0	—	2	1.5
Process-gas-fired	34	350	48	400
Fluidized-bed catalytic cracking unit	1	1700	1	1.4
Blowdown system vapor recovery/flare	1	31	1	0.1
Process gas flares	2	0.2	4	0.4
Fluid coking	1	3.9	0	—
Total	57	2300	73	600

Source: CARB (1994).

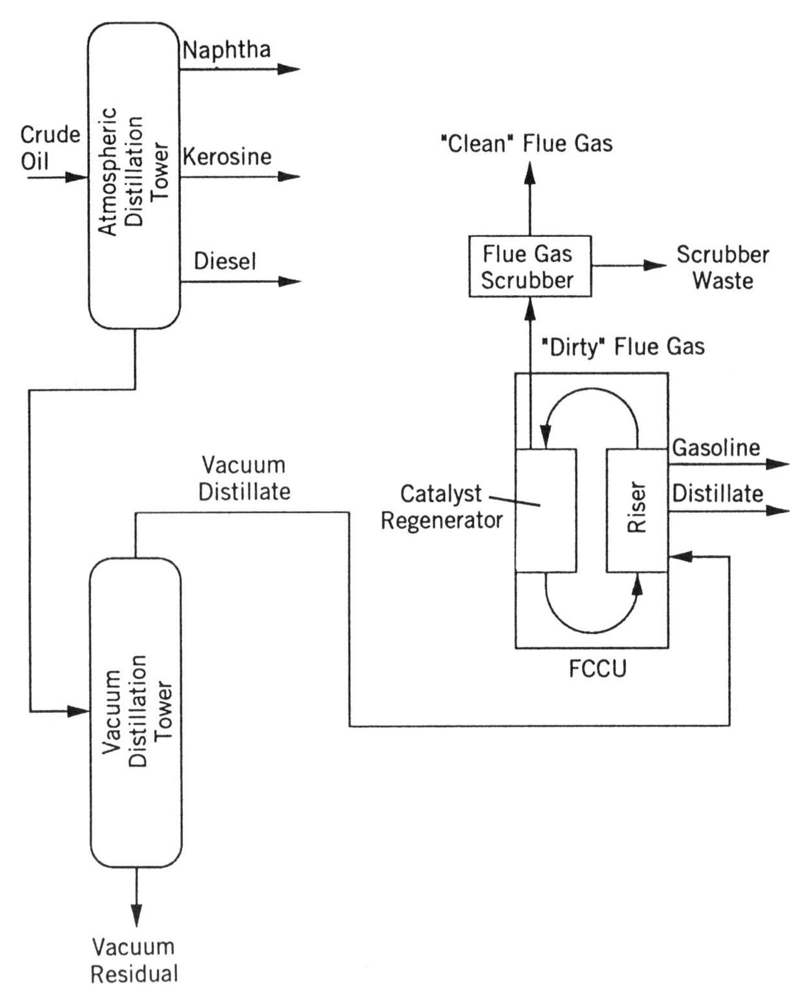

Figure 10-2. A simplified flow diagram showing addition of flue-gas scrubber to remove SO_x from fluidized-bed catalytic cracking unit flue gas (Allen and Rosselot, 1994).

by contacting the gas stream with aqueous or solid adsorbents. However, the scrubber generates either a solid, slurry, or aqueous byproduct or waste stream that must be managed.

Hydrotreating is the largest volume secondary processing step used in petroleum refining. It is used on all boiling ranges of hydrocarbons from naphtha through heavy-boiling vacuum distillates such as FCCU feed, and is sometimes even applied to residual oils. Whatever the feed, the process is the same: organic sulfur compounds in the oil are reacted with hydrogen in the presence of a catalyst at elevated temperatures and pressures to convert the organic sulfur to hydrogen sulfide. The hydrogen sulfide can then be

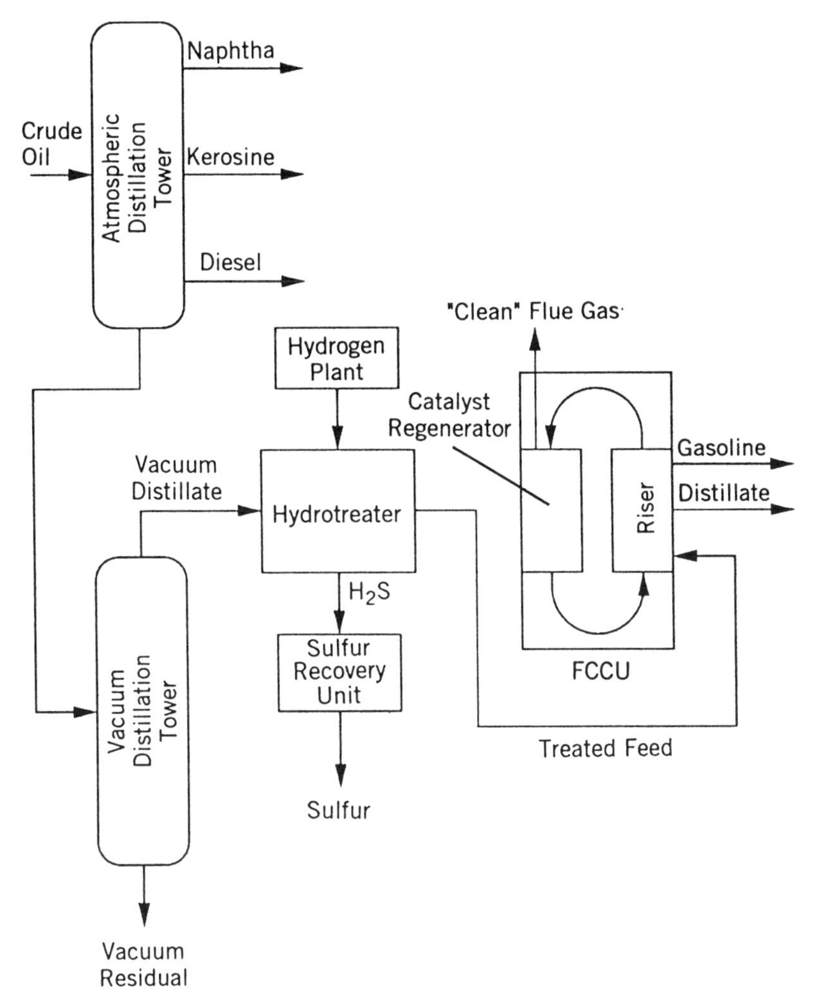

Figure 10-3. A simplified flow diagram showing addition of a hydrotreater to prevent SO_x emission from a fluidized-bed catalytic cracking unit (Allen and Rosselot, 1994).

removed from the process and safely converted to elemental sulfur in a Claus sulfur-recovery unit. The elemental sulfur is a salable byproduct of petroleum refining.

Figure 10-3 shows that adding a hydrotreater to the feed for an FCCU is a much more extensive project than adding a flue-gas scrubber. The main addition to the refinery is the hydrotreater process, but because the hydrotreater consumes a large quantity of hydrogen, many refineries would also need to add a hydrogen plant to support the hydrotreater. The hydrogen sulfide generated in the hydrotreater will likely require installation of more sulfur recovery capacity in the refinery as well, as a large hydrotreater can produce 200–300 tons/day of sulfur.

Question for Discussion

10-3. What are the difficulties in using substitute raw materials to reduce SO_x emissions from refineries?

Problem Statement

10-4. A hydrotreater might contain tens of tons of catalyst that during operation becomes covered with coke and loses its activity. Instead of constantly circulating between hydrotreatment and regeneration, however, hydrotreater catalysts are typically removed from the unit during shutdown and regenerated by burning off the coke. There is a limit to the number of times the catalyst can be regenerated because metals in the feedstock permanently poison the catalyst and because high temperatures during the regeneration process can cause damage to the catalyst support. Because large amounts of catalyst are involved, it is important that the regeneration process be carefully controlled so that damage to the catalyst, particularly sintering, is minimized. Therefore, it is desirable to allow the catalyst to reach only as high a temperature as is required to burn off the necessary coke. In this problem, you are to find the range of temperature at which the weight percent of coke left on the catalyst is 1–3%. The catalyst is restored to nearly all of its original activity with this amount of coke residual, but has not begun to sinter. The fraction of coke on the catalyst is 20% (by weight) when regeneration is begun, and remains constant until 300°C. Above 300°C the weight percent of coke is modeled by the following equation:

$$W = 20e^{-2\times10^{-4}(T-300)^2},$$

where T is temperature in °C. Keeping this range of temperature in mind, describe what will happen if regeneration is accomplished by charging a large fluidized-bed reactor with spent catalyst (the most common method for on-site regeneration). If large-scale off-site regeneration methods can increase the number of times a catalyst can be regenerated, and therefore reduce catalyst waste, should they be pursued? Discuss this in terms of the waste management hierarchy described in Chapter 1.

10-5. Organic sulfur contained in crude oil is often found in dibenzothiophenes, one of which is pictured in Figure 10-4. The phenyl rings on these compounds tend to be highly methylated. If the No. 1 and/or No. 8 carbons have a methyl group attached to them, the sulfur is shielded from contact with catalytic surfaces during hydro-

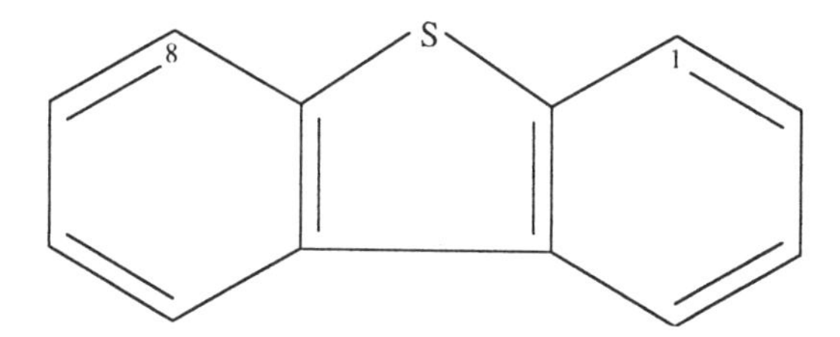

Figure 10-4. Dibenzothiophene or diphenylene sulfide.

treatment. If both the No. 1 and the No. 8 carbons are methylated, the steric hindrance is so strong that the rate of reaction for reducing the sulfur is two orders of magnitude less than the reaction rate when only carbon No. 1 is methylated. If a hydrotreater can be modeled as an ideal plug-flow reactor, and a volume of reactor V would be required to reduce the sulfur in 1-methylated dibenzothiophenes, what volume of reactor would be required to reduce the sulfur in the 1,8-methylated dibenzothiophenes? (Your answer will be in terms of V.) Assume that the time required for a plug-flow reactor to process one reactor volume of feed to a given conversion is

$$\tau = \frac{\Lambda}{\nu} \propto \frac{1}{k},$$

where τ is the space–time of the reactor, Λ is the volume of the reactor, ν is the volumetric feed rate to the reactor, and k is the reaction rate constant. What are the implications of this finding if the sulfur concentration in gasoline is required by regulations to be lower than the concentration of 1,8-methylated dibenzothiophenes in atmospheric distillate?

10.1.1.3 ***(Chapter 8) Flowsheet Boundaries and SO_x Emissions*** One benefit of hydrotreating the feed to the FCCU that will become more important as the sulfur in gasoline provisions of the Clean Air Act Amendments of 1990 are enacted is that the finished products made by a refinery have much less sulfur when the FCCU processes hydrotreated feed. This is because removing sulfur from the feed reduces the sulfur not only in the FCCU coke, but in all the FCCU products. When consumers burn fuels, such as gasoline, that are produced in an FCCU, the sulfur contained in the fuels is converted into SO_x. Therefore, pollution from the product enduser is reduced when the feed to the FCCU is hydrotreated. Sulfur in gasoline is of further concern because it lowers the activity of catalytic converters in automobiles and reduces their efficacy.

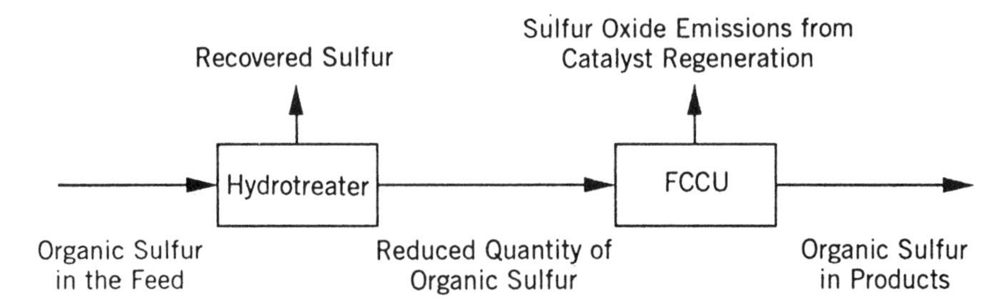

Figure 10-5. Flow diagram of sulfur in a hydrotreater and FCCU.

Problem Statement

10-6. Given that 12% of the sulfur in the FCCU feed for the refinery with no hydrotreater from Table 10-2 is deposited in the coke laydown on the catalyst (Lane and Latimer, 1992), and a hydrotreater would remove 90% of the organic sulfur from the feed to the FCCU, find the reduction in SO_x emissions in tons per year to be achieved from addition of a hydrotreater. Neglect H_2S emissions from the riser. A flow diagram of sulfur for a hydrotreater and FCCU is given in Figure 10-5. Provide separate estimates for reductions from both the FCCU and from the combustion of the fuel products from the FCCU during enduse. Assume that during combustion, all the organic sulfurs are converted to SO_x. What impact does the placement of the boundaries for a hydrotreater flowsheet have on the perceived SO_x reduction due to installation of a hydrotreater?

10.1.1.4 ***(Chapter 9) The Economics of FCCU SO_x Reduction Alternatives*** The feed hydrotreater, even though it requires a much larger capital investment than a scrubber, can produce a superior return on investment for the refiner. Although the primary reason for adding the hydrotreater may be to remove sulfur from the feed, hydrotreating has a number of effects. One is that the quality of the feed, in terms of the product yields from the FCCU, is substantially enhanced by hydrotreating. Therefore, the value of the products from the FCCU are much higher when processing hydrotreated feed. This is due mainly to the higher yield of gasoline, but can be compounded by allowing the refinery to process lower quality, less expensive crude oil while maintaining or increasing production of premium products.

The approximate costs and revenues for FCCU feed hydrotreaters and flue-gas scrubbers are given in Table 10-3. In addition to being approximate, the numbers in this table are not generally applicable to all refineries. Equipment needs and unit yields are specific to each refinery, crude slate, and product market. For example, a refinery with the hardware to fully utilize the hydrotreater, as well as a product market that favors high-octane gaso-

Table 10-3. Economics of adding a SO_x scrubber versus a 40,000-barrel/day hydrotreater to reduce FCCU SO_x emissions at a hypothetical refinery

	I	II—Low-Nitrogen Crude Hydrotreater		III—High-Nitrogen Crude Hydrotreater	
Parameters	SO_x Scrubber	Crude a: Low Sulfur and Metals	Crude b: High Sulfur and Metals	Crude a: Low Sulfur and Metals	Crude b: High Sulfur and Metals
Capital, MM$[a]	20	100	125	200	225
Net annual revenue, MM$	−2	10	50	35	75
Improved FCCU yields, MM$		3	3	21	21
Octane increase, MM$		4	4	11	11
Residual upgrading, MM$		3	3	3	3
Reduced crude cost, MM$		0	40	0	40
Hydrogen requirements, MMSCFD[b]		15	25	30	40
Operating pressure, psig		900	900	1500	1500

[a]MM$: million dollars.
[b]MMSCFD: millions of standard cubic feet per day.
Source: Mobil Oil (1992).

line, will get more benefit from addition of a hydrotreater than a refinery without these conditions.

The costs and revenues in this table were generated assuming that the FCCU feed hydrotreater treats all the feed for a 40,000-barrel/day FCCU (a unit of moderate size). Various cases are included to show the sensitivity of the results to the crude charge. Case I is for installation of flue-gas scrubbing treatment, and cases II and III are for addition of a hydrotreater for low- and high-nitrogen crude slates, respectively. (Nitrogen is found in carbazole structures similar to the dibenzothiophene structures for sulfur shown in Figure 10-4. It is sometimes, but not always, a cocontaminant with sulfur and the nitrogen species are also removed by hydrotreatment.) Cases IIa and IIIa are for low-sulfur, low-metals crude, and cases IIb and IIIb are for high-sulfur, high-metals crude. Note that the revenue from adding a hydrotreater tends to increase as nitrogen, sulfur, and metals content in the crude increase. However, hydrogen requirements and operating pressure also increase.

Question for Discussion

10-7. Would the economic returns of adding a hydrotreater shown in Table 10-3 occur in every refinery?

Problem Statement

10-8. If the scrubber and all four hydrotreaters from Table 10-3 are assumed to have the same expected lifetime, it is appropriate to calculate the net present worth of all the options for economic comparison. The present worth of an option is the present worth of the revenue it generates less its capital cost. Calculate the net present worth of all five options assuming that the effective annual interest rate is 8% and the effective annual inflation rate is 3%. Assume an equipment life of 30 years. The present worth (P) of a uniform annual amount (A) is equal to

$$P = A\,\frac{(1+i)^n - 1}{i(1+i)^n}\,,$$

where n is the number of periods and i is the effective interest rate. The effective interest rate can be corrected for inflation using the formula

$$i' = i + e + ie\,,$$

where e is the constant inflation rate and i' is the effective interest corrected for inflation.

10.1.1.5 *(Chapter 9) Ranking SO_x Reduction Alternatives* There are drawbacks to the use of hydrotreaters. Plant safety is a concern because of the hydrogen requirements and the high operating pressures. Hydrogen production requires a substantial amount of energy and thus has substantial air emissions of its own, and Claus units (for recovering elemental sulfur from the H_2S generated in the hydrotreater) generate their own air emissions, which must be managed. Also, installation of a hydrotreater requires more time than does installation of a scrubber and faces greater regulatory permitting hurdles.

Questions for Discussion

10-9. Is it clear which option for arriving at lower SO_x emissions (flue-gas scrubbing or hydrotreating the FCCU feed) is of greater benefit to the environment?

10-10. Which option would be favored if a refinery were required by regulations to reduce SO_x emissions within one year?

10-11. Where in the waste management hierarchy does installation of a scrubber fall? Where does addition of a hydrotreater fall?

Open-Ended Problem Statement

10-12. Suppose that you are charged with recommending either a scrubber or a hydrotreater to reduce SO_x emissions from the hypothetical refinery of Table 10-3. This refinery has the option to process any of the four crude-oil types in the table, so your recommendation may include a recommended crude slate. Justify your recommendation, and include considerations of future liabilities. Rank the options according to weighted evaluations of the criteria on which you have based your decision. Assume that SO_x releases for the high-sulfur-crude slates are the same as the releases from the refinery in Problem Statement 10-6, and that emissions for the low-sulfur-crude slates are half the values obtained in Problem Statement 10-6.

10.1.2 Comparing Claus Plant Tail-Gas Management Strategies

As described in the background material for this problem, Claus plants are used in the petroleum refining industry to convert hydrogen sulfide, which originates in the hydroprocessing of crude oil, to salable elemental sulfur. Claus plants themselves generate a tail gas containing pollutants. In fact, a substantial portion of the sulfur that enters the Claus plant escapes in the tail gas (Allen et al., 1990), and these residual sulfur compounds must be removed before the gas is discharged to the atmosphere. Several processes are available for managing the tail gas from Claus plants. In this module,

Table 10-4. Typical composition of Claus plant tail gas

Compound	% by Volume
H_2S	0.85
SO_2	0.42
S_8	0.05
COS	0.05
CS_2	0.04
CO	0.22
CO_2	2.37
H_2	1.60
H_2O	33.10
N_2	61.30

Source: Allen et al. (1990).

two such processes are compared: the Beavon sulfur removal process (BSRP) and the Shell Claus off-gas treatment (SCOT) process.

10.1.2.1 ***(Chapter 5) Claus Plant Tail-Gas Wastes and Emissions*** The composition of the tail gas generated by a typical Claus plant is given in Table 10-4. More than 90% of this stream is made up of water and nitrogen, but small amounts of sulfur-containing compounds are also present, as well as carbon monoxide, carbon dioxide, and hydrogen. Hydrogen sulfide, besides causing eye and respiratory irritation, smells like rotten eggs and is therefore a particularly important community relations issue for refineries in populated areas.

The SCOT process and the BSRP both create waste hydrogenation catalyst, as shown in the summary of waste streams given in Table 10-5. However, there are significant differences in the waste streams from the two processes. In particular, the sulfur generated by the BSRP is contaminated with va-

Table 10-5. Wastes generated by the SCOT process and by the BSRP; quantities given are per tons of sulfur recovered by the units

SCOT Process	BSRP
Spent hydrogenation catalyst (4.8 lb of catalyst waste/ton of sulfur)	Spent hydrogenation catalyst (4.8 lb of catalyst/ton of sulfur)
Spent amine solution (27 lb of spent solution/ton of sulfur)	Spent Stretford solution (1400 lb spent solution/ton of sulfur)
Off-gas from absorber	Off-gas from absorber (minor)
	Sulfur contaminated with vanadium (1 ton/ton of sulfur)

Source: Allen et al. 1990).

nadium, and would in general require purification before sale. This vanadium-contaminated sulfur is classified as a hazardous waste in some areas of the country because of its vanadium content, creating regulatory hurdles to purification steps.

Problem Statement

10-13. For a Claus unit processing 100 tons/day of sulfur whose tail gas stream has a flow rate of 110 tons/day, calculate the mass fraction of sulfur entering the Claus plant that makes its way into the tail gas. Use the data in Table 10-4, and assume that the tail-gas stream behaves as an ideal gas.

10-14. All of the compounds present in the tail gas are also released as fugitive emissions from the Claus plant and the tail-gas management process. Assuming that the fugitive emissions from a Claus unit are on the order of 500 g/Mg of sulfur produced, estimate the fugitive emissions in tons per year of hydrogen sulfide from a Claus plant producing 100 tons/day of sulfur. In generating this rough estimate, assume that there are 352 operating days per year and that the fugitive-emission profile is the same as the tail gas profile given in Table 10-4.

10-15. Use the data given in Table 10-5 to calculate the emission rates for each type of waste from a BSRP and a SCOT process that manage the tail gas from a Claus plant that produces 100 tons/day of sulfur. (*Hint*: The result of Problem Statement 10-13 is used in the solution to this problem.)

10.1.2.2 *(Chapter 6) Reducing Waste from Tail-Gas Management* Simplified flow diagrams for the SCOT process and the BSRP are given in Figures 10-6 and 10-7. Both processes begin by hydrogenating the Claus plant tail gas so that all the incoming sulfur is converted to H_2S. In this acidic state, the sulfur can be readily separated from the air stream by contacting it with a basic absorbing solution, and it is here that the two processes differ. In the SCOT process, the H_2S is absorbed into a solution containing an amine such as diethanolamine. Then, in an amine regenerator, the H_2S is released and recycled back to the Claus plant. In the BSRP, the hydrogen sulfide is separated from the air stream by contacting it with an alkaline solution containing vanadium. This solution is called *Stretford solution*. Air passed through this solution converts the H_2S to elemental sulfur using vanadium as a catalyst. In both processes, the absorbing solution becomes contaminated and a purge stream is required to maintain its effect.

Problem Statement

10-16. The amine solution in the SCOT process loses its effectiveness because of contaminant loading. Because the contaminants

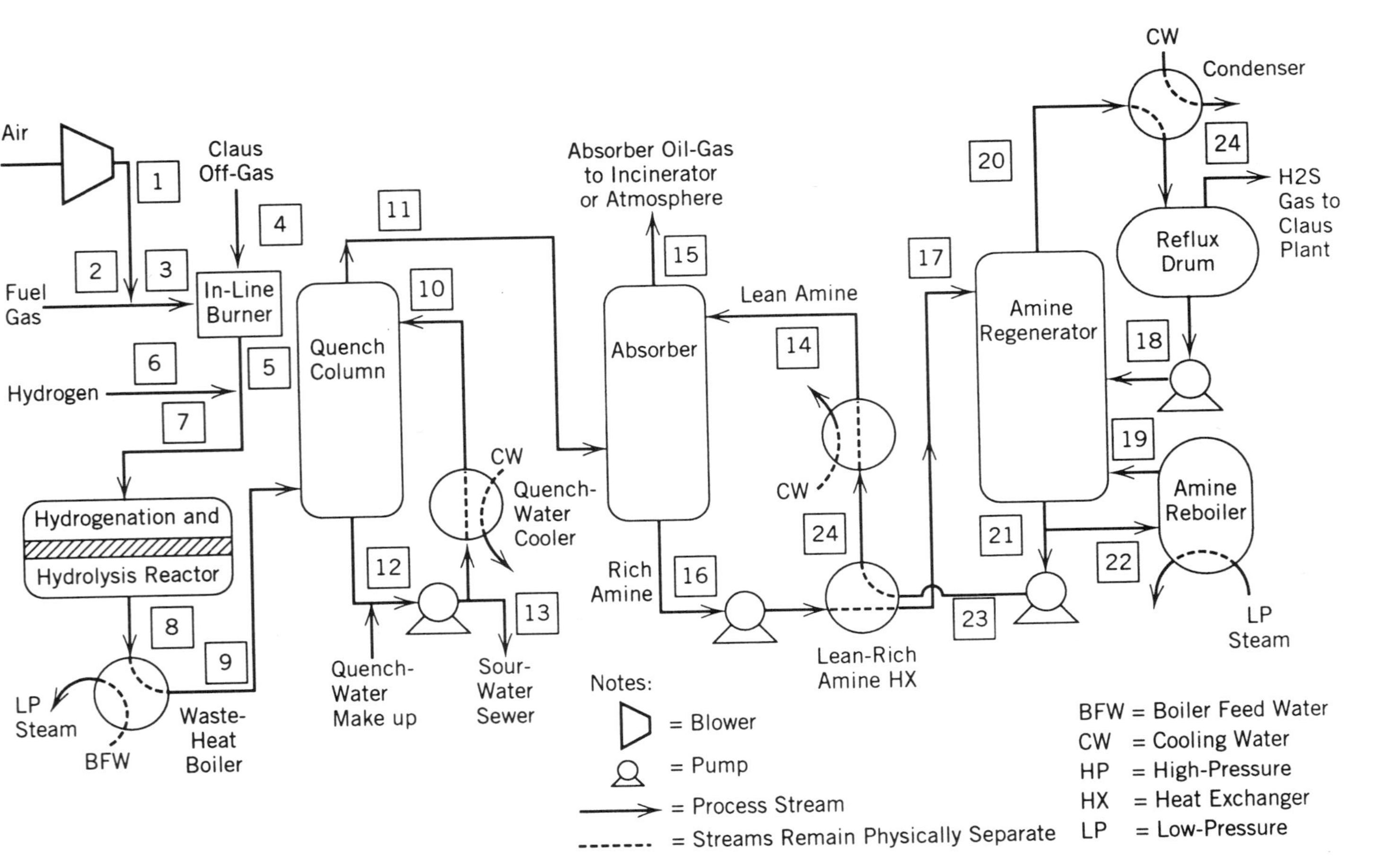

Figure 10-6. Shell Claus off-gas treating (SCOT) process (Allen et al., 1990).

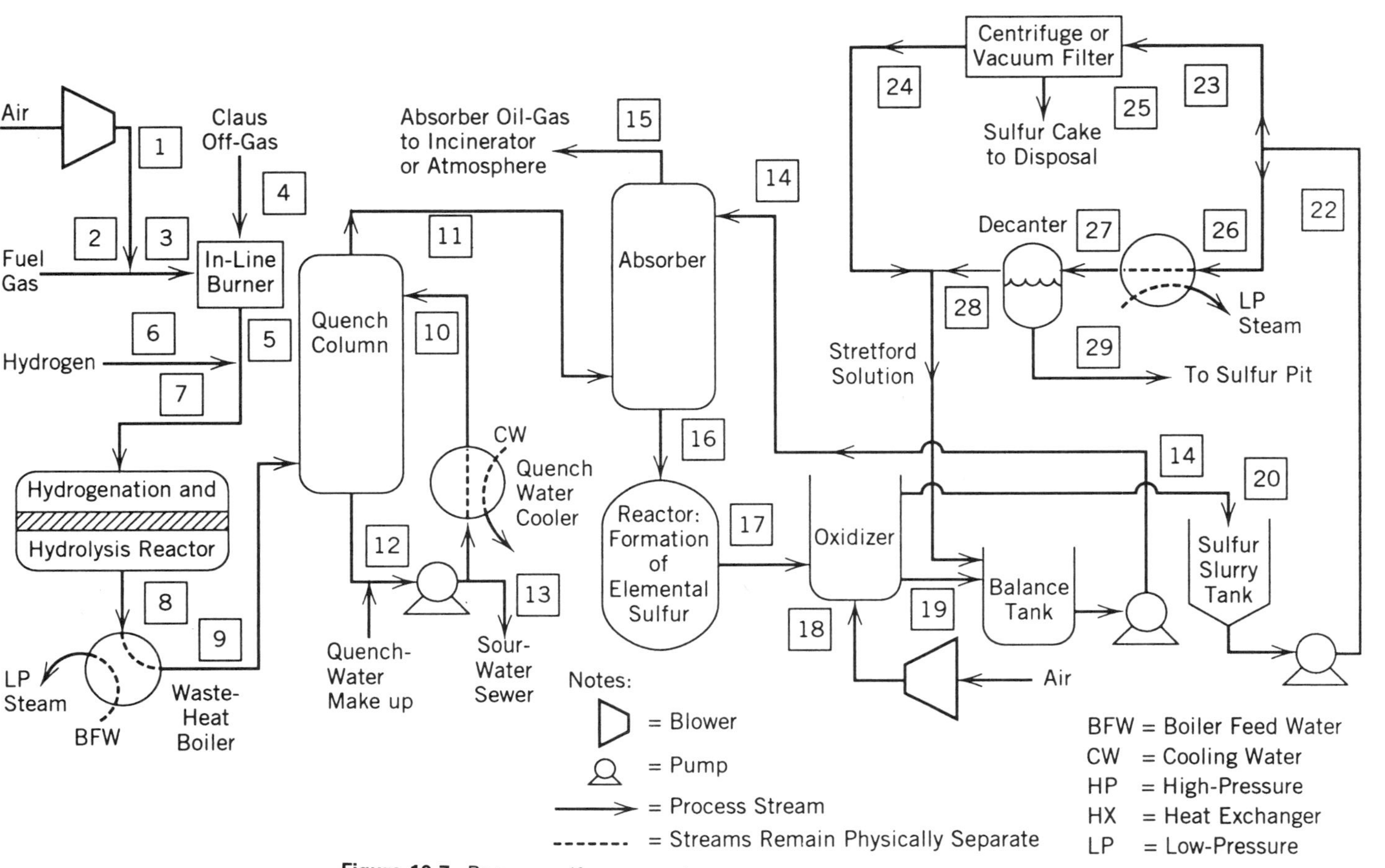

Figure 10-7. Beavon sulfur removal process (BSRP) (Allen et al., 1990).

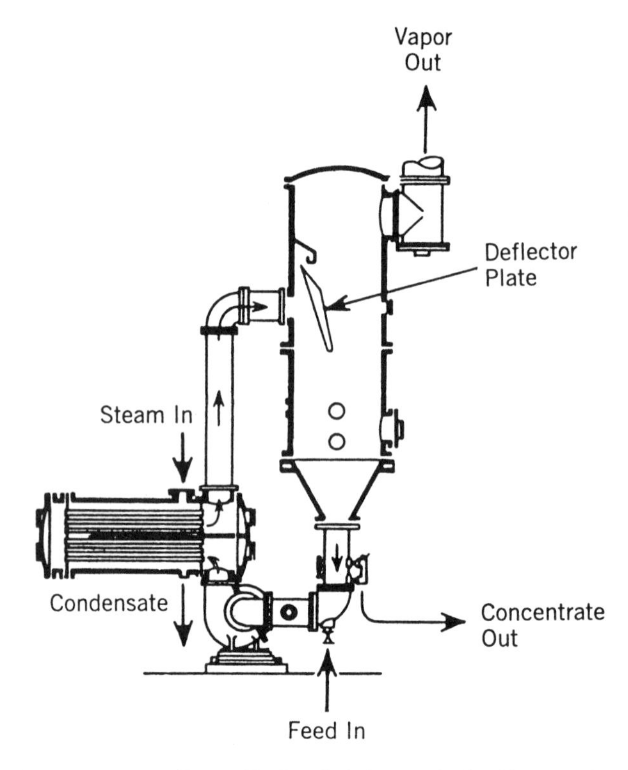

Figure 10-8. An evaporator. (From W. L. McCabe, J. C. Smith, and P. Harriott, *Unit Operations in Chemical Engineering*, 4th ed., McGraw-Hill, New York, 1985. Reproduced with permission of The McGraw-Hill Companies.)

generally have a higher boiling point than the amine solution, one way to significantly reduce the waste generated by the SCOT process is to continuously purify the amine solution in an evaporator like the one pictured in Figure 10-8. In the evaporator, the contaminated solution is boiled to remove the amine/water solution. The residue from the evaporator, consisting of concentrated contaminants carried in the amine solution, is generated at a much lower rate than the amine stream purge required when no purification step is used. Calculate the reduction in waste generated per year by the SCOT process described in Table 10-5 and Problem Statement 10-13, assuming that the residue from the evaporator is generated at $\frac{1}{20}$ the rate of the purge stream required when there is no amine purification step.

10-17. Where in the flowsheet pictured in Figure 10-6 would an evaporator be placed?

Table 10-6. Deviations in conditions by guideword for pollution prevention flowsheet analysis

Guideword	Deviation
NONE	No flow for an input stream
MORE OF	More of any relevant physical property: higher flow rate, higher temperature, higher pressure, higher concentrations, etc.
LESS OF	Less of any relevant physical property: lower flow rate, lower temperature, lower pressure, lower concentrations, etc.
PART OF	Composition of system different: change in ratio of components, component missing, etc.
MORE THAN	More components present in the system: extra phase, impurities, etc.
OTHER THAN	Off-normal operation: startup, shutdown, uprating, low-rate running, alternative operation mode, failure of plant services, maintenance, catalyst change, etc.

10.1.2.3 ***(Chapter 8) Flowsheet Analysis for Tail-Gas Management*** Flowsheets like those pictured in Figures 10-6 and 10-7 can be examined for pollution prevention opportunities by posing a series of hypothetical changes in conditions for each stream in the flowsheet. This is similar to hazard and operability (HAZ-OP) analysis, where potential scenarios are systematically considered. Instead of focusing on the potential hazards caused by a change in conditions, however, the potential impact on waste generation is considered. Table 10-6 is a list of deviations from normal considered in HAZ-OP analysis, adapted for use in examining flowsheets for pollution prevention opportunities.

Question for Discussion

10-18. Would a mass-balance approach be appropriate for estimating the quantity of alkaline purge stream (either Stretford or amine solution)? Would a mass-balance approach be appropriate for estimating the emission rate of the absorber off-gas stream?

Problem Statement

10-19. Find the number of possible deviations that could potentially impact the generation of absorber off-gas using the key words more of, less of, part of, and more than for the process variables temperature, flow, and concentration on the streams flowing into the absorber for the SCOT process (Figure 10-6). Keep in mind that the absorber is only a portion of a process that manages the off-

Table 10-7. Capital costs, waste disposal costs, and utility and chemical requirements for two Claus plant tail-gas management strategies used to handle tail gas from a Claus plant that produces 100 tons/day of sulfur, excluding recycle from SCOT process

Cost Factor	SCOT Process	BSRP
Capital costs, $	$2.9 \times 10^{6,a,c}$–4.0×10^6	$4.9 \times 10^{6,a}$
Utilities and chemicals, $/year[a]		0.16×10^6
Electricity, kW · hr/day	2,800	
Steam, lb/day	80,000	
Fuel gas, MMBtu/day	53	
Chemicals, gal/day	8	
Waste disposal costs[b]	$0.10/lb amine solution	$0.20/lb Stretford solution, $0.06/lb Stretford sulfur

[a]Allen et al. (1990).
[b]WSPA (1993).
[c]This value does not reflect the additional (~2%) capacity required in the Claus plant when the SCOT process is used.

gas from an auxiliary unit to a single process at a refinery, and comment on the time and effort required to analyze an entire refinery flowsheet in this manner.

10.1.2.4 ***(Chapter 9) The Economics of Tail-Gas Management Alternatives*** While the mass of waste generated by the SCOT process is considerably less than that generated by the BSRP, there are factors that favor the use of a BSRP as a tail-gas management strategy. For example, the SCOT process requires the use of an amine absorption solution that is considerably more expensive than Stretford solution. It also requires that the Claus plant have sufficient capacity to recycle the hydrogen sulfide that is generated. Economic characteristics for the two processes are given in Table 10-7.

Problem Statement

10-20. To determine which of the two tail-gas management strategies is economically superior, calculate the net present worth of the SCOT process and the BSRP described in Table 10-7. Assume there are 352 operating days per year and that the expected equipment lifetime is 30 years. Perform your calculations using effective annual interest rates of 5%, 10%, and 15% and neglect inflation. First, consider only capital, utility, and chemical costs. This is what might be done at a facility whose waste disposal costs are charged to overhead. For the SCOT process use the high value ($\$4.0 \times 10^6$) for capital cost. Use values for electricity, steam, fuel gas, and the SCOT amine solution of $0.050/kW·hr, $4.70/1000 lb steam, $3.30/MMBtu, and $7.60/gal, respectively. The net present worth of a project is equal to the present worth of the revenue it generates, less the present worth of the operating costs, less capital costs. The present worth (P) of a uniform annual series (A) is given by

$$P = A\frac{(1+i)^n - 1}{i(1+i)^n},$$

where i is the effective annual interest rate.

10-21. Now perform the same calculations as in Problem Statement 10-20, this time including waste disposal costs and the revenue from the sale of sulfur from the SCOT process, using data in Table 10-5. The SCOT process returns almost 100% of the sulfur to the Claus plant and sulfur is valued at \$100/ton. Use the results of Problem Statement 10-15 for waste quantity values.

10-22. Compare the results of Problem Statements 10-20 and 10-21.

10.1.2.5 *(Chapter 9) Ranking Tail-Gas Management Alternatives*

The BSRP and the SCOT process are two processes that perform the same function (manage the tail gas from a Claus unit) but have very different economic and waste generation characteristics. There are many options for reducing the waste generation from each of the two processes, as well as options for which management strategy to use. In this part of this module, you are asked to make a decision about either replacing a BSRP with a SCOT process or making improvements to the existing BSRP in order to reduce the amount of waste it generates.

Questions for Discussion

10-23. Is it obvious which tail-gas management strategy is the best choice?

10-24. Compare the two processes for waste generation characteristics. How might the environmental and health impacts of the very different waste streams from the two processes be compared?

10-25. Where do these processes fall in the waste management hierarchy? Note that your answer to this question depends on whether you regard the process as the tail-gas management plant, the Claus plant coupled with the tail-gas management plant, or the entire refinery.

Open-Ended Problem Statement

10-26. Suppose that you work for a refinery that manages their Claus plant tail gas with the BSRP. The Claus plant produces 100 tons/day of sulfur. Someone has suggested that the refinery will benefit economically if the BSRP is replaced with a SCOT process. Another suggestion has been to lease a process that would regenerate the Stretford solution and reduce the purge by 20% for a savings of $20,000/yr in waste disposal and raw material costs. The regeneration process uses sulfuric acid and generates sulfur dioxide as well as a small quantity of Glauber salts, which are sometimes classified as a hazardous waste. Develop a recommendation for one of these processes over the other. Rank the two options based on weighted evaluations of the criteria on which you have based your decision.

10.2 PREVENTION OF NITROGEN OXIDE EMISSIONS FROM PROCESS HEATERS

As mentioned in Chapter 2, there are six criteria pollutants for which there are National Ambient Air Quality Standards (NAAQS) established through the Clean Air Act: (1) particulate matter $<10\,\mu m$ in diameter (PM_{10}), (2) sulfur dioxide (SO_2), (3) nitrogen oxides (NO_x), (4) carbon monoxide (CO), (5) ozone (O_3), and (6) lead (Pb). NAAQSs are time-averaged concentrations that cannot be exceeded in the ambient air more than a specified number of times in a year. Attainment areas are those that are in compliance with the NAAQSs; nonattainment areas are those that are not. In nonattainment areas, permitting for a new source or for a major modification of an existing source requires that lowest achievable emission rate (LAER) technology be used. Also, any increase in emissions in a nonattainment area must be offset by decreases in emissions from other sources in the same region. Existing sources in nonattainment areas are also subject to emission control measures. Although volatile organic compounds (VOCs) are not on the list of criteria pollutants, facilities emitting VOCs can be affected by the NAAQS for ozone because of the role VOCs play in the creation of ozone in the lower atmosphere. The same holds true for NO_x: in a region that is an attainment area for NO_x but a nonattainment area for ozone there may be strict controls placed on NO_x emissions.

NO_x emissions are primarily combustion-related and are emitted from a wide variety of sources. Seven oxides of nitrogen are known to occur naturally, but only NO and NO_2 are emitted in large quantities by combustion devices (EPA, 1993). These emissions are a large concern partly because there is a growing body of evidence that ozone formation in some areas of the country is NO_x-limited. Where this is the case, ambient ozone concentrations would respond to levels of NO_x emissions. NO_x emissions are also of concern because, as with sulfur oxides, they are associated with acid precipitation and acid deposition, which over time can cause damage to forests and aquatic environments.

Emissions of NO_x from process heaters are the focus of this section. Process fluids must sometimes be heated, either to raise their temperature before additional processing or so that chemical reactions can occur.

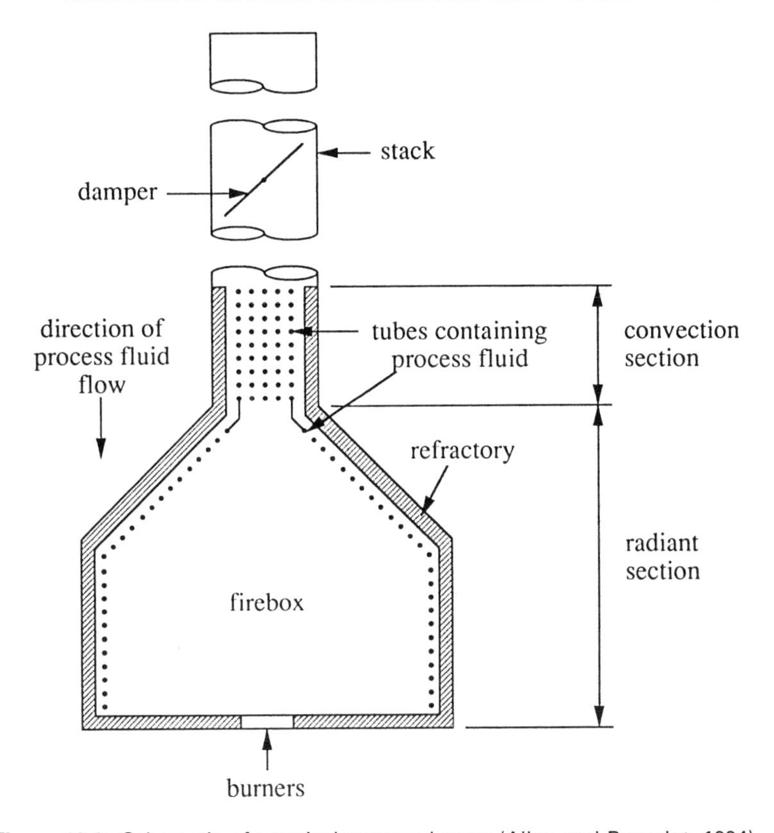

Figure 10-9. Schematic of a typical process heater (Allen and Rosselot, 1994).

Generally, when desired temperatures are above the practical range of steam heat (90–204°C), process heaters are used for heating process fluids. These heaters, which may burn gas, oil, or a combination of gas and oil, are found mostly in the petroleum refining and chemical processing industries.

A diagram of a typical process heater is given in Figure 10-9. Heat is transferred to process fluids primarily by radiation and secondarily by convection. There are many process heater configurations, but most contain the elements shown in this figure: burners, a firebox where radiant heat transfer to process fluid in tubes occurs, and a region where hot flue gases transfer heat to the process fluid convectively. There may be anywhere from one to 100 burners in a process heater (EPA, 1993). Combustion air can be supplied by a natural draft that occurs due to low gas density in the heater, or by a mechanical draft system where fans blow generally preheated air to the burners. Natural draft heaters are simpler, but do not allow precise control of combustion-air flow. Mechanical draft heaters have an efficiency advantage when they use preheated combustion air.

NO_x is formed during combustion through thermal fixation of atmospheric nitrogen introduced in the combustion air (thermal NO_x) or through oxidation of nitrogen introduced with the fuel (fuel NO_x). There is also prompt NO, which is formed near the flame zone in a process heater, but this is not generally considered to be a major contributor to NO_x emissions (Wood, 1994).

A simplified mechanism for the formation of thermal NO_x is as follows (EPA, 1993):

$$N_2 + O \rightleftharpoons NO + N \tag{1}$$

$$N + O_2 \rightleftharpoons NO + O \tag{2}$$

$$\text{Net:} \quad N_2 + O_2 \rightarrow 2NO$$

Reaction 1 is the rate-determining step and has a high activation energy, which requires high temperatures. At temperatures above 2800°F, the rate of formation of NO is given by

$$[NO] = k_1 \exp\left(-\frac{k_2}{T}\right)[N_2][O_2]^{1/2} t,$$

where brackets indicate mole fraction, k_1 and k_2 are constants, T is the peak flame temperature in K, and t is the residence time of reactants at the peak flame temperature (EPA, 1993). Note that the rate of thermal NO_x formation increases exponentially with increasing flame temperature but is only weakly dependent on the concentration of oxygen. Thermal NO_x can therefore be reduced by reducing combustion temperatures.

Fuel NO_x can comprise the majority of total NO_x emissions when high-nitrogen fuels are burned (EPA, 1993). Typically, nitrogen in fuel is limited to liquid and solid fuels. The nitrogen to nitrogen bond in nitrogen molecules is much stronger than the nitrogen to carbon bonds in fuels containing nitrogen. During combustion, the nitrogen atoms in fuels are readily released and oxidized to NO, with conversion rates ranging from 15% to 100% (EPA, 1993). It has been proposed that the formation of fuel NO_x from liquid fuels is a gas-phase reaction involving volatilized nitrogen-containing compounds. The gas-phase reactions are strongly dependent on stoichiometry (the air : fuel ratio), with increasing levels of excess air leading to higher rates of formation. The rate of formation of fuel NO_x is only weakly dependent on temperature, in contrast to the rate of formation of thermal NO_x.

10.2.1 Comparison of NO_x Emission Management Strategies

10.2.1.1 ***(Chapter 5) Process Heater Emissions*** There are approximately 3200 process heaters in use at nearly 200 refineries in the United States (EPA, 1993). A large, integrated refinery may have as many as 100 process heaters, while a small refinery can have as few as four. Table 10-8 shows that the refining processes with the highest process heat requirements per barrel of feed are thermal processes such as thermal cracking and coking and hydroconversion processes such as alkylation and catalytic reforming. It has been estimated that in 1991 the total energy consumption for process heaters at refineries was 2.3×10^{15} Btu (EPA, 1993). In 1985, the known process heater requirements in the chemical manufacturing industry were 6.8×10^{14} Btu (EPA, 1993). Major chemical industry process heater requirements are shown in Table 10-9. The data in this table indicate that pyrolysis furnaces used in the production of ethylene and propylene account for approximately half of the total process heat requirements in the chemical industry.

Problem Statement

10-27. Emission factors for NO_x are expressed in terms of pounds per unit of heat generated in the burner. For process heaters, these factors must be multiplied by the thermal flow rate of the heaters to arrive at NO_x emissions, according to the equation

$$E_\mathrm{i} = f_\mathrm{i} V_i \mathrm{HHV}_i C_i \ ,$$

where E_i is the NO_x emissions from heater i, f_i is the appropriate emission factor for that heater, V_i is the volumetric flow rate of the fuel, HHV is the higher heating value of the fuel, and C_i is the efficiency of reduction and control techniques applied to the heater. If the uncontrolled emission factors for NO_x from natural-gas-fired process heaters are 0.098 and 0.197 lb/MMBtu for natural draft and mechanical draft heaters, respectively (EPA, 1993), what are the NO_x emissions in pounds per year from the process heaters described in Table 10-10? These process heaters burn refinery fuel gas (note that the emission factors given above were for natural-gas-fired process heaters) with an HHV of 1380 Btu/SCF (standard cubic foot). The fuel gas contains on average 9 mol% hydrogen, which results in a higher flame temperature and therefore higher NO_x emissions than natural gas. To correct for the hydrogen content, assume that process gas containing 50 mol% hydrogen results in NO_x emissions that are 20% higher than when natural gas is burned (EPA, 1993), and that the increase in NO_x emissions depends linearly on the mol% hydrogen. There are 352 operating days per year. Efficiencies for various reduction and control techniques are given in Table 10-11.

Table 10-8. Major refinery processes requiring a fired heater

Process	Process Description	Heaters Used	Process Heat Requirements, 10^3 Btu/bbl feed	Feedstock Temperature Outlet of Heater, °F
	Distillation			
Atmospheric Column	Separates light hydrocarbons from crude in a distillation column under atmospheric conditions	preheater, reboiler	89	700
Vacuum Column	Separates heavy gas oils from atmospheric distillation bottoms under vacuum	preheater, reboiler	63	750–830
	Thermal Processes			
Thermal Cracking	Thermal decomposition of large molecules into lighter, more valuable products	fired reactor	700	850–1000
Coking	Cracking reactions allowed to go to completion; lighter products and coke produced	preheater	230	900–975
Visbreaking	Mild cracking of residuals to improve their viscosity and produce lighter gas oils	fired reactor	145	850–950

	Catalytic Cracking			
Fluidized	Cracking of heavy petroleum products with the aid of a catalyst	preheater	100	600–885
Hydrocracking	Cracking of heavy feedstocks to produce lighter products in the presence of hydrogen and a catalyst	preheater	195	400–850
	Hydroprocessing			
Hydrodesulfurization	Remove contaminating metals, sulfur, and nitrogen from the feedstock in the presence of hydrogen and a catalyst	preheater	65[a]	390–850
Hydrotreating	Same as hydrodesulfurization, except for lighter feedstocks	preheater	75[b]	600–800
	Hydroconversion			
Alkylation	Combination of two hydrocarbons to produce a higher-molecular-weight hydrocarbon	reboiler	377[c]	400
Catalytic reforming	Low-octane naphthas are converted to high-octane, aromatic naphthas in the presence of hydrogen and a catalyst	preheater	270	850–1000

[a]Heavy gas oils and middle distillates.
[b]Light distillate.
[c]Btu/bbl (barrel) of total alkylate.
Source: EPA (1993).

Table 10-9. Energy requirements of major fired heater applications in the chemical industry in 1985

Chemical	Process	Heater Type	Firebox Temperature, °F	Energy Requirement, 10^{12} Btu/yr	% of Known Chemical Industry Requirements
		Low- and Medium-Temperature Applications			
Benzene	Reformate extraction	Reboiler	700	65	9.9
Styrene	Ethylbenzene dehydrogenation	Steam superheater	1500–1600	32	4.9
Vinyl chloride monomer	Ethylene dichloride cracking	Cracking furnace		13	1.9
p-Xylene	Xylene isomerization	Reactor fired preheater		13	2.0
Dimethyl terephthalate	Reaction of *p*-xylene and methanol	Preheater, hot-oil furnace	480–540	11	1.7
Butadiene	Butylene dehydrogenation	Preheater, reboiler	1100	2.6	0.4
Ethanol (synthetic)	Ethylene hydration	Preheater	750	1.3	0.2
Acetone	Various	Hot-oil furnace		0.8	0.1
		High-Temperature Applications			
Ethylene/propylene	Thermal cracking	Pyrolysis furnace	1900–2300	340	52
Ammonia	Natural-gas reforming	Steam hydrocarbon reformer	1500–1600	150	23
Methanol	Hydrocarbon reforming	Steam hydrocarbon	1000–2000	26	4.0
Totals				650	100

Table 10-10. Process heater profile at a petroleum refinery

Heater No.	Heater Rating, MMBtu/hr[a]	Actual Fuel Flow Rate, MSCF/d[b]	Control/Reduction Technology	Draft Type[c]
1	457	5400	LNB	ND
2	161	2100	LNB	ND
3	108	1200	LNB	ND
4	17.2	120	none	ND
5	17.2	130	none	ND
6	21.1	190	none	ND
7	21.1	170	none	ND
8	129	570	LNB	MD
9	73	860	LNB	MD
10	527	4400	none	ND
11	39.6	530	none	ND
12	64	250	none	ND
13	288	2400	LNB	ND
14	220	2000	LNB	ND
15	91	1200	LNB	ND
16	91	990	none	ND
17	12	170	none	ND

[a]MMBtu: million Btu.
[b]MSCF: thousand standard cubic feet.
[c]ND: natural draft; MD: mechanical draft.

Table 10-11. Reduction efficiencies for NO_x reduction and control techniques applied to refinery fuel-gas-fired process heaters

Technique	Control Efficiency (controlled emissions ÷ uncontrolled emissions)
Low NO_x burner (LNB)	0.5
Ultra-low NO_x burner (ULNB)	0.25
Selective noncatalytic reduction (SNCR)	0.4
Selective catalytic reduction (SCR)	0.25
Low NO_x burner with flue-gas recirculation (LNB + FGR)	0.45

Source: EPA (1993).

10.2.1.2 (Chapter 6) Methods for Preventing NO_x Emissions from Process Heaters There are two general ways in which NO_x emissions can be prevented from process heaters: (1) operational improvements and (2) heater modifications. Many control methods, such as selective catalytic reduction, are also available.

Combustion conditions markedly affect NO_x and other criteria pollutant emissions. For example, excess air is needed for complete combustion and low hydrocarbon and carbon monoxide emissions. However, at stack oxygen levels up to 6%, NO_x emissions increase by 6–9% for every 1% increase in stack oxygen levels. Above stack oxygen levels of 6%, NO_x levels decrease, but thermal efficiency is lost. Burner adjustments also affect NO_x emissions: long, low-intensity flames emit less NO_x, but some heaters require the uniform heat flux from compact flames.

By determining the key parameters that influence emissions and using equations that predict emissions based on these parameters, advanced control of combustion can be achieved. In addition to providing the ability to continually optimize combustion conditions as variables change, these systems provide an accurate estimate of emissions, and are called *predictive emissions monitoring systems*. They allow for horizon control of combustion emissions instead of reactionary control, and promote energy conservation as well as prevention of criteria pollutant emissions.

Conventional process heaters were designed to maximize thermal efficiency while minimizing hydrocarbon and carbon monoxide emissions. Now, they are designed or modified to minimize NO_x emissions as well. Heater modifications that alter the stoichiometry of combustion affect fuel NO_x and would not apply to gas burners, unless the alteration in stoichiometry resulted in lower flame temperatures as well. Heater modifications for reducing NO_x emissions include low-excess-air burners, staged-combustion burners, and flue-gas recirculation (Wood, 1994). Brief descriptions of these modifications are given in Table 10-12. Modifications that reduce NO_x emissions but that can reduce combustion efficiency significantly are water/steam injection and reduced air preheat, neither of which is included in the table.

Questions for Discussion

10-28. Considering the effect of operating conditions and other parameters on NO_x emissions from process heaters, comment on the accuracy of the emission factor method of Problem Statement 10-27 used to estimate these emissions.

10-29. In some parts of the country, emissions from process heaters must be continuously monitored for reporting and permitting purposes. Predictive emission monitoring systems (PEMS) have been shown to be as accurate as the more traditional chemical emissions monitoring systems (CEMS). Comment on the advantages of the predictive systems over the chemical systems. In many cases, in order to use PEMS instead of CEMS, a facility would have to go through a laborious and expensive process of demonstrating the equivalency of PEMS and CEMS. How is this a barrier to pollution prevention?

Problem Statement

10-30. Quantitatively describe the effect of halving the variables T, $[N_2]$, $[O_2]$, and t one at a time on NO formation in a process heater at temperatures above 2800°F. Discuss the limits to making these

Table 10-12. Combustion modifications for NO_x emission reduction

Technique	Description	Advantages	Disadvantages	Impacts to Consider	Applicability	NO_x Reduction (Garg, 1994)
Low excess air (LEA)	Reduces oxygen availability	Easy operation modification	Low NO_x reduction potential	High carbon monoxide emissions; flame length; flame stability	All fuels	1–15% (20–25%)
Off-stoichiometric combustion (OSC) Burners out-of-service (BOOS)[a] Overfire air (OFA)[b] Air Lances[c]	Staged combustion, creating fuel-rich and fuel-lean zones	Low operating cost; no capital equipment required for BOOS	—[a–c]	Flame length; forced-draft fan capacity; burner header pressure	All fuels; multiple burner devices	30–60% (25–50%)
Low-NO_x burners (LNB)	Provides internal staged combustion, thus reducing peak flame temperatures and oxygen availability	Low operating cost; compatible with FGR as a combination technology to maximize NO_x reduction	Moderately high capital cost; Applicability depends on combustion device and fuels, design characteristics, waste streams, etc.	Forced-draft fan capacity; flame length; design compatibility; turndown flame stability	All fuels	30–50% (55–80%)
Flue-gas recirculation (FGR)	Up to 20–30% of the flue gas is recirculated and mixed with the combustion air, thus decreasing peak flame temperatures	High NO_x reduction potential for natural-gas and low-nitrogen fuels	Moderately high capital cost; moderately high operating cost; affects heat transfer and system pressures	Forced-draft fan capacity; furnace pressure; burner pressure drop; turndown flame stability	Gas fuels and low-nitrogen fuels	40–80% (40–60%)

[a]Typically requires higher air flow to control carbon monoxide.
[b]Relatively high capital cost.
[c]Moderate capital cost.
Source: S. C. Wood, "Select the Right NO_x Control Technology," *Chem. Eng. Prog.*, **90**(1), 32–38, Jan. 1994. Reproduced with permission of the American Institute of Chemical Engineers.

adjustments, and describe how they might affect the formation of CO due to incomplete combustion.

10.2.1.3 *(Chapter 8) Heat Integration* As discussed in Chapter 8, heat-exchange network (HEN) synthesis, also called *pinch analysis*, can be used to systematically examine all the heating and cooling requirements in a flowsheet and determine the extent to which streams that need to have their temperature raised can be heated by streams that need to be cooled. Heat transfer between streams, called *heat integration*, conserves fuel and reduces combustion-related emissions, including NO_x.

Heat integration is limited economically and practically by the size of the exchangers that are required and the piping modifications that must be made. An additional constraint is found at refineries that fuel their process heaters with process gas: the process gas is generally a waste if it is not burned as fuel at the refinery. Therefore, heat integration that eliminates the need for heating from process gas is not desirable.

Questions for Discussion

10-31. To what extent do permits on individual process units inhibit the use of pollution prevention alternatives such as heat integration?

10-32. If permits for process heaters require that NO_x emissions fall below a certain level in units of mass per energy released by combustion, do they provide any incentive to pursue heat integration?

10-33. Industrial energy consumption in the United States in 1991 has been estimated to be 19.4×10^{15} Btu. If refineries could reduce the energy required from their process heaters by 10% through heat integration, would it affect this value for industrial energy consumption in the nation?

Problem Statement

10-34. It has been suggested that a heat exchanger contacting the crude charge with hot heavy vacuum gas oil (HVGO) be added to the crude preheat train at a refinery so that the atmospheric distillation column reboiler duty may be reduced. Because of space limitations, the exchanger is limited to a heat exchange surface area of $\leq 1100\ m^2$. Stream data are given in Table 10-13. At the given temperatures, both streams have a specific heat capacity of approximately 1.9 J/g/°C. The proposed heat exchanger has an overall heat-transfer coefficient of $U = 230\ W/m^2/°C$ and a temperature correction factor (necessary because the exchanger is not a counterflow double-pipe arrangement) of $F = 0.85$. Find the outlet temperatures of the two streams when the largest possible heat exchanger is installed. Remember that heat transfer is equal to

Table 10-13. Data for heat-exchange streams

Property	Crude Charge (Cold Stream)	HVGO (Hot Stream)
Inlet temperature, °F	152	332
Flow rate, barrels/day	175,000	26,000
Density, g/cm^3	0.85	0.95

$$q = UAF\,\Delta T_{\mathrm{lm}}\,,$$

where A is heat exchange surface area and ΔT_{lm} is the log mean temperature difference for a counterflow double-pipe exchanger, given by

$$\Delta T_{\mathrm{lm}} = \frac{(T_{\mathrm{h2}} - T_{\mathrm{c2}}) - (T_{\mathrm{h1}} - T_{\mathrm{c1}})}{\ln[(T_{\mathrm{h2}} - T_{\mathrm{c2}})/(T_{\mathrm{h1}} - T_{\mathrm{c1}})]}\,.$$

The subscripts for the temperatures in this equation are as shown in Figure 10-10. If heat-transfer coefficients are approximated as constants, heat transfer can also be given by

$$q = -\dot{m}_{\mathrm{h}} c_{\mathrm{ph}} \Delta T_{\mathrm{h}} = \dot{m}_{\mathrm{c}} c_{\mathrm{pc}} \Delta T_{\mathrm{c}}\,,$$

where $\dot{m}$ is mass flow rate, c_{p} is the specific heat capacity, ΔT is the outlet temperature less the inlet temperature, and the subscripts c and h denote the cold and hot streams, respectively. (*Note*: there are 2.64×10^{-4} gal cm^3 and 42 gal in a barrel.) Calculate the heat flow rate in this exchanger.

10-35. Given that the reboiler is a mechanical-draft natural-gas-fired process heater, use the heat flow rate calculated in Problem Statement 10-34 to estimate the reduction in NO_x emissions due to addition of the heat exchanger. The NO_x emission factor for this type of heater is 0.197 lb/MMBtu.

Figure 10-10. Representative counterflow heat exchanger for refinery in Problem Statement 10-34.

10.2.1.4 *(Chapter 9) Heat-Integration Economics*

Studies have indicated that additional heat integration at existing refineries can be profitable. In one study, it was estimated that a refinery could achieve an 11% reduction in NO_x emissions at a profit through heat integration (Rossiter and Kumana, 1994). In another study, three additional heat exchangers on the crude feed were estimated to reduce the atmospheric distillation tower reboiler duty by 20% (API, 1993). These additional heat exchangers had a projected payback period of 16 months.

Problem Statement

10-36. If the capital costs of the heat exchanger of Problem Statements 10-34 and 10-35 are approximately $350/m^2 and the cost of the fuel firing the reboiler is $3/MMBtu, what is the payback period for the heat exchanger? Neglect interest, so that payback period is equal to

$$t = \frac{\text{capital cost}}{\text{revenue}}.$$

10.2.1.5 *(Chapter 9) Choosing between Mutually Exclusive Pollution Prevention Options*

Sometimes the pollution prevention alternatives available to a facility are mutually exclusive. One cannot, for example, both eliminate a pond to reduce emissions and cover it to reduce emissions. In this portion of this problem module, you are asked to consider a problem where there are mutually exclusive options that must be ranked.

Question for Discussion

10-37. How many burner modifications can one process heater have? How many control strategies?

Open-Ended Problem

10-38. Develop a strategy for reducing emissions from the heaters described in Problem Statement 10-27 by at least 65%. You may have to rely on control as well as prevention strategies to accomplish this goal. Control strategies are those designed to remove NO_x from stack gases after it is formed but before it reaches ambient air. One such control strategy is selective noncatalytic reduction (SNCR), where NO_x-reducing chemicals such as ammonia or urea are injected into the flue gas. In selective catalytic reduction (SCR), ammonia is mixed with flue gases that are then fed to a catalytic reactor where NO_x is reduced to N_2. High reduction of NO_x emissions can be achieved with these controls, but ammonia, nitrous oxide, carbon monoxide, and particulate matter can be produced by SNCR, and ammonia and particulate matter emissions are pro-

Table 10-14. Costs of reduction and control techniques for natural-draft natural-gas-fired heaters in 1991 dollars

Heater Rating, MMBtu/hr	Technique	Total Annual Cost at Various Fractions of Operating Capacity, $/yr		
		0.1	0.5	0.9
17	LNB	9,250	9,250	9,250
	ULNB	9,940	9,940	9,940
	SNCR	24,800	25,700	26,700
	SCR[a]	155,000	158,000	160,000
	LNB + FGR[a]	40,400	40,900	41,400
36	LNB	14,700	14,700	14,700
	ULNB	15,400	15,400	15,400
	SNCR	39,000	41,000	43,000
	SCR[a]	247,000	252,000	257,000
	LNB + FGR[a]	63,700	64,800	66,000
77	LNB	21,200	21,200	21,200
	ULNB	21,900	21,900	21,900
	SNCR	61,900	66,100	70,400
	SCR[a]	399,000	410,000	420,000
	LNB + FGR[a]	97,600	100,000	103,000
121	LNB	36,900	36,900	36,900
	ULNB	37,600	37,600	37,600
	SNCR	81,500	88,100	94,800
	SCR[a]	532,000	548,000	565,000
	LNB + FGR[a]	142,000	146,000	150,000
186	LNB	55,000	55,000	55,000
	ULNB	55,700	55,700	55,700
	SNCR	106,000	116,000	126,000
	SCR[a]	700,000	726,000	752,000
	LNB + FGR[a]	195,000	201,000	207,000

[a]Requires conversion to mechanical draft; annual cost figures include the cost of conversion.
Source: EPA (1993).

duced by SCR, along with spent catalyst waste (EPA, 1993). SCR is also very expensive, and the increased complexity of the process leads to increased down time during which emissions are uncontrolled. Use the approximate total annual cost data for reduction and control strategies that are provided in Tables 10-14 and 10-15 for a range of process heater sizes to develop your strategy. Remember that the reduction efficiencies are multiplicative, not additive, when used in combination. Develop your proposal only on the basis of costs and describe the effects on reliability and environ-

Table 10-15. Costs of reduction and control techniques for mechanical-draft natural-gas-fired heaters in 1991 dollars

Heater Rating, MMBtu/hr	Technique	Total Annual Cost at Various Fractions of Operating Capacity, \$/yr 0.1	0.5	0.9
40	LNB	20,700	20,700	20,700
	ULNB	21,700	21,700	21,700
	SNCR	42,000	45,500	49,100
	SCR	237,000	242,000	248,000
	LNB + FGR	37,500	38,700	40,000
77	LNB	44,800	44,800	44,800
	ULNB	45,800	45,800	45,800
	SNCR	62,600	69,400	76,300
	SCR	358,000	369,000	380,000
	LNB + FGR	69,900	72,300	74,700
114	LNB	80,700	80,700	80,700
	ULNB	81,700	81,700	81,700
	SNCR	79,500	89,700	99,900
	SCR	460,000	476,000	492,000
	LNB + FGR	113,000	116,000	120,000
174	LNB	86,100	86,100	86,100
	ULNB	87,100	87,100	87,100
	SNCR	103,000	119,000	134,000
	SCR	604,000	629,000	653,000
	LNB + FGR	127,000	133,000	138,000
263	LNB	123,000	123,000	123,000
	ULNB	124,000	124,000	124,000
	SNCR	133,000	157,000	180,000
	SCR	791,000	828,000	864,000
	LNB + FGR	177,000	185,000	193,000

Source: US EPA (1993).

mental impacts your choices will have. Do the costs of NO_x reduction increase linearly with increasing reduction?

REFERENCES

Allen, D. T. and K. S. Rosselot, *Pollution Prevention for Chemical Processes: A Handbook with Solved Problems from the Refining and Chemical Processing Industries*, prepared for the Hazardous Waste Research and Information Center, Champaign, IL, HWRIC TR-022, Aug. 1994.

Allen, P., A. Jackman, and R. Powell, "Petroleum Refining Industry Waste Audit," report submitted to the California Department of Health Services, May 1990.

American Petroleum Institute (API), "Environmental Design Considerations for Petroleum Refining Crude Processing Units," Publication 311, Feb. 1993.

California Air Resources Board (CARB), "1991 Emission Inventory: Emissions by Device within Facility," March 1994.

Garg, A., "Specify Better Low-NO_x Burners for Furnaces," *Chem. Eng. Prog.*, **90**(1), 46–49, Jan. 1994.

Lane, P. A. and J. A. Latimer, "Controlling FCC SO_x Emissions with DESOX: Effective Performance in Partial Combustion," in *Advanced Fluid Catalytic Cracking Technology*, AIChE Symposium Series, Vol. 88, New York, 1992, p. 291.

McCabe, W. L., J. C. Smith, and P. Harriott, *Unit Operations in Chemical Engineering*, 4th ed., McGraw-Hill, New York, 1985.

Mobil Oil Corp., personal communication, Paulsboro, NJ, 1992.

Rossiter, A. P. and J. D. Kumana, "Rank Pollution Prevention and Control Options," *Chem. Eng. Prog.*, **90**(2), 39–44, Feb. 1994.

United States Environmental Protection Agency (EPA), "Alternative Control Techniques Document—NO_x Emissions from Process Heaters (Revised)," available through NTIS as PB94-120235, Sep. 1993.

United States Environmental Protection Agency (EPA), AIRS Graphics, "United States Facilities with SO_2 Emissions GE 100 TPY, Year of Record: 1989–1993, SIC 2911," March 1994a.

United States Environmental Protection Agency (EPA), "AFS Emissions by SIC Report: Emissions for SIC 2911 by Region," April 1994b.

Western States Petroleum Association (WSPA), "Waste Minimization Compilation: Western States Petroleum Association Refineries," Project 251-A, prepared by England, Shahin & Associates, Irvine, CA, March 1993.

Wood, S. C., "Select the Right NO_x Control Technology," *Chem. Eng. Prog.*, **90**(1), 32–38, Jan. 1994.

MICROSCALE POLLUTION PREVENTION

In previous chapters, pollution prevention strategies at the process level and on the scale of entire industry sectors were outlined. While these process changes have been characterized as macro- or mesoscale, many of the approaches that have been described rely on a molecular-level understanding of chemical and physical processes. For example, the synthesis of new catalysts that result in higher reaction yields and less wastes relies on an understanding of surface chemistry. The design of highly selective separation technologies relies on an understanding of adsorption and thermodynamics. The minimization of nitrogen oxide pollutants relies on an understanding of combustion physics and chemistry. These and a large number of other general engineering principles are employed in pollution prevention, but it is impossible to treat them comprehensively in a single volume. Instead, two case studies of pollution prevention at the microscale are discussed in this section: systematic design of substitute solvents and molecular-level reaction pathway synthesis. These case studies illustrate how general engineering principles are used as a foundation on which pollution prevention approaches are built.

11

MICROSCALE POLLUTION PREVENTION

In this chapter, two of many potential case studies of pollution prevention at a microscale are presented. These are the systematic design of substitute solvents and the systematic evaluation of reaction pathways. Both case studies employ optimization tools similar in concept to those presented in Chapter 8 and illustrate the power of emerging computational tools. However, they represent only a very limited subset of the engineering principles and analysis tools that can be used in preventing pollution. Thus, this final chapter is intended to be more of an introduction than a conclusion.

11.1 SYSTEMATIC DESIGN OF SUBSTITUTE MATERIALS

Group contribution theory and quantitative structure–activity relationships (QSARs) can aid in the systematic design of substitutes for environmentally harmful chemicals. The basic premise of group contribution methods is that a molecule can be considered to be a collection of functional groups, each of which contributes in a definable manner to the properties of the molecule. [Sources for further information on group contribution theory include Reid et al. (1987) and Kier and Hall (1976).] The first step in using group contribution theory to design chemical substitutes is to define the requirements of the new chemical, specifically, to set upper and/or lower bounds for flammability, volatility, toxicity, environmental fate, solubility, and other properties. At this point, an attempt could be made to find a chemical with acceptable characteristics by searching property tables, but such a chemical might not exist. Even if it does, its properties might not have been measured. To design a chemical to fit a particular use, equations that model the effect

Table 11-1. Selected solubility parameter group contributions

Group	Δ_d	Δ_h	Δ_p
$—CH_3$	0.34	−0.85	−0.59
—Cl	−0.15	0.26	3.11
—Br	2.23	−0.69	1.61
—OH	−0.65	10.63	5.55
$>CH_2$	0.34	−0.85	−0.59
>C=O	−1.14	4.85	4.67

Source: Joback (1992).

of different functional groups on the desired properties can be solved, adding functional groups, until a chemical is postulated that falls within the set boundaries.

A simplified example for finding a substitute solvent for toluene illustrates the systematic design of substitute materials. In identifying solvent substitutes, properties such as vapor pressure, flammability, toxicity, and solvating capability (the ability to dissolve solutes) can all be important. To keep the analysis for this example simple, only solvating capability is considered here. This property can be characterized by solubility parameters that are estimated using the following equations:

$$\delta_d = 13.29 + \Sigma\Delta_{i,\delta_d},$$

$$\delta_p = 5.07 + \Sigma\Delta_{i,\delta_p},$$

and

$$\delta_h = 7.23 + \Sigma\Delta_{i,\delta_h}.$$

Here, δ_d, δ_p, and δ_h are the solubility parameters that characterize dispersive, polar, and hydrogen bonding forces, and Δ_i are the contributions of functional group i to the dispersive, polar, and hydrogen bonding parameters. Table 11-1 gives Δ_i for several functional groups.

The use of the group contributions and the equations relating group contributions to properties can be illustrated by the estimation of the solubility parameters for toluene. Toluene contains five aromatic =CH— groups, one aromatic =C< group, and one methyl group ($—CH_3$). The group contributions for the solubility parameter describing hydrogen bonding are −0.4, −2.8, and −0.85 for these three groups, respectively (see Figure 11-1). The solubility parameter δ_h is therefore given by

$$\delta_h = 1.6 = 7.23 + 5(-0.4) - 2.8 - 0.85.$$

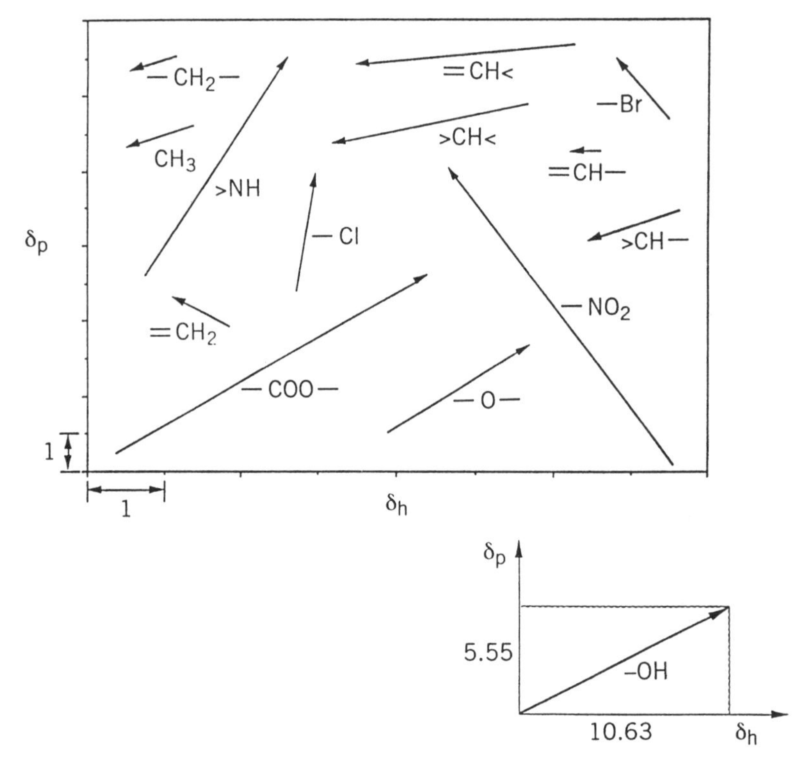

Figure 11-1. Functional group vectors for two solubility parameters (Joback, 1992).

The other solubility parameters for toluene, as well as any other molecules that can be constructed out of functional groups whose group contributions to solubility are known, can also be estimated.

The problem of finding a substitute for toluene that matches its solvating abilities can be reduced to finding a material that matches toluene's solubility parameters, which are 16.4, 8.0, and 1.6 for the dispersive, polar, and hydrogen bonding parameters, respectively. Group contribution methods can be used to systematically build chemical structures that match these solubility parameters. The process of systematic design is most easily visualized for a problem in which only two properties are matched, as shown in Figure 11-1, where the polar and hydrogen bonding parameters form the axes. The vectors in this figure represent the group contribution values, such as those in Table 11-1, for the functional groups. For example, the vector for the methyl group can be represented as $(\delta_h, \delta_p) = (-0.85, -0.59)$, so it points downward and to the left with a slope of $-0.59/-0.85$ in Figure 11-1. The length of the vector is $[(-0.59)^2 + (-0.85)^2]^{1/2}$. The placement of the vector in Figure 11-1 is arbitrary. In Figure 11-2, the axes are the same and the area within the circle is a polar and hydrogen bonding solubility parameter

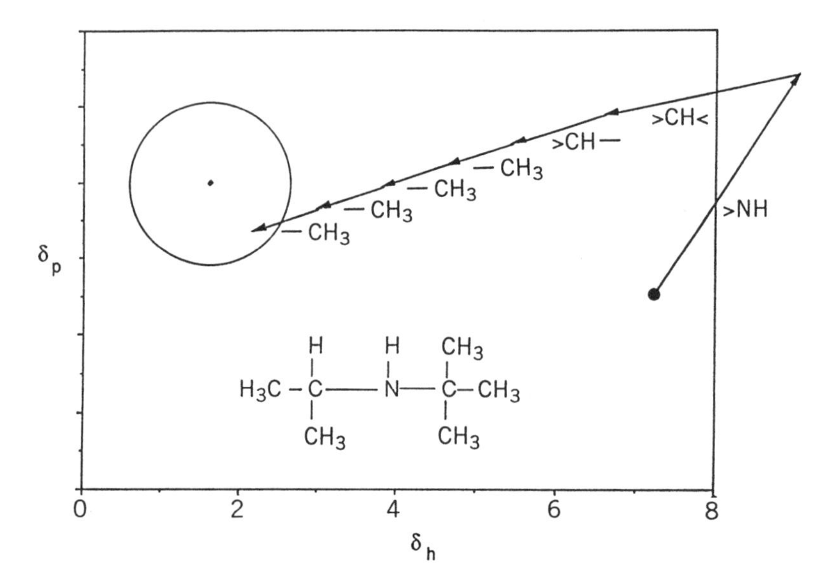

Figure 11-2. Diagram for designing a material with hydrogen bonding and polar solubility parameters similar to those of toluene (Joback, 1992).

target area for a toluene substitute. Possible candidates for the substitute chemical are found by adding vectors to the coordinates (7.23, 5.07) until a vector ends inside the circle. These starting coordinates are the values for the solubility parameters from the group contribution equations when no functional groups are present. Figure 11-2 shows one possible substitute for toluene.

While this simple example of molecular design has focused on only two parameters (the polar and hydrogen bonding solubility parameters), there is, in principle, no limit to the number of physical properties that could be considered. Groups must be chosen so that a feasible molecule is constructed (e.g., the choice of three methyl groups and no other functionalities is not a feasible molecule), but otherwise, this approach is quite general, and is already in use in applications such as pharmaceutical design.

The most serious limitations of a group contribution approach to the molecular design of chemical products are as follows:

- The optimization may lead to structures that, while feasible, are difficult to synthesize.
- The optimization may not be able to consider molecules containing functional groups for which no group contributions are available.
- The group contributions are based on regressions from data on a limited

set of compounds, and these group contributions may not be accurate for structures generated by the optimization. For example, the group contribution of a hydroxyl group to a property such as vapor pressure in an alcohol can be quite different from that in a glycol. Thus, care must be taken in applying group contribution methods to new structures.

- Group contribution methods may not be available for all relevant properties. The development of group contribution or molecularly based approaches to estimating properties of environmental interest is the major impediment to the use of these methods for pollution prevention.

Each of these difficulties could be addressed in the optimization. For example, it is possible to estimate the potential for stratospheric ozone depletion in a new material. Stratospheric ozone depletion potentials have only been calculated for a few compounds, but it is clear that the property depends strongly on two parameters: (1) atmospheric lifetime and (2) the number of chlorine and bromine atoms in the molecule. Both of these parameters can be estimated using a group contribution approach as discussed in Chapter 6, making it possible to include at least a crude estimate of stratospheric ozone depletion potential in the design of a substitute material. Other environmental factors could be considered in a similar manner.

While group contribution methods have great potential in pollution prevention, many revolutionary design innovations are outside their scope. For example, in the selection of replacement solvents, the use of supercritical fluids or eliminating the need for the solvent entirely might be overlooked. Further, caution must be exercised in labeling any material as environmentally benign. The complexity of the environment makes it impossible to predict all the potential consequences of a new substance. For example, chlorinated fluorocarbons (CFCs) were initially regarded as environmentally beneficial since they replaced flammable or explosive refrigerants and it was not until 40 years after their invention that their potential to severely impact the environment was discovered.

11.2 MOLECULAR-LEVEL REACTION PATHWAY SYNTHESIS

The selection of chemical synthesis pathways that minimize the formation of undesired byproducts is a central concept in pollution prevention. A number of detailed case studies have been presented that describe how, once the reaction pathways are known, reaction conditions can be selected to minimize pollutants. For example, Hopper et al. (1992) examined the production of acrylonitrile and allyl chloride. Senkan used a detailed chemical kinetic model to describe the high temperature oxidation and pyrolysis of C_2HCl_3, a prototypical chlorinated hydrocarbon (Senkan, 1992; Chang and Senkan,

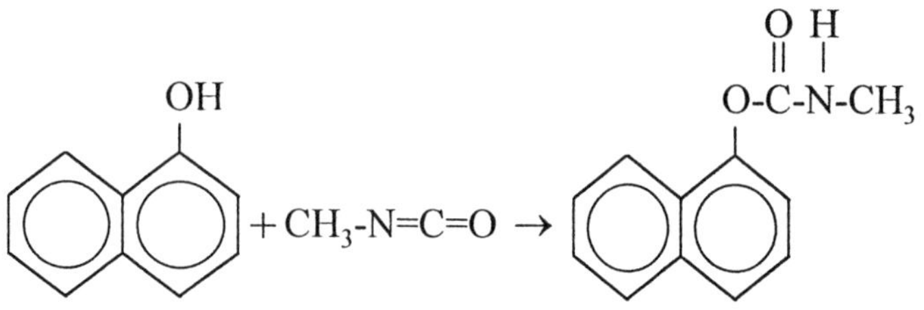

OH

+ CH_3-N=C=O →

O H
|| |
O-C-N-CH_3

α-naphthol | methyl isocyanate | 1-naphthyl-*N*-methylcarbamate (Sevin)

Figure 11-3. Traditional reactants for producing Sevin.

1988, 1989). The high-temperature reactions of chlorinated hydrocarbons are of interest in the production of commodity chemicals such as ethylene dichloride and vinyl chloride as well as in the formation of undesired byproducts, such as chlorinated dibenzodioxins and dibenzofurans. A group of approximately 150 elementary reactions was used to follow the formation of byproducts such as phosgene and carbon tetrachloride in the oxidation of C_2HCl_3, revealing major formation pathways and reaction conditions that minimize the formation of the byproducts. A number of additional examples of reaction pathway synthesis for pollution prevention are available. Chapman and Tsou (1993) have proposed a synthesis of styrene that avoids the use of benzene; Gladfelter (1993) has proposed a route to isocyanates that avoids phosgene. Generalizing these successes in reaction pathway synthesis is a substantial challenge, however. A number of tools are available for predicting equilibrium product distributions (see, e.g., Govind, 1985), but general methods for predicting pathways and rates are in their infancy.

In this section, the optimization methods that are emerging for the systematic design of reaction pathways are briefly described. The methods are illustrated by a case study of the synthesis of 1-naphthyl-N-methylcarbamate (Crabtree and El-Halwagi, 1995), manufactured until 1984 by Union Carbide and known by the trade name Sevin. In the Union Carbide synthesis, α-naphthol and methyl isocyanate were used as reactants, as shown in Figure 11-3. The objective in this case study is to examine the potential for other synthesis pathways, particularly those that could be accomplished at less cost or that would avoid the use of methyl isocyanate, a highly toxic intermediate.

The first step in identifying alternate synthesis routes is to select a set of alternative reactants. A list of potential reactants is shown in Figure 11-4. This list could be expanded, but is limited here in order to keep the equations tractable. In deciding which of these reactants to use, a series of constraints must be satisfied, including

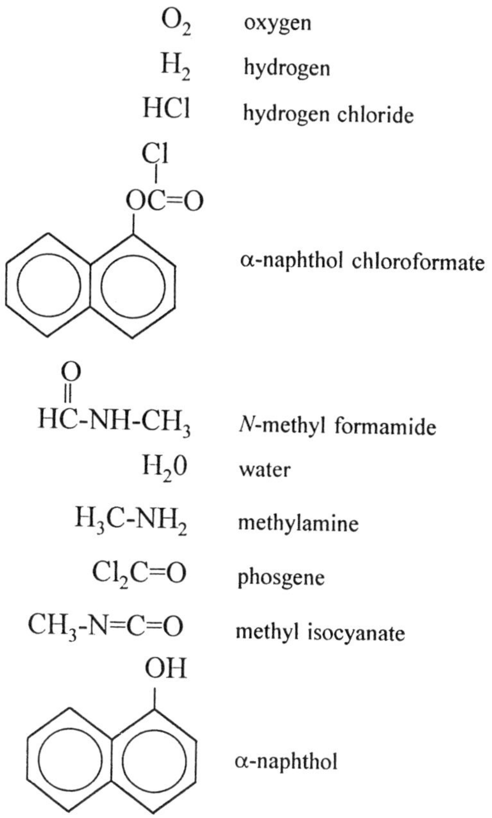

O_2	oxygen
H_2	hydrogen
HCl	hydrogen chloride
$C_{10}H_7OC(O)Cl$	α-naphthol chloroformate
$HC(=O)\text{-}NH\text{-}CH_3$	*N*-methyl formamide
H_2O	water
$H_3C\text{-}NH_2$	methylamine
$Cl_2C{=}O$	phosgene
$CH_3\text{-}N{=}C{=}O$	methyl isocyanate
$C_{10}H_7OH$	α-naphthol

Figure 11-4. Potential reactant candidates for production of Sevin. (From E. W. Crabtree and M. M. El-Halwagi, "Synthesis of Environmentally Acceptable Reactions," in *Pollution Prevention via Process and Product Modifications*, M. M. El-Halwagi and D. P. Petrides, eds., AIChE Symposium Series, **90**(303), 117–127, 1995. Reproduced with permission of the American Institute of Chemical Engineers, Copyright © 1995 AIChE. All rights reserved.)

- Stoichiometric constraints
- Thermodynamic constraints
- Environmental constraints

The goal is to then find the most economical solution that satisfies the constraints.

At a minimum, atomic balances must be satisfied when considering stoichiometric constraints. Sevin contains 12 carbon atoms, 11 hydrogen atoms, two oxygen atoms, and one nitrogen atom. The reactants must have these atoms in the proper ratio. This constraint can be formalized in matrix form as follows:

$$\begin{bmatrix} 12 \\ 11 \\ 1 \\ 2 \\ 0 \end{bmatrix} \begin{matrix} \text{carbon atoms} \\ \text{hydrogen atoms} \\ \text{nitrogen atoms} \\ \text{oxygen atoms} \\ \text{chlorine atoms} \end{matrix} = \begin{bmatrix} 0 & 0 & 0 & 11 & 2 & 0 & 1 & 1 & 2 & 10 \\ 0 & 2 & 1 & 7 & 5 & 2 & 5 & 0 & 3 & 8 \\ 0 & 0 & 0 & 0 & 1 & 0 & 1 & 0 & 1 & 0 \\ 2 & 0 & 0 & 2 & 1 & 1 & 0 & 1 & 1 & 1 \\ 0 & 0 & 1 & 1 & 0 & 0 & 0 & 2 & 0 & 0 \end{bmatrix} \times \begin{bmatrix} \nu_{O_2} \\ \nu_{H_2} \\ \nu_{HCl} \\ \nu_{C_{11}H_7O_2Cl} \\ \nu_{C_2H_5NO} \\ \nu_{H_2O} \\ \nu_{CH_5N} \\ \nu_{COCl_2} \\ \nu_{C_2H_3NO} \\ \nu_{C_{10}H_8O} \end{bmatrix}$$

where the variables ν_i represent the stoichiometric coefficients in the synthesis reaction. For example, in the synthesis of Sevin from α-naphthol and methyl isocyanate,

$$\nu_{C_{10}H_8O} = 1,$$

and

$$\nu_{C_2H_3NO} = 1.$$

For the reaction

$$0.5\, O_2 + C_2H_5NO + C_{10}H_8O \rightarrow \text{Sevin} + H_2O$$

the values for ν are

$$\nu_{O_2} = 0.5,$$

$$\nu_{C_2H_5NO} = 1,$$

$$\nu_{C_{10}H_8O} = 1,$$

$$\nu_{H_2O} = -1.$$

Note that the value of ν for water is negative because water appears as a product.

In addition to the stoichiometric constraints, thermodynamic constraints must be considered. Thermodynamic constraints limit the maximum extent of the reactions that can occur between potential reactants. The equilibrium constant, defined as the ratio of the product of the chemical activities of the products to the product of the chemical activity of the reactants, sets the limits for allowed concentrations. A detailed derivation of the form of the equilibrium constant, K, for the reaction pathway selection is beyond the scope of this text [see Crabtree and El-Halwagi (1995)] but is given by

$$K \propto \Pi(\text{concentration of species } i)^{-\nu_i},$$

where ν_i is the stoichiometric coefficient. (Using the previous notation, ν is positive for a reactant and negative for a product.) K is determined by evaluating Gibb's free energies of reactants and products and can be found for all possible combinations of reactants and products.

Finally, there is a set of environmental considerations that define the maximum allowable concentrations for hazardous constituents. They can be expressed in the form

$$\text{Concentration} \leq \text{maximum limit value}.$$

These limiting values may be set using data such as the permissible exposure limit (PEL) established through the Occupational Safety and Health Act (OSHA). PELs establish limits for workplace concentrations of hazardous chemicals, and values for the chemicals relevant to this example are shown in Table 11-2.

After the stoichiometric, thermodynamic, and environmental constraints are satisfied, economics are considered. The economic goal may be to maximize the value of the products minus the value of the reactants.

The constraints in this analysis impose linear and non-linear constraints on the types and concentrations of reactants. The economic evaluation is a

Table 11-2. Permissible exposure limits for selected compounds

Compound	Permissible Exposure Limit, $(kmol/m^3) \times 10^9$
Oxygen	—
Hydrogen	—
Hydrogen chloride	191.9860
α-Naphthol chloroformate	—
Methyl formamide	—
Water	—
Methylamine	386.3739
Phosgene	0.4044
Methyl isocyanate	0.8179
α-Naphthol	—

Source: Crabtree and El-Halwagi (1995).

linear function of the reactant and product stoichiometric coefficients and the values of the stoichiometric coefficients are in some cases limited to integer values. Taken together, this results in a mixed integer nonlinear programming problem that can be solved using optimization techniques (Floudas, 1995).

For the Sevin synthesis problem, the preferred synthesis route was found to be (Crabtree and El-Halwagi, 1995)

$$N\text{-methyl formamide} + \alpha\text{-naphthol} \rightarrow \text{Sevin} + \text{hydrogen}$$

A solution that is only slightly less desirable is

$$0.5\ \text{oxygen} + N\text{-methyl formamide} + \alpha\text{-naphthol} \rightarrow \text{Sevin} + \text{water}$$

Although these reactions satisfy stoichiometric, thermodynamic, and environmental constraints, it cannot be assumed that the reaction rates are fast enough to make the synthesis practical. In addition, the optimization considered only a limited set of potential reactants. Nevertheless, this type of optimization allows for the systematic evaluation of a large number of potential reaction pathways and will likely see more and more use as techniques for evaluating chemical synthesis routes develop.

11.3 SUMMARY

Wide-ranging studies are contributing to a molecular-level redesign of some products and processes. In this chapter, a procedure for systematically designing a substitute material (in this case a solvent) is discussed. The approach that was described makes use of group contribution theory, where molecules are systematically designed to contain a distribution of functional groups that result in desired properties. The use of such a procedure relies on an understanding of the ways in which the building blocks of a molecule (its functional groups) affect its properties. Molecular-level reaction pathway synthesis for reducing the use of toxic precursors and reducing the formation of unwanted byproducts is also discussed in this chapter. While there are currently significant limitations to this procedure, it nonetheless shows promise in the systematic identification and evaluation of possible synthesis routes.

QUESTION FOR DISCUSSION

1. Suggest alternatives to OSHA's permissible exposure limit for establishing environmental constraints in the synthesis of Sevin.

OPEN-ENDED PROBLEM STATEMENTS

1. Using the data in Table 11-1 and Figure 11-1, identify three structures that have solubility parameters within a value of ±1 of the values for toluene (i.e., $7 \leq \delta_p \leq 9$ and $0.6 \leq \delta_h \leq 2.6$). Note that the structures must contain appropriate groups and the groups must combine to make a physically reasonable molecule. For example, a structure calling for two aromatic carbons is unreasonable since the two carbons cannot make an aromatic ring.
2. Find three materials that have solubility parameters δ_h and δ_p within a distance

$$d = \sqrt{2}$$

of benzene where

$$d = \sqrt{(\delta_h - \delta_h^{benzene})^2 + (\delta_p - \delta_p^{benzene})^2}\,.$$

3. For the synthesis of Sevin, suggest several additional reactants and complete the stoichiometric matrix for these reactants.

REFERENCES

Chang, W. D. and S. M. Senkan, "Chemical Structure of Fuel-Rich, Premixed, Laminar Flames of Trichloroethylene," 22nd International Symposium on Combustion, 1988.

Chang, W. D. and S. M. Senkan, "Detailed Chemical Kinetic Modeling of the Fuel Rich Flames of Trichloroethylene," *Environ. Sci. Technol.*, **223**, 442, 1989.

Chapman, J. L. and E. Tsou, "The UCLA Styrene Process," *Preprints, Am. Chem. Soc., Environ. Div.*, **33**(2), 308–309, 1993.

Crabtree, E. W. and M. M. El-Halwagi, "Synthesis of Environmentally Acceptable Reactions," in *Pollution Prevention via Process and Product Modifications*, M. M. El-Halwagi and D. P. Petrides, eds., *AIChE Symp. Ser.*, **90**(303), 117–127, 1995.

Floudas, C., *Non-Linear and Mixed Integer Optimization*, Oxford University Press, New York, 1995.

Gladfelter, W. L., "Homogeneous Catalytic Carbonylation of Nitroaromatics: An Alternative to Phosgene Use," *Preprints, Am. Chem. Soc. Environ. Div.*, **33**(2), 323–325, 1993.

Govind, R., "Controlling Hazardous Waste: Computer Assisted Prediction of Reaction Byproducts," *Hazardous Substances*, 26–32, Nov. 1985.

Hopper, J. R., C. L. Yaws, T. C. Ho, M. Vichailak, and A. Muninnimit, "Waste Minimization by Process Modification," in *Industrial Environmental Chemistry: Waste Minimization in Industrial Processes and Remediation of Hazardous Waste*, D. T. Sawyer and A. E. Martell, eds., Plenum, New York, 1992.

Joback, K., Molecular Knowledge Systems, personal communication, 1992.

Kier, L. B. and L. H. Hall, *Molecular Connectivity in Chemistry and Drug Research*, Academic Press, New York, 1976.

Reid, R. C., J. M. Prausnitz, and B. Poling, *The Properties of Gases and Liquids*, 4th ed., McGraw-Hill, New York, 1987.

Senkan, S. M., "Kinetic Models to Predict and Control Minor Constituents in Process Reactions," in *Industrial Environmental Chemistry: Waste Minimization in Industrial Processes and Remediation of Hazardous Waste*, D. T. Sawyer and A. E. Martell, eds., Plenum, New York, 1992.

APPENDIX A

TRI REPORTING FORM R FOR 1994

(IMPORTANT: Type or print; read instructions before completing form)

Form Approved OMB Number: 2070-0093
Approval Expires: 11/92

EPA United States Environmental Protection Agency

FORM R TOXIC CHEMICAL RELEASE INVENTORY REPORTING FORM

Section 313 of the Emergency Planning and Community Right-to-Know Act of 1986, also known as Title III of the Superfund Amendments and Reauthorization Act

TRI FACILITY ID NUMBER

Toxic Chemical, Category, or Generic Name

WHERE TO SEND COMPLETED FORMS:

1. EPCRA Reporting Center
P.O. Box 3348
Merrifield, VA 22116-3348
ATTN: TOXIC CHEMICAL RELEASE INVENTORY

2. APPROPRIATE STATE OFFICE
(See instructions in Appendix F)

Enter "X" here if this is a revision

For EPA use only

IMPORTANT: See instructions to determine when "Not Applicable (NA)" boxes should be checked.

PART I. FACILITY IDENTIFICATION INFORMATION

SECTION 1. REPORTING YEAR

19 ____

SECTION 2. TRADE SECRET INFORMATION

2.1 Are you claiming the toxic chemical identified on page 3 trade secret?

☐ Yes (Answer question 2.2; Attach substantiation forms) ☐ No (Do not answer 2.2; Go to Section 3)

2.2 If yes in 2.1, is this copy: ☐ Sanitized ☐ Unsanitized

SECTION 3. CERTIFICATION (Important: Read and sign after completing all form sections.)

I hereby certify that I have reviewed the attached documents and that, to the best of my knowledge and belief, the submitted information is true and complete and that the amounts and values in this report are accurate based on reasonable estimates using data available to the preparers of this report.

Name and official title of owner/operator or senior management official

Signature

Date Signed

SECTION 4. FACILITY IDENTIFICATION

4.1

Facility or Establishment Name

TRI Facility ID Number

Street Address

City

County

State

Zip Code

Mailing Address (if different from street address)

City

State

Zip Code

PUT LABEL HERE

EPA Form 9350-1(Rev. 12/4/92) - Previous editions are obsolete

EPA
United States Environmental Protection Agency

EPA FORM R

PART I. FACILITY IDENTIFICATION INFORMATION (CONTINUED)

TRI FACILITY ID NUMBER
Toxic Chemical, Category, or Generic Name

SECTION 4. FACILITY IDENTIFICATION (Continued)

4.2	This report contains information for: (Important: check only one)	a. ☐ An entire facility		b. ☐ Part of a facility		
4.3	Technical Contact	Name			Telephone Number (include area code)	
4.4	Public Contact	Name			Telephone Number (include area code)	
4.5	SIC Code (4-digit)	a.	b.	c.	d.	e.	f.
4.6	Latitude and Longitude	Latitude			Longitude		
		Degrees	Minutes	Seconds	Degrees	Minutes	Seconds
4.7	Dun & Bradstreet Number(s) (9 digits)	a.					
		b.					
4.8	EPA Identification Number(s) (RCRA I.D. No.) (12 characters)	a.					
		b.					
4.9	Facility NPDES Permit Number(s) (9 characters)	a.					
		b.					
4.10	Underground Injection Well Code (UIC) I.D. Number(s) (12 digits)	a.					
		b.					

SECTION 5. PARENT COMPANY INFORMATION

5.1	Name of Parent Company	
	☐ NA	
5.2	Parent Company's Dun & Bradstreet Number	
	☐ NA	(9 digits)

EPA
United States Environmental Protection Agency

EPA FORM R

PART II. CHEMICAL-SPECIFIC INFORMATION

TRI FACILITY ID NUMBER

Toxic Chemical, Category, or Generic Name

SECTION 1. TOXIC CHEMICAL IDENTITY

(Important: DO NOT complete this section if you complete Section 2 below.)

1.1	CAS Number (Important: Enter only one number exactly as it appears on the Section 313 list. Enter category code if reporting a chemical category.)
1.2	Toxic Chemical or Chemical Category Name (Important: Enter only one name exactly as it appears on the Section 313 list.)
1.3	Generic Chemical Name (Important: Complete **only** if Part I, Section 2.1 is checked "yes." Generic Name must be structurally descriptive.)

SECTION 2. MIXTURE COMPONENT IDENTITY

(Important: DO NOT complete this section if you complete Section 1 above.)

2.1	Generic Chemical Name Provided by Supplier (Important: Maximum of 70 characters, including numbers,letters, spaces, and punctuation.)

SECTION 3. ACTIVITIES AND USES OF THE TOXIC CHEMICAL AT THE FACILITY

(Important: Check all that apply.)

3.1	**Manufacture the toxic chemical:**	a. ☐ Produce b. ☐ Import	If produce or import: c. ☐ For on-site use/processing d. ☐ For sale/distribution e. ☐ As a byproduct f. ☐ As an impurity
3.2	**Process the toxic chemical:**	a. ☐ As a reactant b. ☐ As a formulation component	c. ☐ As an article component d. ☐ Repackaging
3.3	**Otherwise use the toxic chemical:**	a. ☐ As a chemical processing aid b. ☐ As a manufacturing aid	c. ☐ Ancillary or other use

SECTION 4. MAXIMUM AMOUNT OF THE TOXIC CHEMICAL ON-SITE AT ANY TIME DURING THE CALENDAR YEAR

4.1	☐ (Enter two-digit code from instruction package.)

EPA Form 9350-1(Rev. 12/4/92) - Previous editions are obsolete.

EPA
United States Environmental Protection Agency

EPA FORM R

PART II. CHEMICAL-SPECIFIC INFORMATION (CONTINUED)

TRI FACILITY ID NUMBER
Toxic Chemical, Category, or Generic Name

SECTION 5. RELEASES OF THE TOXIC CHEMICAL TO THE ENVIRONMENT ON-SITE

			A. Total Release (pounds/ year) (enter range code from instructions or estimate)	B. Basis of Estimate (enter code)	C. % From Stormwater
5.1	Fugitive or non-point air emissions	☐ NA			
5.2	Stack or point air emissions	☐ NA			
5.3	Discharges to receiving streams or water bodies (enter one name per box)				
5.3.1	Stream or Water Body Name				
5.3.2	Stream or Water Body Name				
5.3.3	Stream or Water Body Name				
5.4	Underground injections on-site	☐ NA			
5.5	Releases to land on-site				
5.5.1	Landfill	☐ NA			
5.5.2	Land treatment/ application farming	☐ NA			
5.5.3	Surface impoundment	☐ NA			
5.5.4	Other disposal	☐ NA			

☐ Check here only if additional Section 5.3 information is provided on page 5 of this form.

EPA Form 9350-1 (Rev. 12/4/92) - Previous editions are obsolete.

Range Codes: A = 1 - 10 pounds; B = 11 - 499 pounds; C = 500 - 999 pounds.

EPA United States Environmental Protection Agency	**EPA FORM R** **PART II. CHEMICAL-SPECIFIC INFORMATION (CONTINUED)**	TRI FACILITY ID NUMBER Toxic Chemical, Category, or Generic Name

SECTION 5.3 ADDITIONAL INFORMATION ON RELEASES OF THE TOXIC CHEMICAL TO THE ENVIRONMENT ON-SITE

5.3 Discharges to receiving streams or water bodies (enter one name per box)	A. Total Release (pounds/ year) (enter range code from instructions or estimate)	B. Basis of Estimate (enter code)	C. % From Stormwater
5.3.__ Stream or Water Body Name			
5.3.__ Stream or Water Body Name			
5.3.__ Stream or Water Body Name			

SECTION 6. TRANSFERS OF THE TOXIC CHEMICAL IN WASTES TO OFF-SITE LOCATIONS

6.1 DISCHARGES TO PUBLICLY OWNED TREATMENT WORKS (POTW)

6.1.A Total Quantity Transferred to POTWs and Basis of Estimate

6.1.A.1 Total Transfers (pounds/year) (enter range code or estimate)	6.1.A.2 Basis of Estimate (enter code)

6.1.B POTW Name and Location Information

6.1.B.__ POTW Name		6.1.B.__ POTW Name	
Street Address		Street Address	
City	County	City	County
State	Zip Code	State	Zip Code

If additional pages of Part II, Sections 5.3 and/or 6.1 are attached, indicate the total number of pages in this box ☐ and indicate which Part II, Sections 5.3/6.1 page this is, here. ☐
(example: 1, 2, 3, etc.)

EPA Form 9350-1(Rev. 12/4/92) - Previous editions are obsolete

Range Codes: A = 1 - 10 pounds; B = 11 - 499 pounds; C = 500 - 999 pounds.

EPA
United States Environmental Protection Agency

EPA FORM R

PART II. CHEMICAL-SPECIFIC INFORMATION (CONTINUED)

TRI FACILITY ID NUMBER

Toxic Chemical, Category, or Generic Name

SECTION 6.2 TRANSFERS TO OTHER OFF-SITE LOCATIONS

6.2.___ Off-site EPA Identification Number (RCRA ID No.)

Off-Site Location Name

Street Address

City | County

State | Zip Code | Is location under control of reporting facility or parent company? ☐ Yes ☐ No

A. Total Transfers (pounds/year) (enter range code or estimate)	B. Basis of Estimate (enter code)	C. Type of Waste Treatment/Disposal/ Recycling/Energy Recovery (enter code)
1.	1.	1. M
2.	2.	2. M
3.	3.	3. M
4.	4.	4. M

SECTION 6.2 TRANSFERS TO OTHER OFF-SITE LOCATIONS

6.2.___ Off-site EPA Identification Number (RCRA ID No.)

Off-Site Location Name

Street Address

City | County

State | Zip Code | Is location under control of reporting facility or parent company? ☐ Yes ☐ No

A. Total Transfers (pounds/year) (enter range code or estimate)	B. Basis of Estimate (enter code)	C. Type of Waste Treatment/Disposal/ Recycling/Energy Recovery (enter code)
1.	1.	1. M
2.	2.	2. M
3.	3.	3. M
4.	4.	4. M

If additional pages of Part II, Section 6.2 are attached, indicate the total number of pages in this box ☐ and indicate which Part II, Section 6.2 page this is, here. ☐ (example: 1, 2, 3, etc.)

EPA Form 9350-1 (Rev. 12/4/92) - Previous editions are obsolete.

Range Codes: A = 1 - 10 pounds; B = 11 - 499 pounds; C = 500 - 999 pounds.

EPA
United States Environmental Protection Agency

EPA FORM R

PART II. CHEMICAL-SPECIFIC INFORMATION (CONTINUED)

TRI FACILITY ID NUMBER

Toxic Chemical, Category, or Generic Name

SECTION 7A. ON-SITE WASTE TREATMENT METHODS AND EFFICIENCY

☐ **Not Applicable (NA) - Check here if no on-site waste treatment is applied to any waste stream containing the toxic chemical or chemical category.**

a. General Waste Stream (enter code)	b. Waste Treatment Method(s) Sequence [enter 3-character code(s)]	c. Range of Influent Concentration	d. Waste Treatment Efficiency Estimate	e. Based on Operating Data?
7A.1a	7A.1b 1 ☐ 2 ☐ 3 ☐ 4 ☐ 5 ☐ 6 ☐ 7 ☐ 8 ☐	7A.1c	7A.1d %	7A.1e Yes ☐ No ☐
7A.2a	7A.2b 1 ☐ 2 ☐ 3 ☐ 4 ☐ 5 ☐ 6 ☐ 7 ☐ 8 ☐	7A.2c	7A.2d %	7A.2e Yes ☐ No ☐
7A.3a	7A.3b 1 ☐ 2 ☐ 3 ☐ 4 ☐ 5 ☐ 6 ☐ 7 ☐ 8 ☐	7A.3c	7A.3d %	7A.3e Yes ☐ No ☐
7A.4a	7A.4b 1 ☐ 2 ☐ 3 ☐ 4 ☐ 5 ☐ 6 ☐ 7 ☐ 8 ☐	7A.4c	7A.4d %	7A.4e Yes ☐ No ☐
7A.5a	7A.5b 1 ☐ 2 ☐ 3 ☐ 4 ☐ 5 ☐ 6 ☐ 7 ☐ 8 ☐	7A.5c	7A.5d %	7A.5e Yes ☐ No ☐

If additional copies of page 7 are attached, indicate the total number of pages in this box ☐ and indicate which page 7 this is, here. ☐ (example: 1, 2, 3, etc.)

EPA Form 9350-1 (Rev. 12/4/92) - Previous editions are obsolete.

EPA
United States Environmental Protection Agency

EPA FORM R

PART II. CHEMICAL-SPECIFIC INFORMATION (CONTINUED)

TRI FACILITY ID NUMBER

Toxic Chemical, Category, or Generic Name

SECTION 7B. ON-SITE ENERGY RECOVERY PROCESSES

☐ **Not Applicable (NA)** - **Check here if no on-site energy recovery is applied to any waste stream containing the toxic chemical or chemical category.**

Energy Recovery Methods [enter 3-character code(s)]

1 ☐ 2 ☐ 3 ☐ 4 ☐

SECTION 7C. ON-SITE RECYCLING PROCESSES

☐ **Not Applicable (NA)** - **Check here if no on-site recycling is applied to any waste stream containing the toxic chemical or chemical category.**

Recycling Methods [enter 3-character code(s)]

1 ☐ 2 ☐ 3 ☐ 4 ☐ 5 ☐

6 ☐ 7 ☐ 8 ☐ 9 ☐ 10 ☐

EPA
United States
Environmental Protection
Agency

EPA FORM R

PART II. CHEMICAL-SPECIFIC INFORMATION (CONTINUED)

TRI FACILITY ID NUMBER

Chemical, Category, or Generic Name

SECTION 8. SOURCE REDUCTION AND RECYCLING ACTIVITIES

	All quantity estimates can be reported using up to two significant figures.	Column A 1991 (pounds/year)	Column B 1992 (pounds/year)	Column C 1993 (pounds/year)	Column D 1994 (pounds/year)
8.1	Quantity released *				
8.2	Quantity used for energy recovery on-site				
8.3	Quantity used for energy recovery off-site				
8.4	Quantity recycled on-site				
8.5	Quantity recycled off-site				
8.6	Quantity treated on-site				
8.7	Quantity treated off-site				

8.8	Quantity released to the environment as a result of remedial actions, catastrophic events, or one-time events not associated with production processes (pounds/year)	
8.9	Production ratio or activity index	

8.10 Did your facility engage in any source reduction activities for this chemical during the reporting year? If not, enter "NA" in Section 8.10.1 and answer Section 8.11.

	Source Reduction Activities [enter code(s)]	Methods to Identify Activity (enter codes)		
8.10.1		a.	b.	c.
8.10.2		a.	b.	c.
8.10.3		a.	b.	c.
8.10.4		a.	b.	c.

		YES	NO
8.11	Is additional optional information on source reduction, recycling, or pollution control activities included with this report? (Check one box)	☐	☐

* Report releases pursuant to EPCRA Section 329(8) including "any spilling, leaking, pumping, pouring, emitting, emptying, discharging, injecting, escaping, leaching, dumping, or disposing into the environment." Do not include any quantity treated on-site or off-site.

EPA Form 9350-1(Rev. 12/4/92) - Previous editions are obsolete

APPENDIX B

REPORTED HAZARDOUS WASTE MANAGEMENT IN THE UNITED STATES BY WASTE TYPE, INDUSTRIAL SECTOR, AND GEOGRAPHIC REGION FOR EACH MANAGEMENT TECHNOLOGY FOR THE YEAR 1986*

*From R. D. Baker, J. L. Warren, N. Behmanesh, and D. T. Allen, "Management of Hazardous Waste in the United States," *Hazardous Waste & Hazardous Materials*, **9**(1), 37–59, 1992.

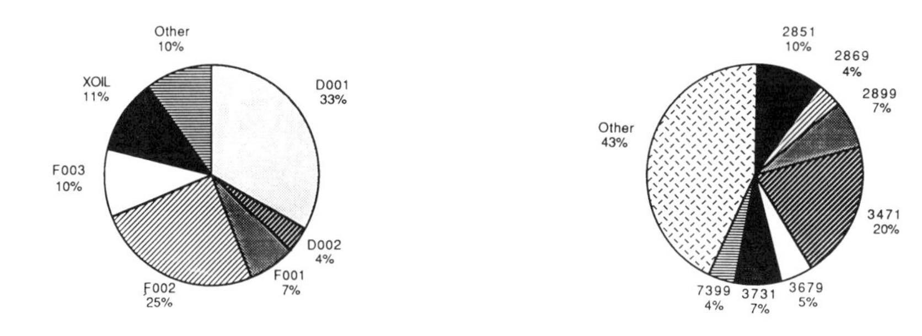

Waste Code	Waste Description
D001	Ignitable Waste
D002	Corrosive Waste
F001	Spent Halogenated Solvents Used In Degreasing
F002	Spent Halogenated Solvents
F003	Spent Nonhalogenated Solvents
XOIL	Waste Oil

(*a*)

Industry Code	Industry Description
2851	Paints and Allied Products
2869	Industrial Organic Chemicals, nec
2899	Chemical Preparations, nec
3471	Plating and Polishing
3679	Electronic Components, nec
3731	Ship Building and Repairing
7399	Business Services, nec

(*b*)

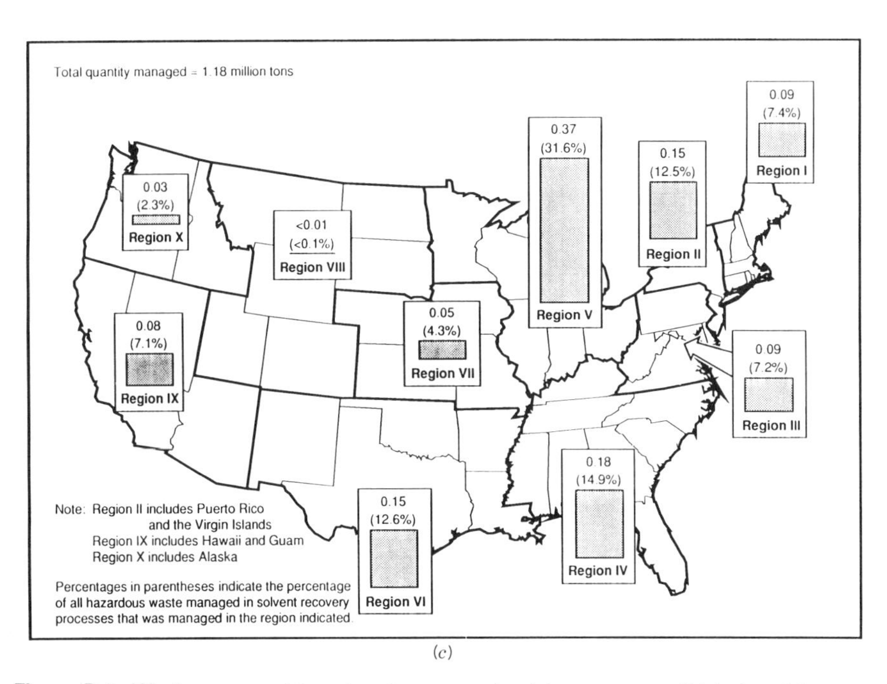

(*c*)

Figure B-1. Waste managed in solvent recovery by (*a*) waste type, (*b*) industrial sector, and (*c*) geographic region. (Reprinted with permission of Mary Ann Liebert, Inc., publishers.)

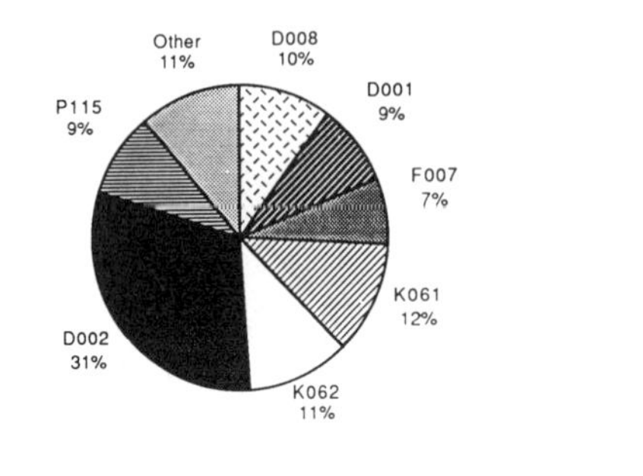

Waste Code	Waste Description
D002	Corrosive Waste
D008	Lead
D011	Silver
F007	Spent Cyanide Plating Bath Solutions From Electroplating Operations
K061	Emission Control Dust/ Sludge From Primary Production of Steel in Electric Furnaces
K062	Spent Pickle Liquor From Steel Finishing Operations That Produce Iron or Steel
P115	Thallium (I) Sulfate

(*a*)

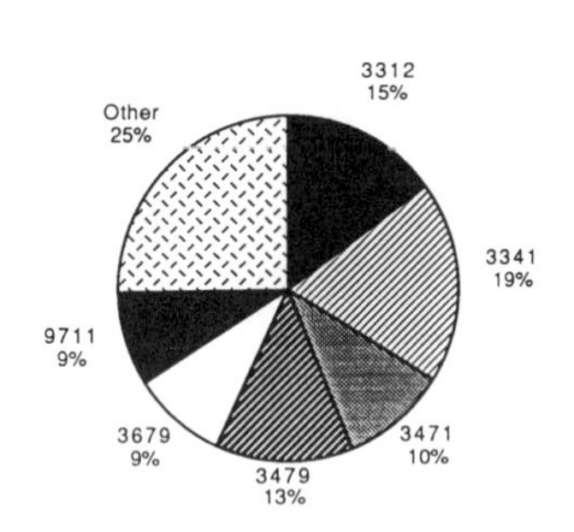

Industry Code	Industry Description
3312	Blast Furnaces and Steel Mills
3341	Secondary Nonferrous Metals
3471	Plating and Polishing
3479	Metal Coating and Allied Services
3679	Electronic Components, nec
9711	National Security

(*b*)

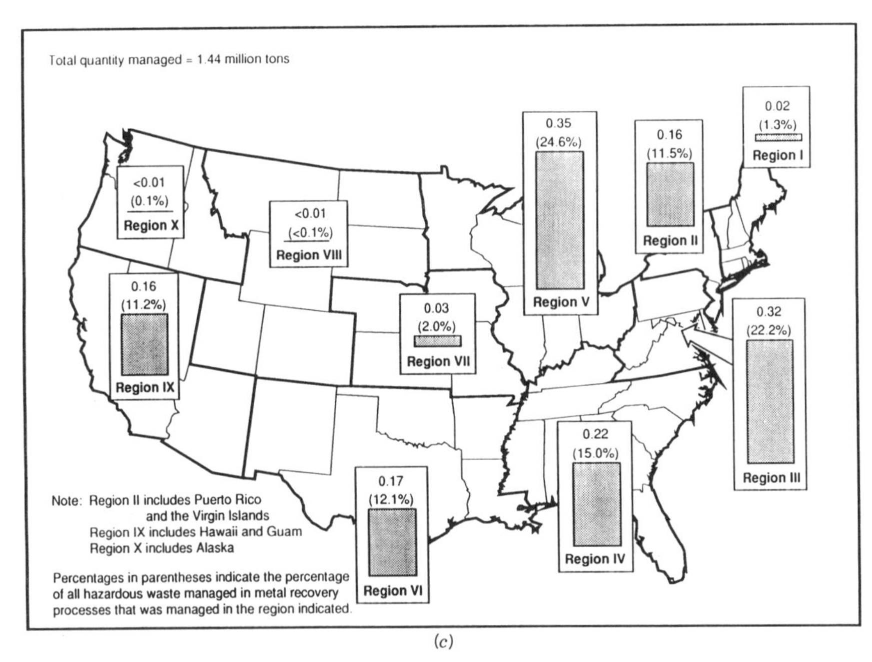

(*c*)

Figure B-2. Waste managed in metal recovery by (*a*) waste type, (*b*) industrial sector, and (*c*) geographic region. (Reprinted with permission of Mary Ann Liebert, Inc., publishers.)

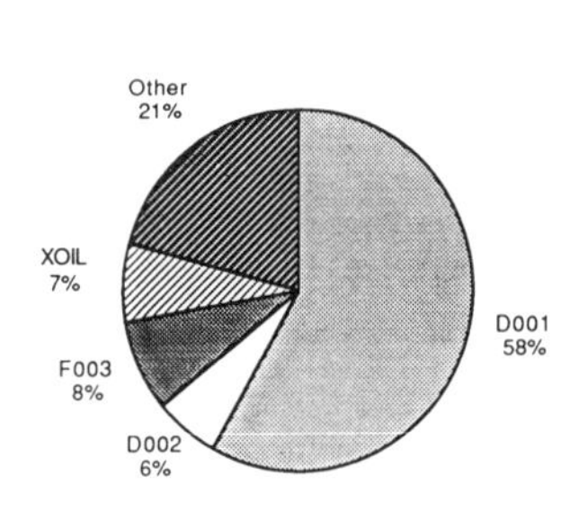

Waste Code	Waste Description
D001	Ignitable Waste
D002	Corrosive Waste
F003	Spent Nonhalogenated Solvents
XOIL	Waste Oil

(*a*)

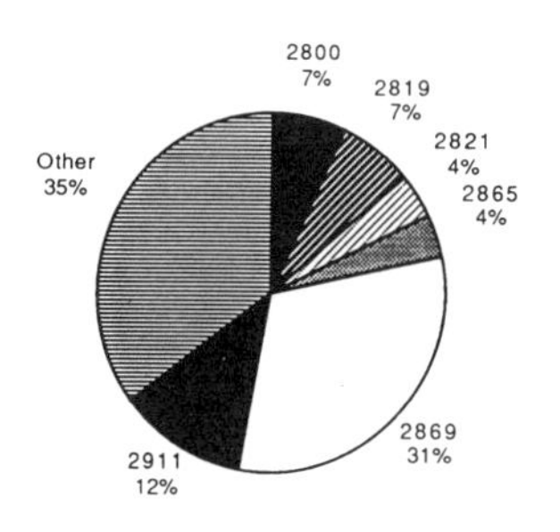

Industry Code	Industry Description
2800	General Chemical Manufacturing
2819	Industrial Inorganic Chemicals, nec
2821	Plastic Materials and Resins
2865	Cyclic Crudes and Intermediates
2869	Industrial Organic Chemicals, nec
2911	Petroleum Refining

(*b*)

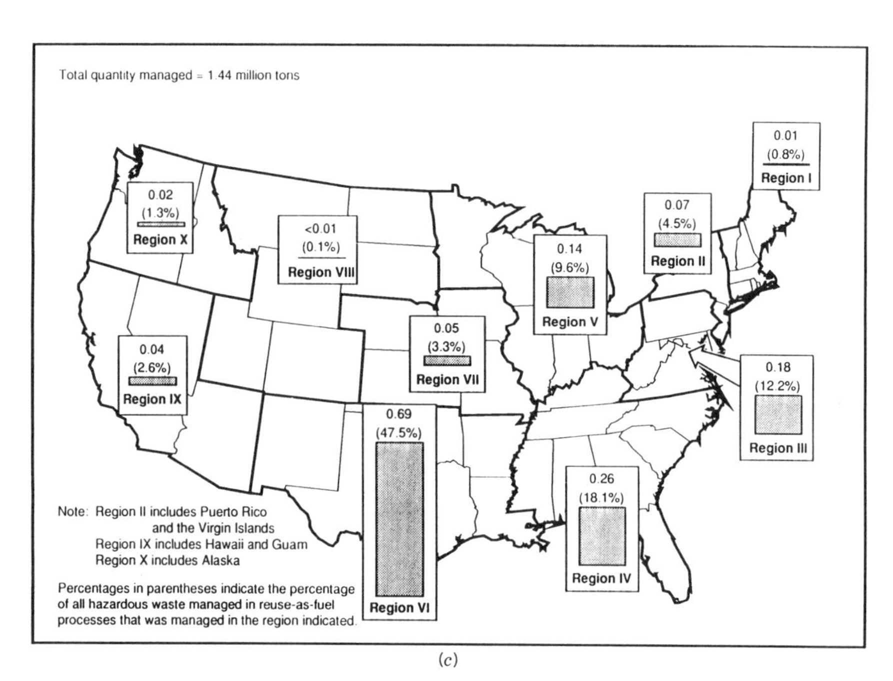

(*c*)

Figure B-3. Waste managed through reuse as fuel by (*a*) waste type, (*b*) industrial sector, and (*c*) geographic region. (Reprinted with permission of Mary Ann Liebert, Inc., publishers.)

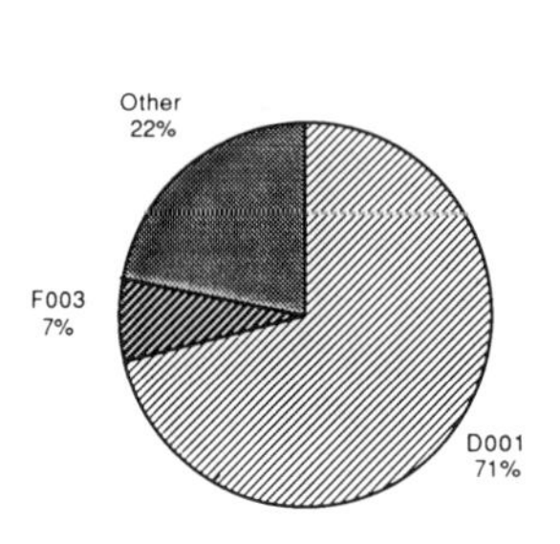

Waste Code	Waste Description
D001	Ignitable Waste
F003	Spent Nonhalogenated Solvents

(*a*)

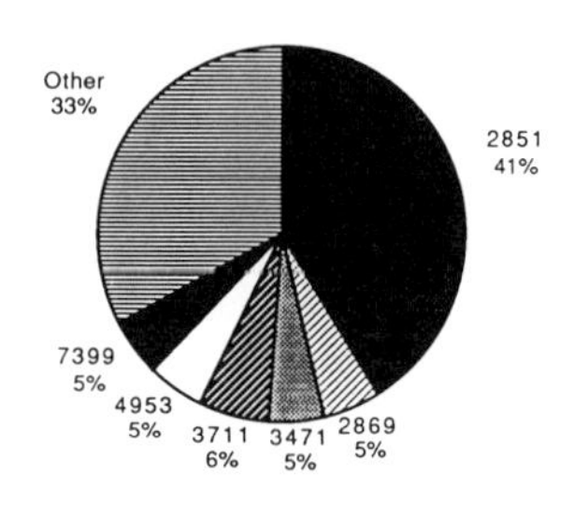

Industry Code	Industry Description
2851	Paints and Allied Products
2869	Industrial Organic Chemicals, nec
3471	Plating and Polishing
3711	Motor Vehicles and Car Bodies
4953	Refuse Systems
7399	Business Services, nec

(*b*)

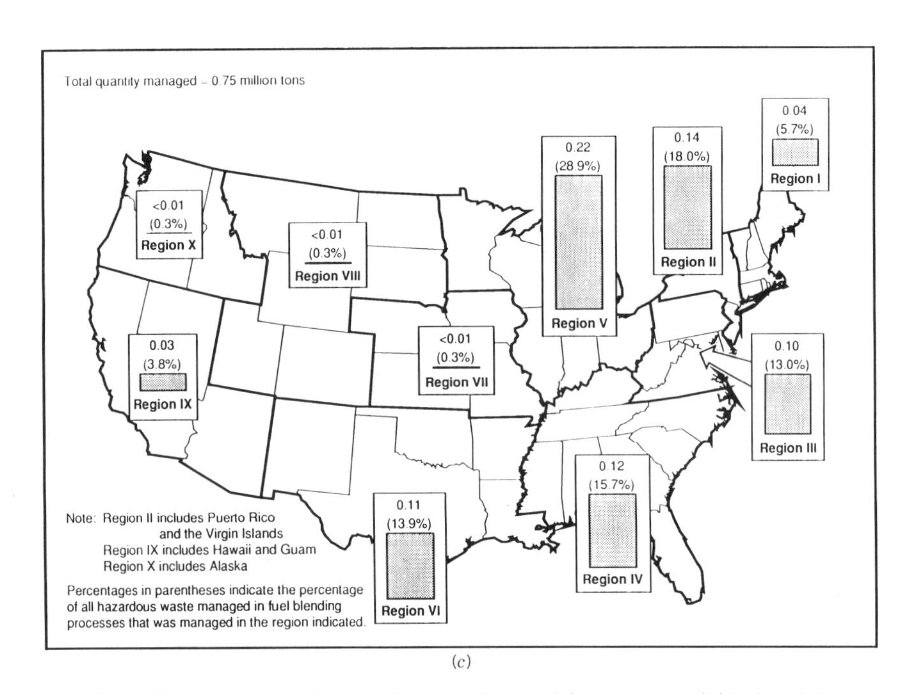

(*c*)

Figure B-4. Waste managed through fuel blending by (*a*) waste type, (*b*) industrial sector, and (*c*) geographic region. (Reprinted with permission of Mary Ann Liebert, Inc., publishers.)

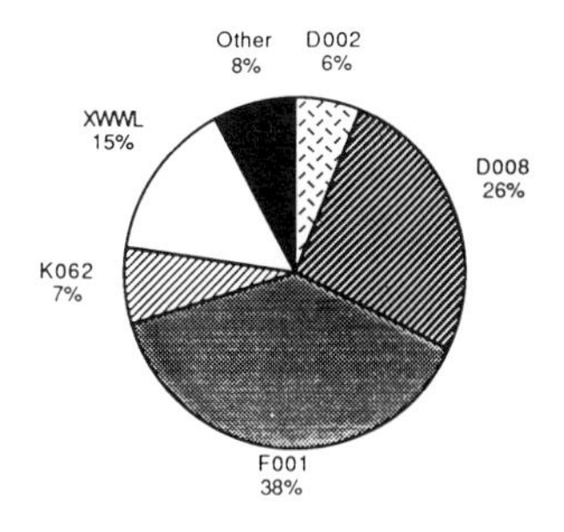

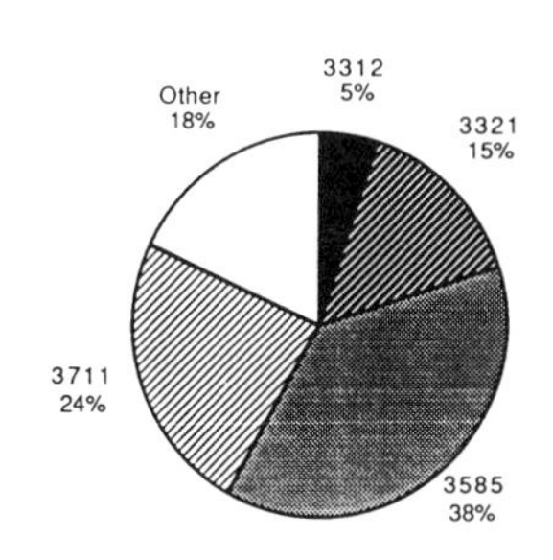

Waste Code	Waste Description
D002	Corrosive Waste
D008	Lead
F001	Spent Halogenated Solvents Used In Degreasing
K062	Spent Pickle Liquor From Steel Finishing Operations That Produce Iron or Steel
XWWL	Hazardous Wastewater Treatment Liquid

(*a*)

Industry Code	Industry Description
3312	Blast Furnaces and Steel Mills
3321	Gray Iron Foundries
3585	Refrigeration and Heating Equipment
3711	Motor Vehicles and Car Bodies

(*b*)

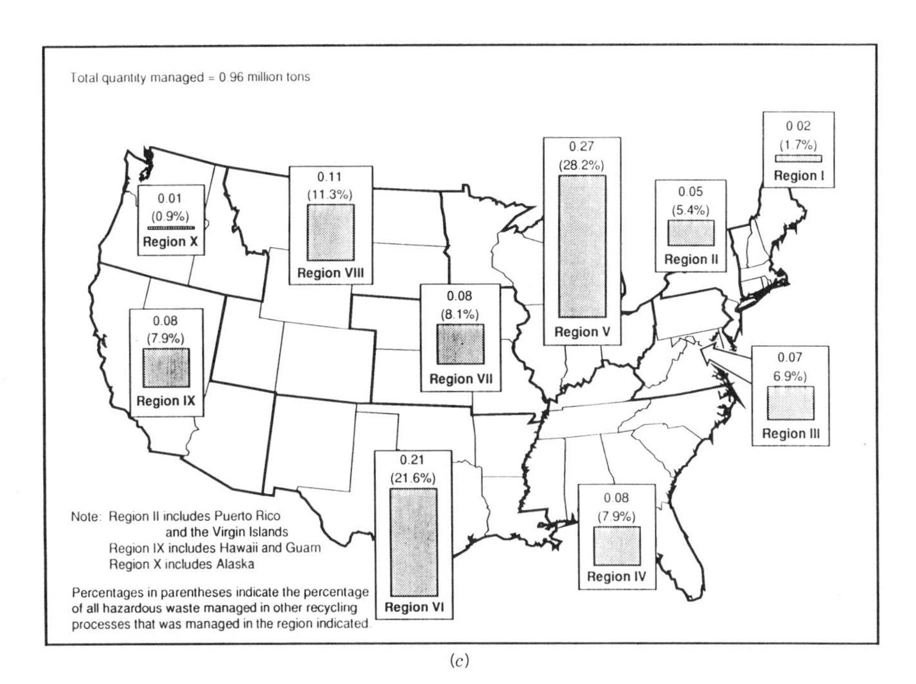

(*c*)

Figure B-5. Waste managed in other recycling by (*a*) waste type, (*b*) industrial sector, and (*c*) geographic region. (Reprinted with permission of Mary Ann Liebert, Inc., publishers.)

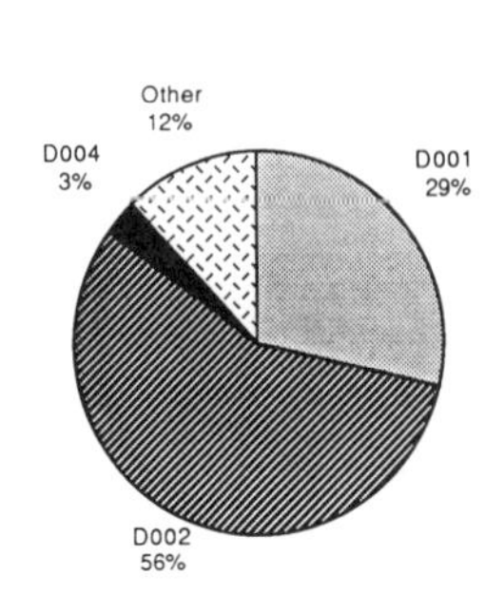

Waste Code	Waste Description
D001	Ignitable Waste
D002	Corrosive Waste
D004	Arsenic

(*a*)

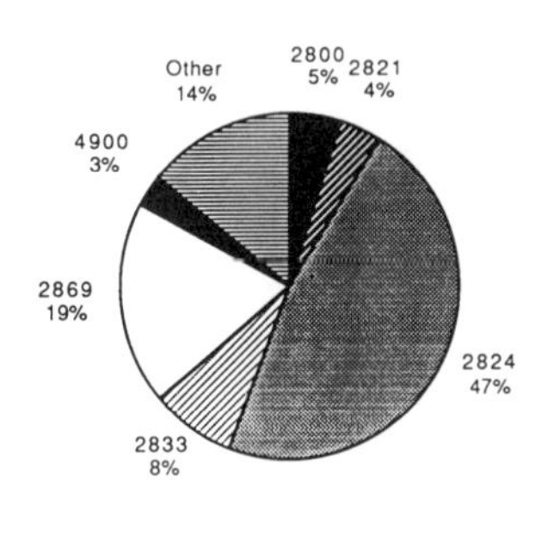

Industry Code	Industry Description
2800	General Chemical Manufacturing
2821	Plastics Materials.and Resins
2824	Organic fibers, noncellulosic
2833	Medicinals and botanicals
2869	Industrial Organic Chemicals, nec
4900	Electrical, Gas, and Sanitary Services

(*b*)

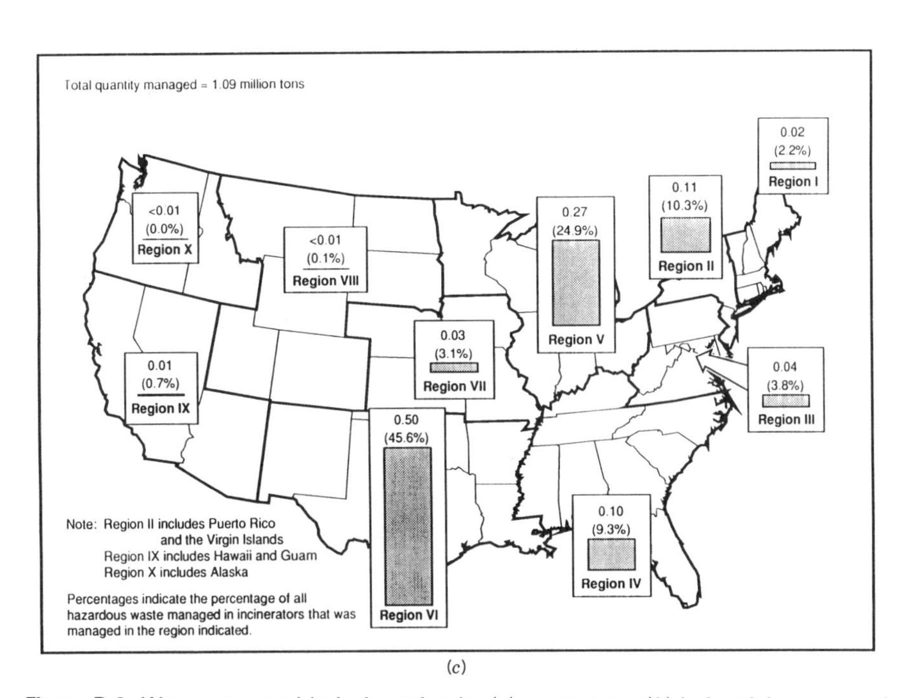

(*c*)

Figure B-6. Waste managed in incineration by (*a*) waste type, (*b*) industrial sector, and (*c*) geographic region. (Reprinted with permission of Mary Ann Liebert, Inc., publishers.)

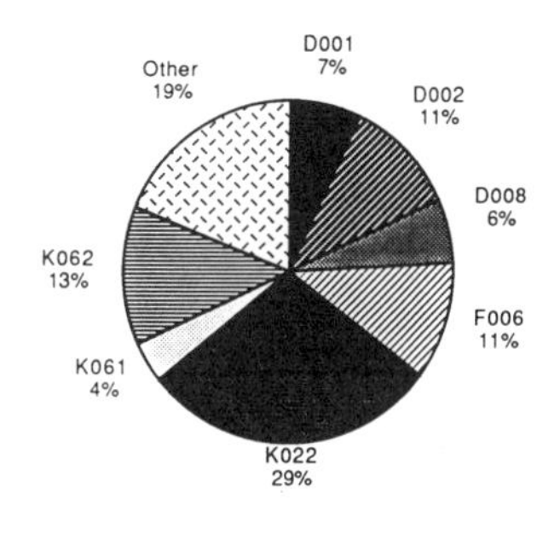

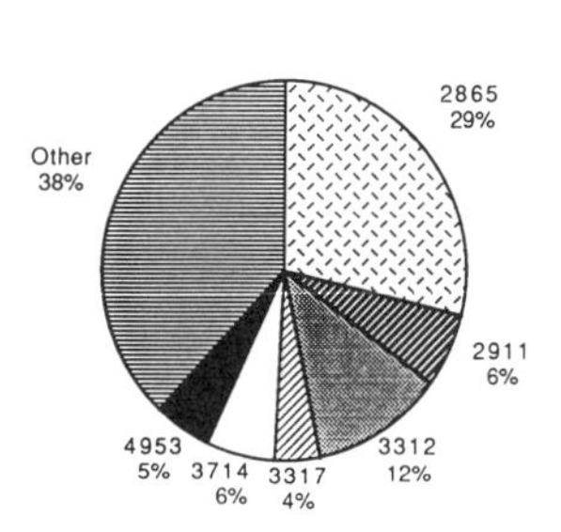

Waste Code	Waste Description
D001	Ignitable Waste
D002	Corrosive Waste
D008	Lead
F006	Wastewater Treatment Sludges From Certain Electroplating Operations
K022	Distillation Bottom Tars From the Production of Phenol/Acetone From Cumene
K061	Emission Control Dust/ Sludge From Primary Production of Steel in Electric Furnaces
K062	Spent Pickle Liquor From Steel Finishing Operations That Produce Iron or Steel

(*a*)

Industry Code	Industry Description
2865	Cyclic Crudes and Intermediates
2911	Petroleum Refining
3312	Blast Furnaces and Steel Mills
3317	Steel Pipe and Tubes
3714	Motor Vehicle Parts and Accessories
4953	Refuse Systems

(*b*)

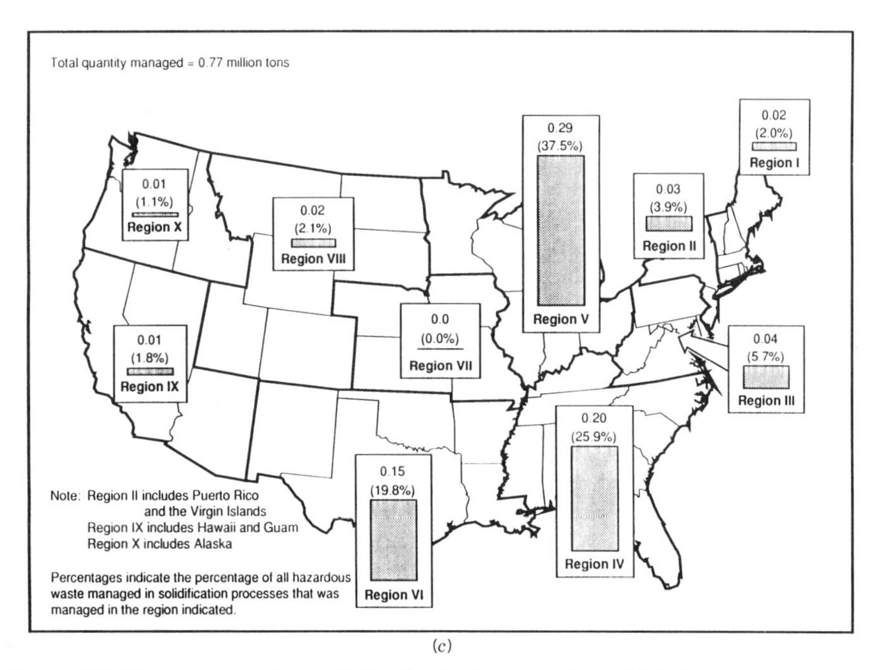

(*c*)

Figure B-7. Waste managed in solidification by (*a*) waste type, (*b*) industrial sector, and (*c*) geographic region. (Reprinted with permission of Mary Ann Liebert, Inc., publishers.)

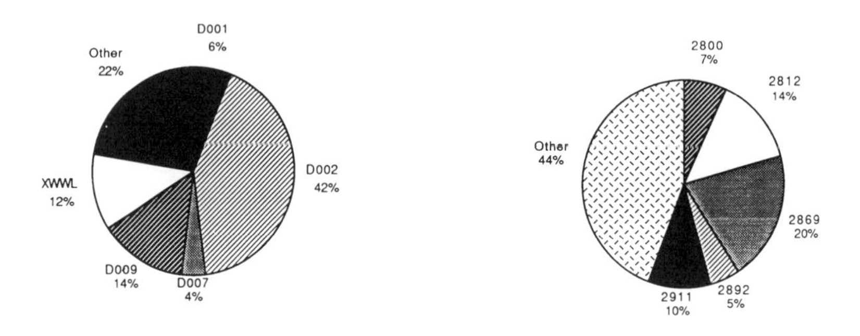

Waste Code	Waste Description
D001	Ignitable Waste
D002	Corrosive Waste
D007	Chromium
D009	Mercury
XWWL	Hazardous Wastewater Treatment Liquid

(*a*)

Industry Code	Industry Description
2800	General Chemical Manufacturing
2812	Alkalies and Chlorine
2869	Industrial Organic Chemicals, nec
2892	Explosives
2911	Petroleum Refining

(*b*)

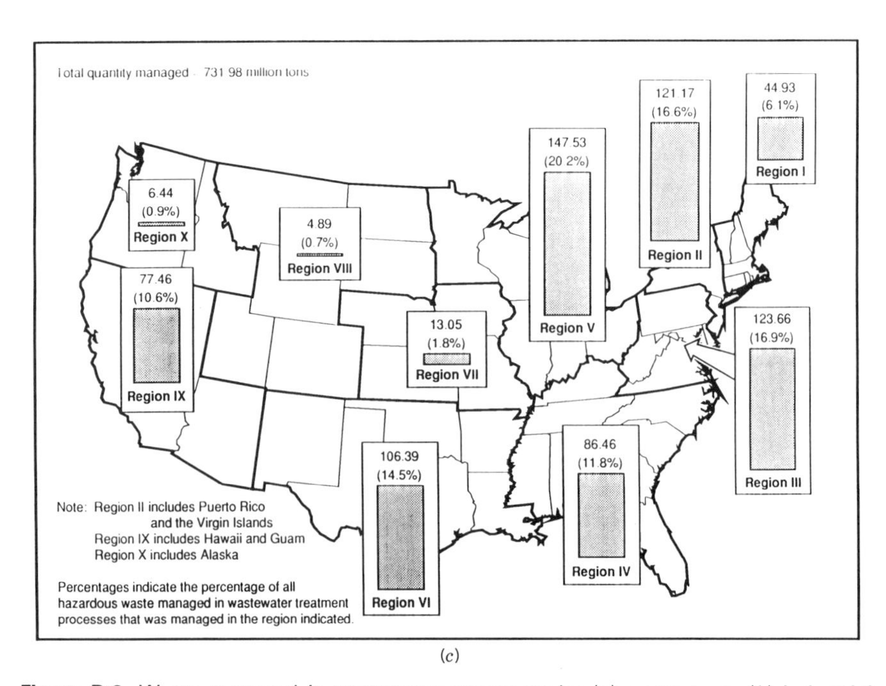

(*c*)

Figure B-8. Waste managed in wastewater treatment by (*a*) waste type, (*b*) industrial sector, and (*c*) geographic region. (Reprinted with permission of Mary Ann Liebert, Inc., publishers.)

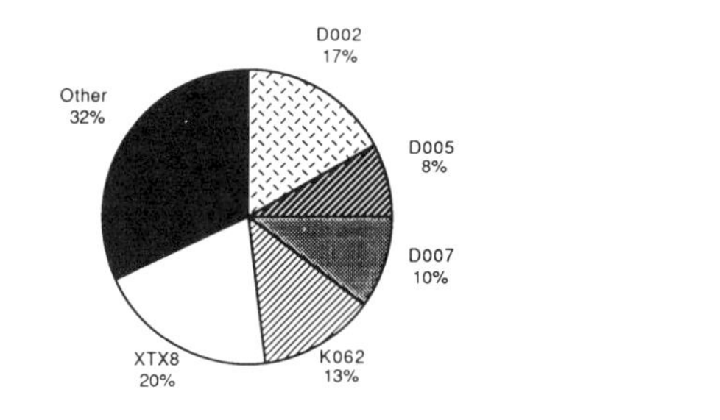

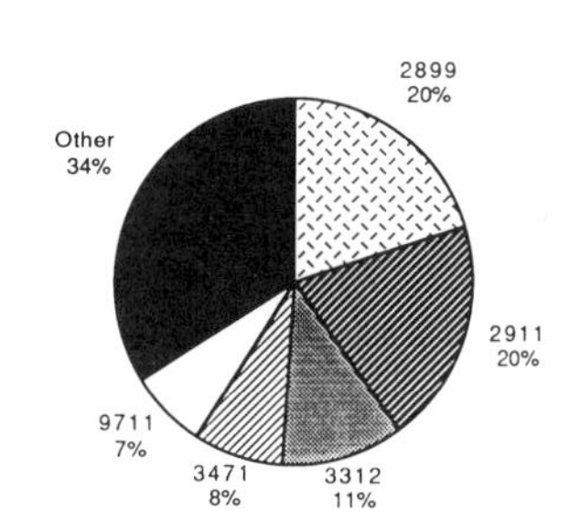

Waste Code	Waste Description
D002	Corrosive Waste
D005	Barium
D007	Chromium
K062	Spent Pickle Liquor From Steel Finishing Operations That Produce Iron or Steel
XTX8	

(*a*)

Industry Code	Industry Description
2899	Chemical Preparations, nec
2911	Petroleum Refining
3312	Blast Furnaces and Steel Mills
3471	Plating and Polishing
9711	National Security

(*b*)

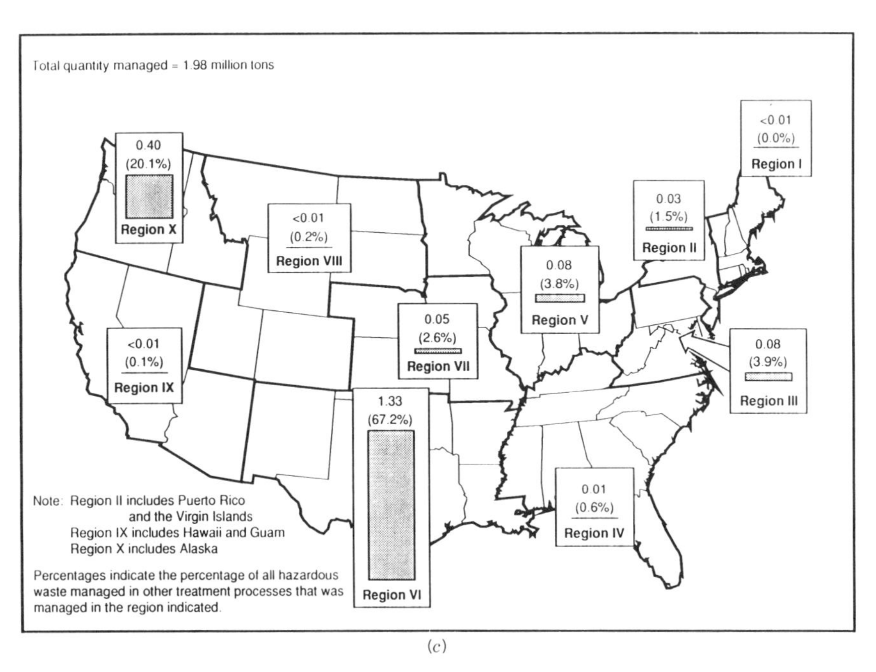

(*c*)

Figure B-9. Waste managed in other treatments by (*a*) waste type, (*b*) industrial sector, and (*c*) geographic region. (Reprinted with permission of Mary Ann Liebert, Inc., publishers.)

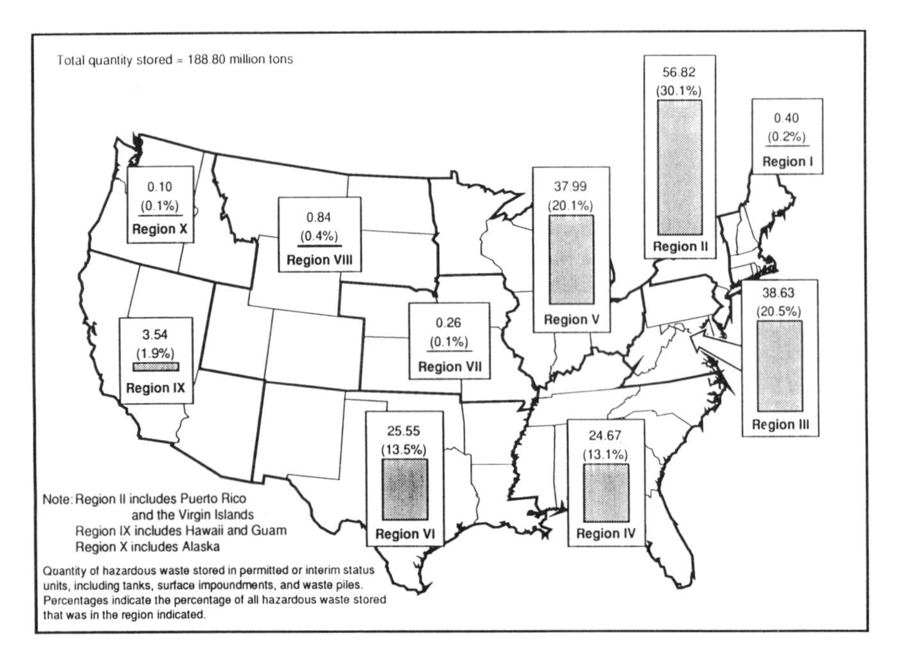

Figure B-10. Waste stored by geographic region. (Reprinted with permission of Mary Ann Liebert, Inc., publishers.)

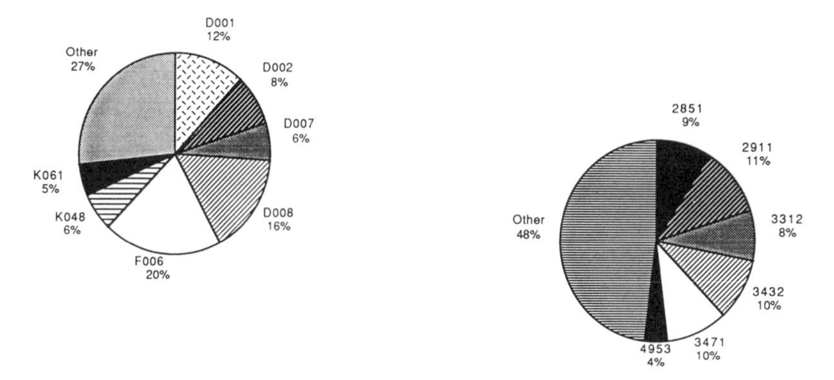

Waste Code	Waste Description
D001	Ignitable Waste
D002	Corrosive Waste
D007	Chromium
D008	Lead
F006	Wastewater Treatment Sludges From Certain Electroplating Operations
K048	Dissolved Air Floatation (DAF) Float From Petroleum Refining Industry
K061	Emission Control Dust/ Sludge From Primary Production of Steel in Electric Furnaces

(*a*)

Industry Code	Industry Description
2851	Paints and Allied Products
2911	Petroleum Refining
3312	Blast Furnaces and Steel Mills
3432	Plumbing Fittings and Brass Goods
3471	Plating and Polishing
4953	Refuse Systems

(*b*)

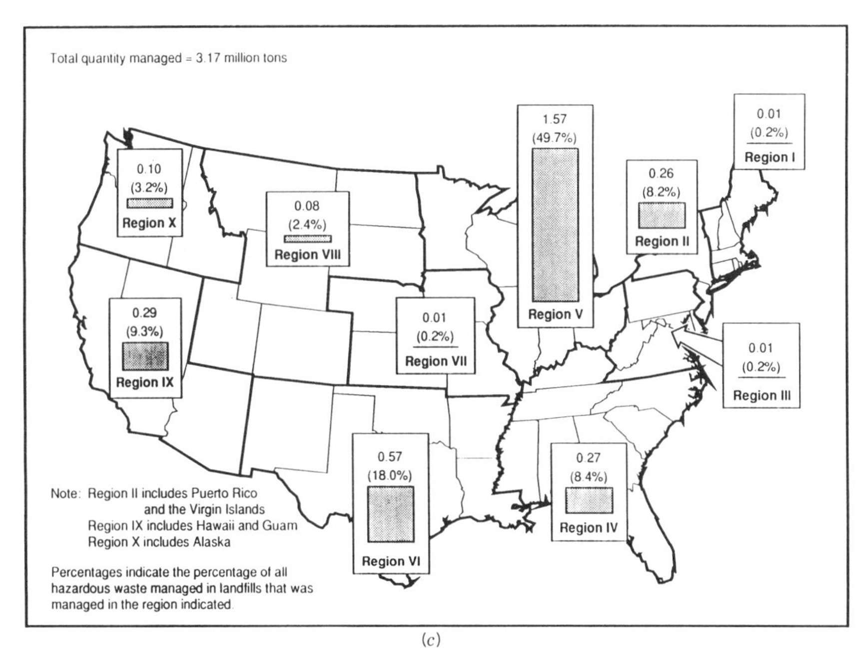

(*c*)

Figure B-11. Waste managed in landfills by (*a*) waste type, (*b*) industrial sector, and (*c*) geographic region. (Reprinted with permission of Mary Ann Liebert, Inc., publishers.)

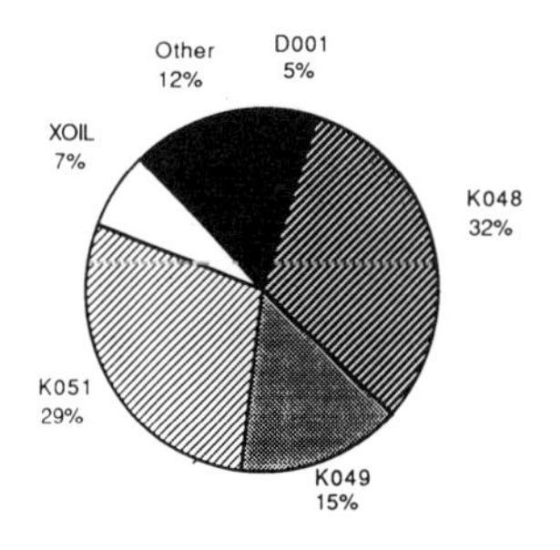

Waste Code	Waste Description
D001	Ignitable Waste
K048	Dissolved Air Floatation (DAF) Float From the Petroleum Refining Industry
K049	Slop Oil Emulsion Solids From the Petroleum Refining Industry
K051	API Separator Sludge From the Petroleum Refining Industry
XOIL	Waste Oil

(*a*)

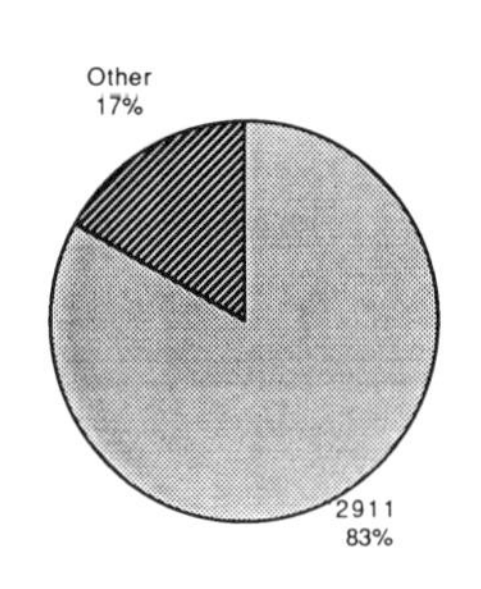

Industry Code	Industry Description
2911	Petroleum Refining

(*b*)

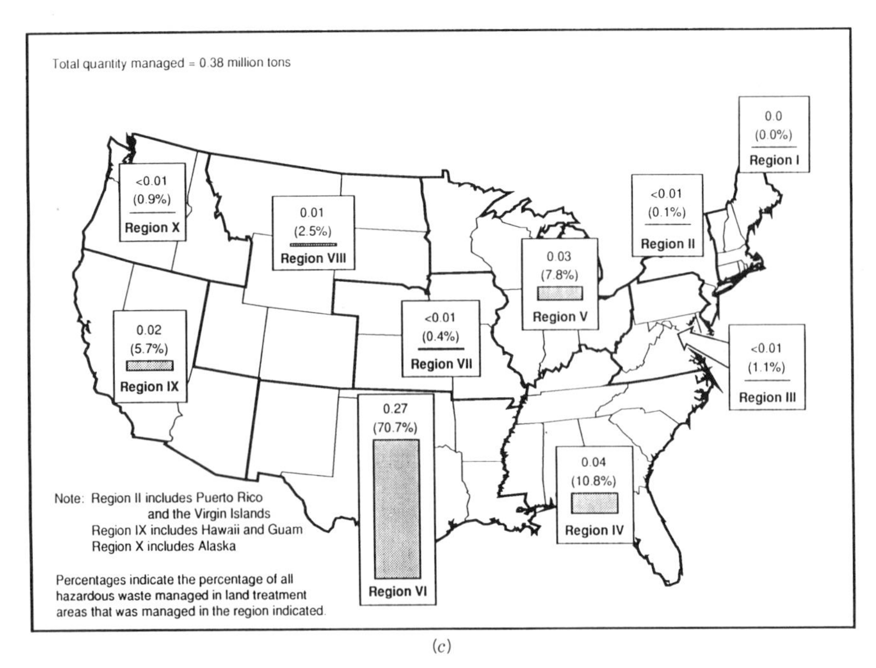

(*c*)

Figure B-12. Waste managed in land treatment by (*a*) waste type, (*b*) industrial sector, and (*c*) geographic region. (Reprinted with permission of Mary Ann Liebert, Inc., publishers.)

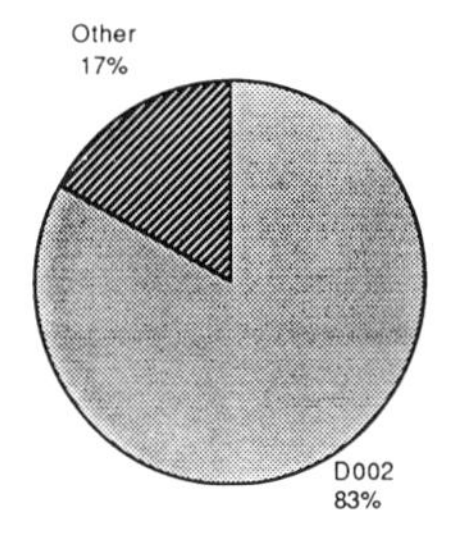

Waste Code	Waste Description
D002	Corrosive Waste

(*a*)

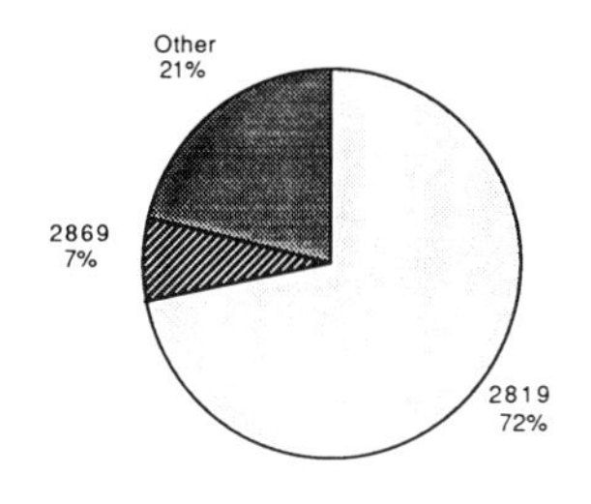

Industry Code	Industry Description
2819	Industrial Inorganic Chemicals, nec
2869	Industrial Organic Chemicals, nec

(*b*)

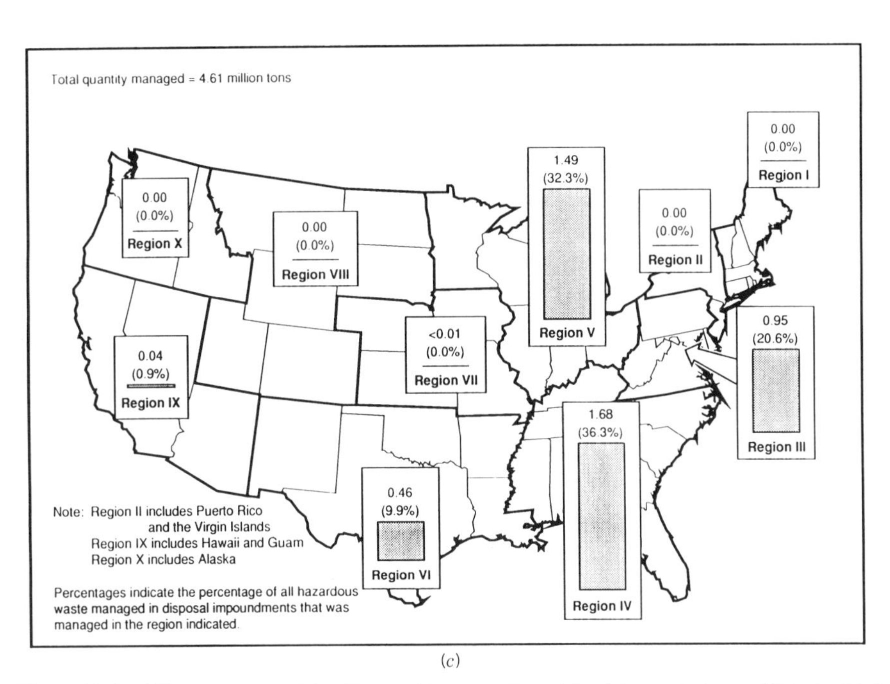

(*c*)

Figure B-13. Waste managed in disposal impoundment by (*a*) waste type, (*b*) industrial sector, and (*c*) geographic region. (Reprinted with permission of Mary Ann Liebert, Inc., publishers.)

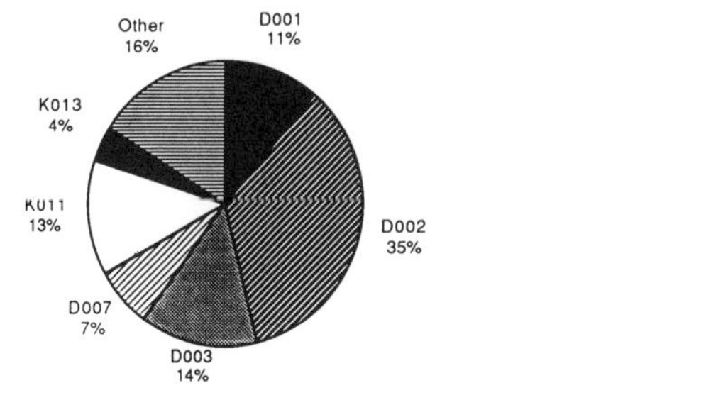

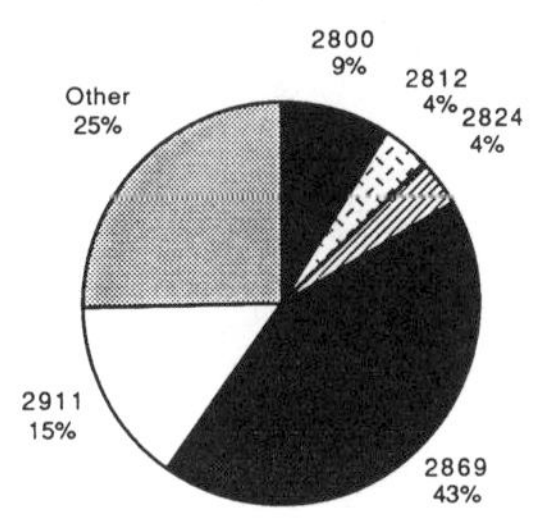

Waste Code	Waste Description
D001	Ignitable Waste
D002	Corrosive Waste
D003	Reactive Waste
D007	Chromium
K011	Bottom Stream From the Wastewater Stripper in Production of Acrylonitrile
K013	Bottom Stream From the Acrylonitrile Column in Production of Acrylonitrile

(*a*)

Industry Code	Industry Description
2800	General Chemical Manufacturing
2812	Alkalies and Chlorines
2824	Organic Fibers, noncellulosic
2869	Industrial Organic Chemicals, nec
2911	Petroleum Refining

(*b*)

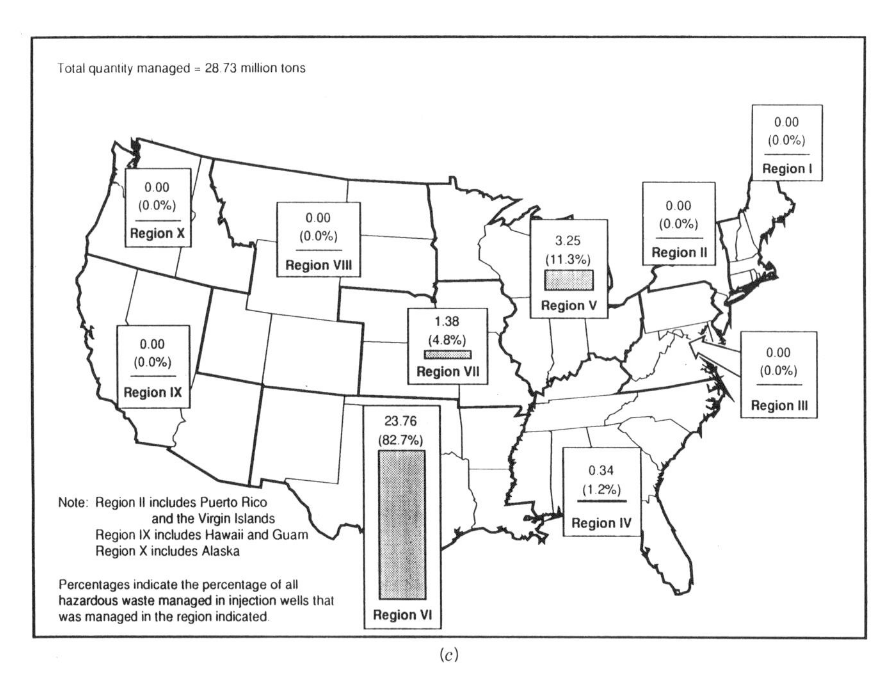

(*c*)

Figure B-14. Waste managed in injection wells by (*a*) waste type, (*b*) industrial sector, and (*c*) geographic region. (Reprinted with permission of Mary Ann Liebert, Inc., publishers.)

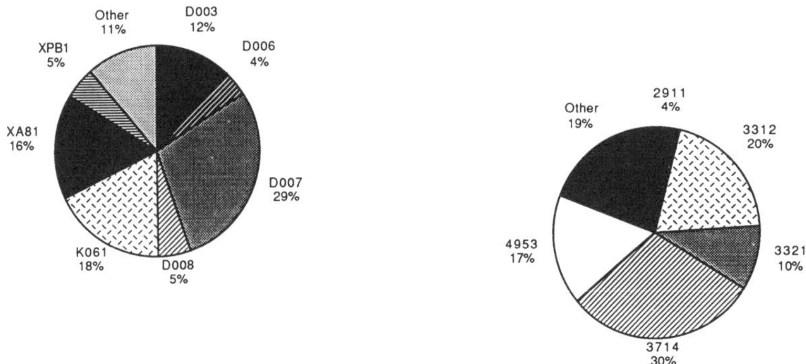

Waste Code	Waste Description
D003	Reactive Waste
D006	Cadmium
D007	Chromium
D008	Lead
K061	Emission Control Dust/ Sludge From Primary Production of Steel in Electric Furnaces
XA81	
XPB1	Waste Which Has Concentration of Polychlorinated Biphenyls Less Than 50 Parts Per Million

(*a*)

Industry Code	Industry Description
2911	Petroleum Refining
3312	Blast Furnaces and Steel Mills
3321	Gray Iron Foundries
3714	Motor Vehicle Parts and Accessories
4953	Refuse Systems

(*b*)

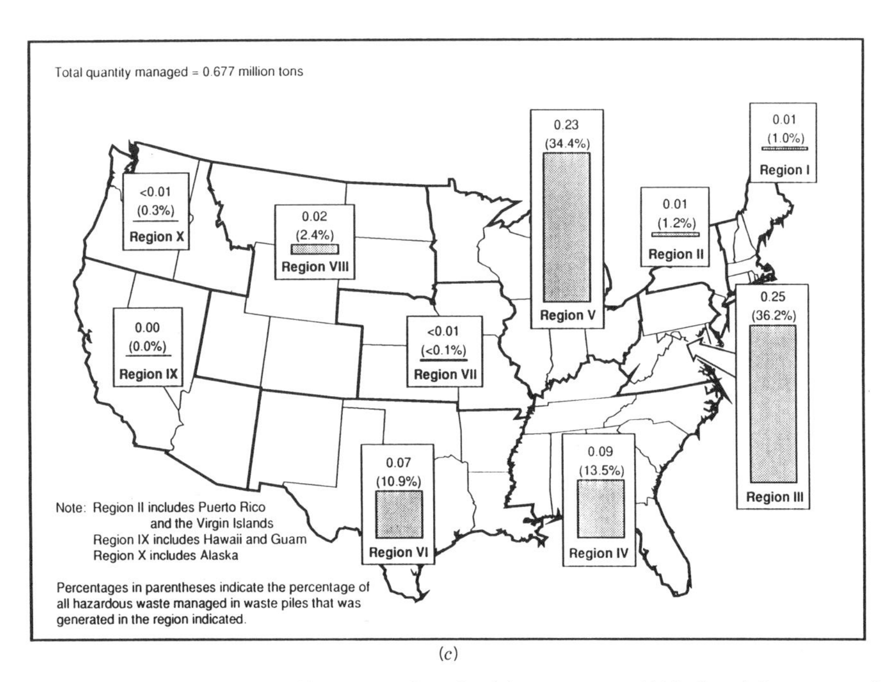

(*c*)

Figure B-15. Waste managed in waste pipes by (*a*) waste type, (*b*) industrial sector, and (*c*) geographic region. (Reprinted with permission of Mary Ann Liebert, Inc., publishers.)

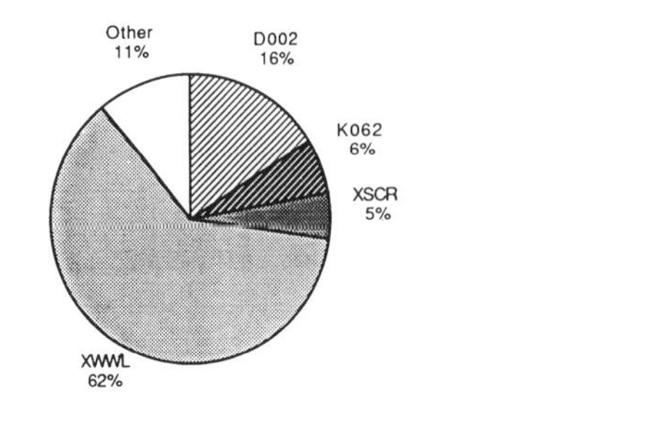

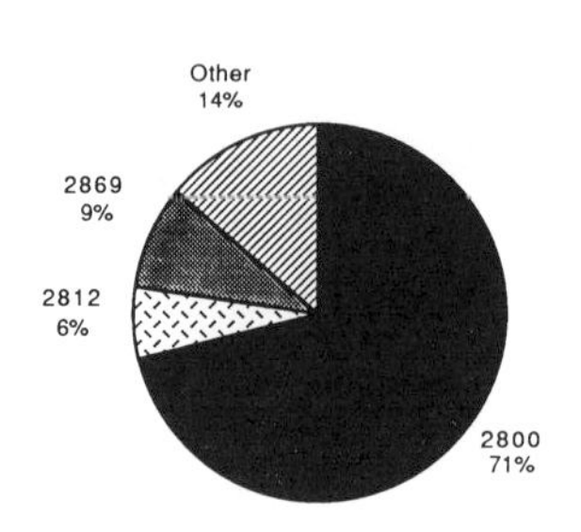

Waste Code	Waste Description
D002	Corrosive Waste
K062	Spent Pickle Liquor From Steel Finishing Operations of Plants That Produce Iron or Steel
XSCR	Hazardous incinerator, boiler, or furnace scrubber water
XWWL	Hazardous wastewater treatment liquid

(*a*).

Industry Code	Industry Description
2800	General Chemical Manufacturing
2812	Alkalies and Chlorine
2869	Industrial Organic Chemicals, nec

(*b*)

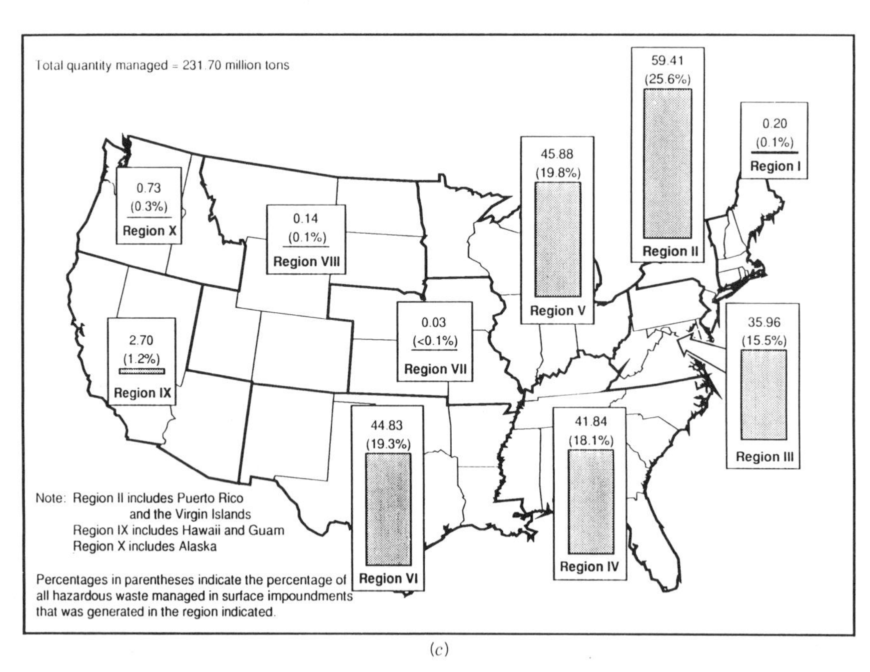

(*c*)

Figure B-16. Waste managed in surface impoundments by (*a*) waste type, (*b*) industrial sector, and (*c*) geographic region. (Reprinted with permission of Mary Ann Liebert, Inc., publishers.)

Index